Methods in Molecular Biology™

Series Editor
John M. Walker
School of Life Sciences
University of Hertfordshire
Hatfield, Hertfordshire, AL10 9AB, UK

For further volumes:
http://www.springer.com/series/7651

The Myc Gene

Methods and Protocols

Edited by

Laura Soucek

Vall d'Hebron Institute of Oncology (VHIO), Barcelona, Spain

Nicole M. Sodir

Department of Biochemistry, University of Cambridge, Cambridge, UK

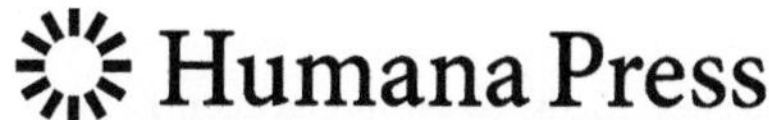

Editors
Laura Soucek
Vall d'Hebron Institute
of Oncology (VHIO)
Barcelona, Spain

Nicole M. Sodir
Department of Biochemistry
University of Cambridge
Cambridge, UK

ISSN 1064-3745 ISSN 1940-6029 (electronic)
ISBN 978-1-62703-428-9 ISBN 978-1-62703-429-6 (eBook)
DOI 10.1007/978-1-62703-429-6
Springer New York Heidelberg Dordrecht London

Library of Congress Control Number: 2013945869

Printed on acid-free paper

Humana Press is a brand of Springer
Springer is part of Springer Science+Business Media (www.springer.com)

Preface

Many brilliant scientists have been working on Myc since its discovery 30 years ago, providing tremendous and invaluable insights into its mechanism of action. However, after a quarter of a century, this infamous pleiotropic transcription factor still represents a challenging but rewarding area of research. Myc controls multiple cellular functions, including cell proliferation, growth, differentiation, and death, both directly and indirectly, through its modulation of downstream transcriptional programs. Such secondary programs not only ramify into all aspects of cell and tissue biology but they also feed back in a context and cell-type specific way to modulate how Myc acts. Despite its wide variety of physiological functions, Myc is mostly known for the role it plays in the development of cancer. Indeed, Myc overexpression or deregulation is associated with more than half of human cancers—a number that is possibly an underestimation. Because of its crucial role in governing intracellular and extracellular aspects of tumorigenesis, Myc represents an obvious and provocative candidate for targeted cancer therapy.

In this book *Myc: Methods and Protocols*, experts in the field summarize the standard and novel techniques that allow the studying of Myc mechanism of action in normal and cancer cells, in vitro and in vivo, in one succinct manual. It also offers a glance at therapeutic approaches for targeting Myc, which will potentially translate soon into clinical applications. This book is directed to biochemists, cell biologists, molecular biologists, medical doctors, and any researcher who is interested in exploring the Myc world. Chapters include introductions to their respective topics, lists of the necessary materials and reagents, step-by-step readily reproducible laboratory protocols, and tips on troubleshooting and avoiding known pitfalls.

This book has been made possible by the wonderful contribution of several colleagues in the Myc field, whose work has been invaluable. To them goes our deepest gratitude for making this project so enjoyable. We would also like to thank our day-to-day laboratory colleagues and friends for their encouragement and moral support. Finally, our warmest welcome goes to those readers who, thanks to this book, will start the fascinating trip into the Myc world. The adventure has just begun!

Barcelona, Spain *Laura Soucek*
Cambridge, UK *Nicole M. Sodir*

Contents

Contributors

BONNIE L. BARRILLEAUX • *Department of Cell Biology and Human Anatomy, Institute of Pediatric Regenerative Medicine, University of California Davis School of Medicine, Shriners Hospital For Children Northern California, Sacramento, CA, USA*

MARIE-EVE BEAULIEU • *Département de Pharmacologie, Faculté de Médecine et des Sciences de la Santé, Université de Sherbrooke, Sherbrooke, QC, Canada*

MIKAËL BÉDARD • *Département de Pharmacologie, Faculté de Médecine et des Sciences de la Santé, Université de Sherbrooke, Sherbrooke, QC, Canada*

KLAUS BISTER • *Center for Chemistry and Biomedicine (CCB), Institute of Biochemistry, University of Innsbruck, Innsbruck, Austria*

MATTEO BOCCI • *Department of Microbiology, Tumor and Cell Biology, Karolinska Institute, Stockholm, Sweden*

CATHERINE A. BURKHART • *Cleveland BioLabs, Inc., Buffalo, NY, USA*

PATRIZIA CAMMARERI • *Beatson Institute for Cancer Research, Glasgow, UK*

REBECCA COTTERMAN • *Department of Cell Biology and Human Anatomy, Institute of Pediatric Regenerative Medicine, University of California Davis School of Medicine, Shriners Hospital For Children Northern California, Sacramento, CA, USA*

JOHN T. CUNNINGHAM • *School of Medicine, University of California San Francisco, San Francisco, CA, USA; Department of Urology, Helen Diller Family Comprehensive Cancer Center, University of California San Francisco, San Francisco, CA, USA*

CHI V. DANG • *Abramson Cancer Center, Abramson Family Cancer Research Institute, Perelman School of Medicine, University of Pennsylvania, Philadelphia, PA, USA*

COLIN J. DANIEL • *Department of Molecular and Medical Genetics, Oregon Health and Science University, Portland, OR, USA*

DEAN W. FELSHER • *Department of Medicine-Oncology and Pathology, Stanford University School of Medicine, Stanford, CA, USA*

CARLA GRANDORI • *Division of Human Biology, Fred Hutchinson Cancer Research Center, Seattle, WA, USA*

ANDREI V. GUDKOV • *Department of Cell Stress Biology, Cleveland BioLabs, Inc., Roswell Park Cancer Institute, Buffalo, NY, USA*

MICHELLE HABER • *Lowy Cancer Research Centre, Children's Cancer Institute Australia for Medical Research, Sydney, Australia*

MARKUS HARTL • *Center for Chemistry and Biomedicine (CCB), Institute of Biochemistry, University of Innsbruck, Innsbruck, Austria*

LIN HE • *Department of Molecular & Cell Biology, University of California at Berkeley, Berkeley, CA, USA*

HEIKO HERMEKING • *Experimental and Molecular Pathology, Institute of Pathology, Ludwig-Maximilians-Universität München, Munich, Germany*

L. ERIC HUANG • *Department of Neurosurgery, University of Utah, Salt Lake City, UT, USA; Department of Oncological Sciences, University of Utah, Salt Lake City, UT, USA*

DAVID J. HUELS • *Beatson Institute for Cancer Research, Glasgow, UK*
RENE JACKSTADT • *Experimental and Molecular Pathology, Institute of Pathology, Ludwig-Maximilians-Universität München, Munich, Germany*
IGOR JURISICA • *Department of Medical Biophysics, Ontario Cancer Institute, The Campbell Family Institute for Cancer Research, University of Toronto, Toronto, ON, Canada; Department of Computer Science, Techna Institute, University Health Network, University of Toronto, Toronto, ON, Canada*
PAUL S. KNOEPFLER • *Department of Cell Biology and Human Anatomy, Institute of Pediatric Regenerative Medicine, University of California Davis School of Medicine, Shriners Hospital For Children Northern California, Sacramento, CA, USA*
LARS-GUNNAR LARSSON • *Department of Microbiology, Tumor and Cell Biology, Karolinska Institute, Stockholm, Sweden*
PIERRE LAVIGNE • *Département de Pharmacologie, Faculté de Médecine et des Sciences de la Santé, Université de Sherbrooke, Sherbrooke, QC, Canada*
ANNE LE • *Department of Pathology, Sol Goldman Pancreatic Cancer Research Center, Johns Hopkins University School of Medicine, Baltimore, MD, USA*
TREVOR D. LITTLEWOOD • *Department of Biochemistry, University of Cambridge, Cambridge, UK*
DAN LU • *Department of Biochemistry, University of Cambridge, Cambridge, UK*
FRANÇOIS-OLIVIER MCDUFF • *Département de Pharmacologie, Faculté de Médecine et des Sciences de la Santé, Université de Sherbrooke, Sherbrooke, QC, Canada*
JAN P. MEDEMA • *Lab of Exp. Oncology and Radiobiology, Academic Medical Center (AMC), Amsterdam, The Netherlands*
ANTJE MENSSEN • *Experimental and Molecular Pathology, Institute of Pathology, Ludwig-Maximilians-Universität München, Munich, Germany*
MARTIN MONTAGNE • *Département de Pharmacologie, Faculté de Médecine et des Sciences de la Santé, Université de Sherbrooke, Sherbrooke, QC, Canada*
MIKHAIL A. NIKIFOROV • *Department of Cell Stress Biology, Roswell Park Cancer Institute, Buffalo, NY, USA*
MURRAY D. NORRIS • *Lowy Cancer Research Centre, Children's Cancer Institute Australia for Medical Research, Sydney, Australia*
LINDA Z. PENN • *Department of Medical Biophysics, Ontario Cancer Institute, The Campbell Family Institute for Cancer Research, University of Toronto, Toronto, ON, Canada*
ROMINA PONZIELLI • *Ontario Cancer Institute, The Campbell Family Institute for Cancer Research, Toronto, ON, Canada*
MICHAEL POURDEHNAD • *Division of Hematology/Oncology, Department of Medicine, Helen Diller Family Comprehensive Cancer Center, University of California San Francisco, San Francisco, CA, USA*
KAVYA RAKHRA • *Department of Medicine-Oncology and Pathology, Stanford University School of Medicine, Stanford, CA, USA*
RACHEL A. RIDGWAY • *Beatson Institute for Cancer Research, Glasgow, UK*
DAVIDE RUGGERO • *School of Medicine, University of California San Francisco, San Francisco, CA, USA; Department of Urology, Helen Diller Family Comprehensive Cancer Center, University of California San Francisco, San Francisco, CA, USA*
OWEN J. SANSOM • *Beatson Institute for Cancer Research, Glasgow, UK*
ROSALIE C. SEARS • *Department of Molecular and Medical Genetics, Oregon Health and Science University, Portland, OR, USA*

NICOLE M. SODIR • *Department of Biochemistry, University of Cambridge, Cambridge, UK*
RUI SONG • *Department of Molecular & Cell Biology, University of California at Berkeley, Berkeley, CA, USA*
LAURA SOUCEK • *Vall d'Hebron Institute of Oncology (VHIO), Barcelona, Spain*
NICOLE SPONER • *Department of Molecular & Cell Biology, University of California at Berkeley, Berkeley, CA, USA*
CRAIG R. STUMPF • *School of Medicine, University of California San Francisco, San Francisco, CA, USA; Department of Urology, Helen Diller Family Comprehensive Cancer Center, University of California San Francisco, San Francisco, CA, USA*
VEDRANA TABOR • *Department of Microbiology, Tumor and Cell Biology, Karolinska Institute, Stockholm, Sweden*
KENNETH K.-W. TO • *School of Pharmacy, Area 39, The Chinese University of Hong Kong, Hong Kong, China*
NATAŠA TODOROVIĆ-RAKOVIĆ • *Department of Experimental Oncology, Institute for Oncology and Radiology of Serbia, Belgrade, Serbia*
WILLIAM B. TU • *Department of Medical Biophysics, Ontario Cancer Institute, The Campbell Family Institute for Cancer Research, University of Toronto, Toronto, ON, Canada*
XIAOLI ZHANG • *Department of Molecular and Medical Genetics, Oregon Health and Science University, Portland, OR, USA*

Chapter 1

The Myc World Within Reach

Nicole M. Sodir and Laura Soucek

Abstract

Myc is a transcriptional coordinator of a wide range of intracellular and extracellular processes required for cell proliferation. These processes are tightly regulated in physiological conditions but hijacked when Myc is oncogenically activated. In fact, aberrantly elevated and/or deregulated activity of Myc is associated with the majority of human cancers. Several switchable mouse transgenic models have been developed and provided insights on the role of Myc in maintaining multiple aspects of the tumor phenotype, indicating that Myc inhibition would constitute an effective and broadly applicable anticancer therapeutic strategy. This issue of "The Myc gene: Methods and Protocols" provides a rich collection of techniques developed or routinely used by Myc investigators and serves as an invaluable resource for exploring the pleiotropic and still puzzling Myc biological functions.

Key words Myc, Transcription factor, Cell proliferation, Cancer, Cell growth, Oncogene, Apoptosis, bHLHZip, Mouse model, Myc protocols

The transcription factor Myc coordinates the diverse intracellular and extracellular transcriptional programs required for the orderly proliferation of somatic cells. These programs include the regulation of cell growth, cell cycle, metabolism, protein biosynthesis, microRNAs expression, invasion, and angiogenesis, as well as a variety of protective mechanisms such as growth arrest and apoptosis [1–6]. Myc belongs to a family of basic helix–loop–helix leucine zipper (bHLHZip) proteins. As a member of this family, Myc heterodimerizes with its partner Max, activating genes by binding at canonical E-Box elements (CACGTG) in target gene promoters, and repressing other genes through binding and inhibition of the transcriptional activator Miz-1 and/or recruitment of the Dnmt3a DNA methyltransferase corepressor [7–9]. Although there are three discrete members of the Myc transcription factor family (c-Myc, N-Myc, and L-Myc), most somatic cells depend solely on c-Myc to coordinate the transcription of their proliferative programs. However, while the function of L-Myc is not well characterized,

Laura Soucek and Nicole M. Sodir (eds.), *The Myc Gene: Methods and Protocols*, Methods in Molecular Biology, vol. 1012, DOI 10.1007/978-1-62703-429-6_1, © Springer Science+Business Media, LLC 2013

N-Myc seems to be able to functionally replace c-Myc in murine development, cellular growth, and differentiation [10].

Expression array, SAGE, chromatin IP, promoter scanning, and whole cell proteomic approaches indicate that Myc modulates an enormous range of gene targets that is estimated to encompass up to a third of the transcriptome [11–20]. Myc expression in normal proliferating somatic cells is highly regulated and tightly dependent upon external growth signals, ensuring that Myc transactivation activities are released only in cells instructed to proliferate. In contrast, in tumor cells, this stringent regulation of Myc expression is compromised, and the same diverse intracellular and extracellular regenerative programs that drive normal somatic cell proliferation are hijacked. Aberrantly elevated and/or deregulated activity of Myc is implicated in the majority of human cancers and often linked to aggressive tumors [21, 22]. In these situations, direct mutational activation of the Myc gene itself (such as amplification or chromosomal translocation) appears to be relatively rare. Instead, Myc is more frequently deregulated due to its constitutive induction by upstream oncogenic signals such as the Wnt/β-catenin, Notch, or RTK/Ras pathways, forcing the cells into an incessant proliferative condition [20, 23]. Studies in vitro have shown that high and deregulated Myc renders cells more sensitive to nutrient deprivation and addicted to continual bioenergetic sources. For instance, glucose withdrawal triggers apoptosis of Myc-overexpressing cells, whereas deprivation of glucose from normal fibroblasts merely causes them to withdraw from the cell cycle into the G1 phase [4, 24].

To explore the impact of overexpressed and deregulated Myc in vivo, various switchable transgenic mouse models with tissue-specific regulatable oncogenic *myc* have been generated. These have shown that induction of Myc drives and maintains multiple aspects of the neoplastic phenotype, such as cell proliferation, dedifferentiation, angiogenesis, and invasiveness, and that subsequent inactivation of Myc triggers dramatic tumor regression, typically involving a variety of mechanisms such as cell death, proliferative arrest, differentiation, shutdown of angiogenesis, and collapse of the tumor microenvironment, which support the "oncogene addiction" concept [25–33]. These observations suggest that Myc could in principle offer an attractive target for cancer therapy.

An ideal cancer drug target must execute an essential function that is continuously required for cancer cell survival but dispensable at least for a short term for maintenance of normal tissues. It should also be functionally nonredundant, rendering it impossible for tumor cells to adapt or evolve independence upon its inhibition [34]. Myc seems to comply with all these criteria. Mouse studies have demonstrated that inhibition of transactivation properties of endogenous Myc by a dominant interfering Myc mutant, Omomyc [35, 36], elicits a therapeutic effect in diverse tumor types even when Myc itself is not the driving oncogene [37, 38]. Indeed, Myc inhibition by Omomyc triggers rapid regression of incipient and established tumors and collapse of the tumor microenvironment [37–40]. Of note, in these studies no tumors resistant to Myc inhibition have been observed, indicating that Myc is an obligate requirement for tumor survival. Although systemic Myc inhibition effectively stalls proliferation in normal proliferating tissues, it is surprisingly well tolerated and does not cause any disruption of normal tissue integrity [37]. In fact, the mild side effects of Myc inhibition observed in continuously regenerating tissues such as skin, intestine, and testis are completely reversible upon restoration of endogenous Myc function. Several groups around the world are using various strategies to target Myc function. These strategies include blockage of Myc expression (such as antisense mRNA, quadruplex-forming oligonucleotides, ribozymes, RNA interference, BET inhibitors, and small molecule drugs), disruption of Myc–Max dimerization, interference with Myc–Max binding to DNA, inhibition of key Myc target genes, and synthetic lethality in Myc-transformed cells [4, 41, 42]. The underlying promise is the same: inhibiting Myc could be the new strategy in cancer therapy and could be applied to most, if not all, human cancers.

This book aims to compile various methods and protocols routinely utilized in the laboratories of Myc experts that focus their efforts on untangling the pleiotropic and puzzling effects of Myc. Chapter 2 by Beaulieu and colleagues takes the reader through the basics of expression, purification, and preparation of the central components of the Myc network, as well as through the analysis and understanding of their dynamic interactions. Chapter 3 by Hartl and Bister offers an interesting journey into the fascinating field of Myc evolution, providing methods to follow Myc and its oncogenic transforming potential in various species. Chapter 4 by Ponzielli and colleagues describes the utilization of two essential techniques, co-immunoprecipitation (co-IP) and in vitro pull-down assays, to evaluate Myc interaction with other proteins besides Max. Chapter 5 by Daniel and colleagues discusses the detection of c-Myc phosphorylation status and its pivotal role in regulating c-Myc activity. Chapter 6 by To and Huang focuses on the dynamics of protein–protein interactions among HIF-1α, c-Myc, and Sp1, exploring the role of Myc in hypoxia, a critical process in cellular biology. Chapter 7 by Lu and Littlewood elaborates on Myc-induced apoptosis in vitro and in vivo, an intrinsic tumor suppressive property embedded in Myc and the most efficient tumor barrier devised by evolution to counteract Myc neoplastic potential. Chapter 8 by Tabor and colleagues discusses senescence, another tumor suppressive mechanism engaged by Myc, and, on a similar note to chapter 7, reviews the most updated protocols to study this aspect of Myc in vitro and in vivo. Chapter 9 by Barrilleaux and colleagues provides the reader with the protocols to analyze the binding and genomic location of Myc in stem and cancer cells.

Given the pleiotropic function of Myc, it is quite challenging to identify bona fide Myc target genes. Chapter 10 by Song and colleagues focuses on the regulation of microRNAs by Myc and describes widely used protocols to assess their expression levels. Chapter 11 by Jackstadt and colleagues provides the reader with comprehensive methodologies for the analysis of Myc-regulated mRNA and miRNA expression, as well as of DNA binding by Myc, allowing the thorough analysis of Myc function in various cellular contexts. Chapter 12 by Grandori applies the siRNA technology to the Myc network and provides protocols to identify Myc synthetic lethal genes to be used as potential therapeutic targets. Chapter 13 by Cunningham and colleagues reveals methods and tips to study translational regulation by Myc, a more recently discovered and fascinating aspect of Myc biology. Chapter 14 by Le and Dang provides precious insights in the study of Myc-dependent metabolism, a potential Achilles' heel in cancer biology. Chapter 15 by Rakhra and Felsher describes the steps involved in the generation of a tetracycline-regulated Myc-dependent mouse model of T-cell acute lymphoblastic leukemia. Chapter 16 by Huels and colleagues reviews the most recent laboratory protocols to study Myc function in intestinal homeostasis, regeneration, and tumorigenesis both in vivo and in vitro. Chapter 17 by Todorovic-Rakovic discusses how to detect Myc amplification in human cancer by chromogenic in situ hybridization (CIHS). Last but not least, Chapter 18 by Burkhart and colleagues offers a glance at methods designed to identify and validate small molecules inhibiting Myc function. Hopefully one day such molecules will be available in the clinic.

In summary, with this book, we hope to inspire more people to approach the Myc field and provide them with a rich collection of tools, insights, and directions to do it effectively. We are confident that Myc will not disappoint your curiosity and expectation for surprise.

References

1. Oster SK, Ho CS, Soucie EL, Penn LZ (2002) The myc oncogene: marvelously complex. Adv Cancer Res 84:81–154
2. Eilers M, Eisenman RN (2008) Myc's broad reach. Genes Dev 22:2755–2766
3. Cole MD, Henriksson M (2006) 25 years of the c-Myc oncogene. Semin Cancer Biol 16:241
4. Dang CV (2012) MYC on the path to cancer. Cell 149:22–35
5. Evan GI, Littlewood TD (1993) The role of c-myc in cell growth. Curr Opin Genet Dev 3:44–49
6. Hoffman B, Liebermann DA (2008) Apoptotic signaling by c-MYC. Oncogene 27:6462–6472
7. Adhikary S, Eilers M (2005) Transcriptional regulation and transformation by Myc proteins. Nat Rev Mol Cell Biol 6:635–645
8. Schneider A, Peukert K, Eilers M, Hanel F (1997) Association of Myc with the zinc-finger protein Miz-1 defines a novel pathway for gene regulation by Myc. Curr Top Microbiol Immunol 224:137–146
9. Brenner C, Deplus R, Didelot C, Loriot A, Vire E, De Smet C, Gutierrez A, Danovi D, Bernard D, Boon T, Pelicci PG, Amati B, Kouzarides T, de Launoit Y, Di Croce L, Fuks F (2005) Myc represses transcription through recruitment of DNA methyltransferase corepressor. EMBO J 24:336–346

10. Malynn BA, de Alboran IM, O'Hagan RC, Bronson R, Davidson L, DePinho RA, Alt FW (2000) N-myc can functionally replace c-myc in murine development, cellular growth, and differentiation. Genes Dev 14:1390–1399
11. Lawlor ER, Soucek L, Brown-Swigart L, Shchors K, Bialucha CU, Evan GI (2006) Reversible kinetic analysis of Myc targets in vivo provides novel insights into Myc-mediated tumorigenesis. Cancer Res 66:4591–4601
12. Fernandez PC, Frank SR, Wang L, Schroeder M, Liu S, Greene J, Cocito A, Amati B (2003) Genomic targets of the human c-Myc protein. Genes Dev 17:1115–1129
13. Shiio Y, Donohoe S, Yi EC, Goodlett DR, Aebersold R, Eisenman RN (2002) Quantitative proteomic analysis of Myc oncoprotein function. EMBO J 21:5088–5096
14. Guo QM, Malek RL, Kim S, Chiao C, He M, Ruffy M, Sanka K, Lee NH, Dang CV, Liu ET (2000) Identification of c-myc responsive genes using rat cDNA microarray. Cancer Res 60:5922–5928
15. Schuhmacher M, Kohlhuber F, Holzel M, Kaiser C, Burtscher H, Jarsch M, Bornkamm GW, Laux G, Polack A, Weidle UH, Eick D (2001) The transcriptional program of a human B cell line in response to Myc. Nucleic Acids Res 29:397–406
16. Watson JD, Oster SK, Shago M, Khosravi F, Penn LZ (2002) Identifying genes regulated in a Myc-dependent manner. J Biol Chem 277:36921–36930
17. O'Connell BC, Cheung AF, Simkevich CP, Tam W, Ren X, Mateyak MK, Sedivy JM (2003) A large scale genetic analysis of c-Myc-regulated gene expression patterns. J Biol Chem 278:12563–12573
18. Menssen A, Hermeking H (2002) Characterization of the c-MYC-regulated transcriptome by SAGE: identification and analysis of c-MYC target genes. Proc Natl Acad Sci USA 99:6274–6279
19. Schuldiner O, Benvenisty N (2001) A DNA microarray screen for genes involved in c-MYC and N-MYC oncogenesis in human tumors. Oncogene 20:4984–4994
20. Meyer N, Penn LZ (2008) Reflecting on 25 years with MYC. Nat Rev Cancer 8:976–990
21. Vita M, Henriksson M (2006) The Myc oncoprotein as a therapeutic target for human cancer. Semin Cancer Biol 16:318–330
22. Cole MD, Nikiforov MA (2006) Transcriptional activation by the Myc oncoprotein. Curr Top Microbiol Immunol 302: 33–50
23. Soucek L, Evan GI (2010) The ups and downs of Myc biology. Curr Opin Genet Dev 20:91–95
24. Shim H, Chun YS, Lewis BC, Dang CV (1998) A unique glucose-dependent apoptotic pathway induced by c-Myc. Proc Natl Acad Sci USA 95:1511–1516
25. Felsher DW, Bishop JM (1999) Reversible tumorigenesis by MYC in hematopoietic lineages. Mol Cell 4:199–207
26. Pelengaris S, Littlewood T, Khan M, Elia G, Evan G (1999) Reversible activation of c-Myc in skin: induction of a complex neoplastic phenotype by a single oncogenic lesion. Mol Cell 3:565–577
27. Pelengaris S, Khan M, Evan GI (2002) Suppression of Myc-induced apoptosis in beta cells exposes multiple oncogenic properties of Myc and triggers carcinogenic progression. Cell 109:321–334
28. Jain M, Arvanitis C, Chu K, Dewey W, Leonhardt E, Trinh M, Sundberg CD, Bishop JM, Felsher DW (2002) Sustained loss of a neoplastic phenotype by brief inactivation of MYC. Science 297:102–104
29. Flores I, Murphy DJ, Swigart LB, Knies U, Evan GI (2004) Defining the temporal requirements for Myc in the progression and maintenance of skin neoplasia. Oncogene 23: 5923–5930
30. Arvanitis C, Felsher DW (2005) Conditionally MYC: insights from novel transgenic models. Cancer Lett 226:95–99
31. D'Cruz CM, Gunther EJ, Boxer RB, Hartman JL, Sintasath L, Moody SE, Cox JD, Ha SI, Belka GK, Golant A, Cardiff RD, Chodosh LA (2001) c-MYC induces mammary tumorigenesis by means of a preferred pathway involving spontaneous Kras2 mutations. Nat Med 7:235–239
32. Podsypanina K, Politi K, Beverly LJ, Varmus HE (2008) Oncogene cooperation in tumor maintenance and tumor recurrence in mouse mammary tumors induced by Myc and mutant Kras. Proc Natl Acad Sci USA 105:5242–5247
33. Shachaf CM, Kopelman AM, Arvanitis C, Karlsson A, Beer S, Mandl S, Bachmann MH, Borowsky AD, Ruebner B, Cardiff RD, Yang Q, Bishop JM, Contag CH, Felsher DW (2004) MYC inactivation uncovers pluripotent differentiation and tumour dormancy in hepatocellular cancer. Nature 431:1112–1117
34. Sodir NM, Evan GI (2011) Finding cancer's weakest link. Oncotarget 2:1307–1313
35. Soucek L, Helmer-Citterich M, Sacco A, Jucker R, Cesareni G, Nasi S (1998) Design and properties of a Myc derivative that

efficiently homodimerizes. Oncogene 17: 2463–2472
36. Soucek L, Jucker R, Panacchia L, Ricordy R, Tato F, Nasi S (2002) Omomyc, a potential Myc dominant negative, enhances Myc-induced apoptosis. Cancer Res 62:3507–3510
37. Soucek L, Whitfield J, Martins CP, Finch AJ, Murphy DJ, Sodir NM, Karnezis AN, Swigart LB, Nasi S, Evan GI (2008) Modelling Myc inhibition as a cancer therapy. Nature 455:679–683
38. Sodir NM, Swigart LB, Karnezis AN, Hanahan D, Evan GI, Soucek L (2011) Endogenous Myc maintains the tumor microenvironment. Genes Dev 25:907–916
39. Soucek L, Whitfield JR, Sodir NM, Massó-Vallés D, Serrano E, Karnezis AN, Swigart LB, Evan GI (2013) Inhibition of Myc family proteins eradicates KRas-driven lung cancer in mice. Genes Dev 27(5):504–513
40. Whitfield JR, Soucek L (2012) Tumor microenvironment: becoming sick of Myc. Cell Mol Life Sci 69:931–934
41. Hermeking H (2003) The MYC oncogene as a cancer drug target. Curr Cancer Drug Targets 3:163–175
42. Evan G (2012) Cancer. Taking a back door to target Myc. Science 335:293–294

Chapter 2

Methods for the Expression, Purification, Preparation, and Biophysical Characterization of Constructs of the c-Myc and Max b-HLH-LZs

Marie-Eve Beaulieu, François-Olivier McDuff, Mikaël Bédard, Martin Montagne, and Pierre Lavigne

Abstract

Specific heterodimerization and DNA binding by the b-HLH-LZ transcription factors c-Myc and Max is central to the activation and repression activities of c-Myc that lead to cell growth, proliferation, and tumorigenesis (Adhikary and Eilers, Nat Rev Mol Cell Biol 6:635–645, 2005; Eilers and Eisenman, Genes Dev 22:2755–2766, 2008; Grandori et al., Annu Rev Cell Dev Biol 16:653–699, 2000; Whitfield and Soucek, Cell Mol Life Sci 69:931–934, 2011). Although many c-Myc-interacting partner proteins are known to interact through their HLH domain (Adhikary and Eilers, Nat Rev Mol Cell Biol 6:635–645, 2005), current knowledge regarding the structure and the determinants of molecular recognition of these complexes is still very limited. Moreover, recent advances in the development and use of b-HLH-LZ dominant negatives (Soucek et al., Nature 455:679–683, 2008) and inhibitors of c-Myc interaction with its protein partners (Bidwell et al., J Control Release 135:2–10, 2009; Mustata et al., J Med Chem 52:1247–1250, 2009; Prochownik and Vogt, Genes Cancer 1:650–659, 2010) or DNA highlight the importance of efficient protocols to prepare such constructs and variants. Here, we provide methods to produce and purify high quantities of pure and untagged b-HLH-LZ constructs of c-Myc and Max as well as specific c-Myc/Max heterodimers for their biophysical and structural characterization by CD, NMR, or crystallography. Moreover, biochemical methods to analyze the homodimers and heterodimers as well as DNA binding of these constructs by native electrophoresis are presented. In addition to enable the investigation of the c-Myc/Max b-HLH-LZ complexes, the protocols described herein can be applied to the biochemical characterization of various mutants of either partner, as well as to ternary complexes with other partner proteins.

Key words b-HLH-LZ, c-Myc, Max, Expression, Purification, Specific heterodimerization, Disulphide oxidation, Native electrophoresis, DNA binding, Electrophoretic mobility shift assay

1 Introduction

The b-HLH-LZ domains of c-Myc and Max are responsible for their heterodimerization, DNA binding and interactions with other proteins [1–5] and hence play crucial roles in the molecular biology of c-Myc. Designed b-HLH-LZ domains are now being

Laura Soucek and Nicole M. Sodir (eds.), *The Myc Gene: Methods and Protocols*, Methods in Molecular Biology, vol. 1012, DOI 10.1007/978-1-62703-429-6_2, © Springer Science+Business Media, LLC 2013

used as dominant negatives and inhibitors of c-Myc [6–8]. Our understanding of the structural determinants for specific heterodimerization and DNA binding as well as the engineering of dominant negatives necessitates the production of large amounts of pure b-HLH-LZ domains of c-Myc, Max and respective mutant forms. The methods presented here were used to understand the determinants of the molecular recognition between the b-HLH-LZ domains of c-Myc and Max (termed c-Myc'SH and Max'SH respectively throughout) as well as the DNA binding properties of these constructs and variants either in their homodimeric or heterodimeric states [9–15]. They encompass efficient biosynthesis in *E. coli*, purification, and preparation of homo- and heterodimeric constructs necessary to carry out biophysical characterization using circular dichroism (CD), solution-state nuclear magnetic resonance (NMR), and electrophoretic mobility shift assays (EMSA).

First, we explain the protocols used to produce and to purify from heterologous *E. coli* expression systems large quantities of c-Myc'SH and Max'SH (Subheadings 3.1–3.3). In contrast to the procedure to express and purify the Max'SH construct from the soluble lysis fraction [13], expression and purification of c-Myc'SH from inclusion bodies (IB) in the absence of tag (e.g., His tag) necessitate additional and critical steps during expression and cationic exchange purification [11] to obtain high yields of pure constructs. The expression protocols in minimal medium to produce isotopically labeled c-Myc'SH and Max'SH for NMR studies are also provided [9, 11, 15]. Although these procedures are derived from standard expression and purification protocols, we describe key modifications to optimize yields and purity of these constructs.

In Subheading 3.4, we also describe how to prepare specific heterodimeric complexes chemically cross-linked with a disulfide bond using these constructs and dimethylsulfoxide (DMSO) as the oxidizing agent [16].

In Subheading 3.5, a native electrophoresis protocol specifically adapted to distinguish the c-Myc'SH and Max'SH monomeric, homodimeric, and heterodimeric species is explained in detail [10]. This protocol is derived from the moving boundary electrophoresis on gels developed by Chrambach and Jovin [17]. Briefly, the method uses buffer systems at various pH values, which define moving boundaries within known leading and trailing ion mobility. It allows concentration and distinction of proteins under native conditions within moving boundaries set up at various pH values. In the specific protocol developed for c-Myc'SH and Max'SH constructs, free base histidine is used as the trailing ion and potassium hydroxide as the leading ion in a HEPES buffer.

Finally, an electrophoretic mobility shift assay (EMSA) method using the purified c-Myc'SH, Max'SH, and fluorescently labeled Alexa-Fluor™ DNA probes is described. The EMSA protocol is derived from the original work of Gardner and Revzin [18] and of

Fried and Crothers [19]. This assay is based on the observation that protein/DNA complexes migrate more slowly than free linear DNA in a non-denaturing polyacrylamide or agarose gel electrophoresis. It can be used in conjunction with mutagenesis to identify important DNA binding sequences or residues or to determine the optimal conditions for stabilization of specific complexes with regard to salt concentration, stoichiometry of the complex, or to estimate binding affinities (K_A) of the constructs to DNA. The resolution depends on the stability of the complex. The low ionic strength of the electrophoresis buffer helps to stabilize transient interactions.

2 Materials

2.1 Expression of the c-Myc and Max b-HLH-LZ (c-Myc'SH and Max'SH) Constructs

1. 2YT medium (rich medium) per liter: 16 g Bacto tryptone, 10 g yeast extract, 5 g NaCl. Autoclave in 2 L baffled erlenmeyers (*see* **Note 1**).
2. 10× M9 salts per 500 mL: 34 g Na_2HPO_4 (or 64 g $Na_2HPO_4 \cdot 7H_2O$), 15 g KH_2PO_4, 2.5 g NaCl. Autoclave.
3. To produce isotopically labeled proteins for NMR structure determination, use M9 minimal medium containing 1 g/L of (U-^{15}N) $^{15}NH_4Cl$ and/or 4 g/L of (U-^{13}C) glucose as the sole sources of nitrogen and carbon. M9 minimal medium per liter: 100 mL of 10× M9 salts, 850 mL H_2O_{dd}, 100 μL of sterile 1 M $CaCl_2$, 100 μL of sterile 1 M $MgSO_4$, 100 μL of sterile 0.1 M $FeSO_4$, 1 g of ^{14}N or $^{15}NH_4Cl$, 4 g of (^{12}C or ^{13}C) glucose. Autoclave.
4. 1,000× ampicillin: 50 mg/mL of ampicillin. Store in 1 mL aliquots at −20 °C.
5. 1,000× chloramphenicol: 34 mg/mL of chloramphenicol in 70 % v/v ethanol. Store at −20 °C.
6. 1,000× IPTG, per 50 mL: 0.716 g of isopropyl β-D-1-thiogalactopyranoside (IPTG). Store in 1 mL aliquots at −20 °C.
7. Arabinose solution: 20 % w/v of L-arabinose in water, fresh.
8. IB Laemmli buffer: 125 mM Tris–HCl pH 6.8, 3 % w/v sodium dodecyl sulfate, 20 % v/v glycerol, 20 % v/v β-mercaptoethanol, 0.02 % w/v bromophenol blue, 4 M urea.

2.2 Purification of the c-Myc and Max b-HLH-LZ (c-Myc'SH and Max'SH) Constructs

1. Lysis buffer: 50 mM KH_2PO_4, 700 mM NaCl, 5 mM $MgCl_2$, pH 7.0. Store at 4 °C.
2. Triton solution: 10 % v/v Triton × 100.
3. DNase I solution: 1 mg/mL of bovine pancreatic DNase I in 5 mM sodium acetate pH 5.0, 1 mM $CaCl_2$, 50 % v/v glycerol. Store at −20 °C.

4. PEI solution: 5 % w/v polyethyleneimine (PEI) in H_2O_{dd}. Store at 4 °C.
5. Bull cracker buffer part A: 50 mM sodium acetate pH 5.0, 6 M urea (ultra pure), 500 mM guanidinium chloride (ultra pure), 25 mM dithiothreitol (DTT) (*see* **Note 2**).
6. Bull cracker buffer part B: 2 M urea. Store at 4 °C for up to 2 weeks.
7. HiTrap™ SP Sepharose HP columns from GE Healthcare or the equivalent.
8. FPLC buffer A: 50 mM sodium acetate pH 5.0.
9. FPLC buffer B: 50 mM sodium acetate pH 5.0, 5 M NaCl.
10. U8 buffer: 50 mM Tris–HCl, 4 M urea, pH 7.4. Store at 4 °C for up to 2 weeks.
11. Desalting trifluoroacetic acid (TFA) buffer: 0.05 % v/v TFA.
12. HiTrap™ SP Sepharose HP columns (GE Healthcare).
13. Amicon®Ultra centrifugal filters Ultracel®-3K (Millipore™) or the equivalent.
14. Acetonitrile for lyophilization.
15. FPLC system (ÄKTAPrime™ or equivalent).

2.3 Specific Heterodimerization and Structural Characterization

1. Buffer C: 10 mM sodium cacodylate, 10 mM glacial acetic acid, 100 mM KCl, pH 5.0.
2. Dithiothreitol (DTT).
3. Amicon®Ultra centrifugal filters Ultracel®-3K (Millipore™) or the equivalent.
4. Dimethyl sulfoxide (DMSO).
5. H_2O_{dd} containing 0.1 % v/v TFA.
6. Acetonitrile.
7. High pressure liquid chromatography (HPLC) equipped with a C18 column.

2.4 Specific Heterodimerization Assessment by Cationic PAGE, His-HEPES-KOH

1. Acrylamide–urea solution: acrylamide 30 %, 6 M urea. Use 7.5 mL of a 40 % acrylamide stock solution (38.67 % w/v acrylamide, 1.33 % w/v *bis*-acrylamide) and 3.6 g urea to obtain a final volume of 10 mL (do not add H_2O_{dd}) (*see* **Note 3**).
2. Resolving gel buffer: 250 mM HEPES-KOH pH 7.75, 6.35 M urea. Adjust pH with 10 N KOH.
3. Stacking gel buffer: 750 mM HEPES-KOH pH 7.0, 6.35 M urea. Adjust pH with 10 N KOH.
4. 10 % Ammonium persulfate (APS) in water, fresh.
5. *N*,*N*,*N*′,*N*′-tetramethylethylenediamine (TEMED).
6. Isopropanol.

7. Anolyte buffer, for 500 mL: 25 mM HEPES, 175 mM histidine free base. Do not adjust the pH; it should be at pH 7.3.
8. Catholyte buffer, for 500 mL: 200 mM HEPES-KOH, pH 7.0. Adjust pH with 10 N KOH.
9. Sample buffer (2×): 62.5 mM HEPES-KOH pH 7.7, 6 M urea (*see* **Note 4**).
10. Protein stain solution: 100 mL methanol, 20 mL glacial acetic acid, 80 mL water, 0.4 g Coomassie Brilliant Blue R250. Dissolve the dye in methanol, followed by addition of the acid and water.
11. Destain solution: same composition as above but the dye is omitted.
12. Additional materials: 0.75 mm × 7.25 cm × 10 cm gel cassette and casting setup (e.g., from Bio-Rad or the equivalent), 10 wells combs, standard electrophoresis apparatus for acrylamide gels, power supply, plastic tank at gel dimensions for staining, 10-mL glass pipet, Kimwipes.

2.5 DNA Binding by Electrophoretic Mobility Shift Assay

1. Alexa-Fluor™-labeled DNA probe (*see* **Note 5**) at ~2 μM in 10 mM Tris–HCl pH 7.5.
2. Protein stocks.
3. Binding buffer 10×: 200 mM Tris–HCl pH 8.0, 750 mM KCl, 25 mM DTT, 250 μg/mL BSA, 10 μg/mL poly dI-dC (*see* **Note 6**), 50 % v/v glycerol (*see* **Note 7**).
4. Stock solution of 50× TA buffer: 2 M Trizma base, 1 M acetic acid, pH 8.0. Dilute in water to prepare 2× TA buffer to be used for the gel and 1× TA buffer for the electrophoretic migration.
5. 30 % w/v acrylamide (29:1 acrylamide:*bis*-acrylamide); *see* **Note 3**.
6. 10 % Ammonium persulfate (APS) in water.
7. *N*,*N*,*N*′,*N*′-tetramethylethylenediamine (TEMED).

3 Methods

3.1 Expression of the c-Myc and Max b-HLH-LZ Constructs (c-Myc'SH and Max'SH) in Rich Medium or in Minimal M9 Medium

1. For c-Myc'SH production in rich medium, inoculate 1 L of 2YT medium containing ampicillin with BL21-AI *Escherichia coli* (*E. coli*) transformed with the c-Myc'SH-pET-3a construct (*see* **Note 8**), grow at 37 °C with agitation at 250 rpm until OD600 reaches 0.8 and induce expression with 10 mL of Arabinose solution, incubating at 37 °C for 12 h. Alternatively, for c-Myc'SH production in M9 minimal medium, inoculate 1 L of M9 minimal medium containing chloramphenicol and ampicillin with BL21-CodonPlus (Strategene) *E. coli* transformed with the c-MycSH'-pET3a construct and grow at 37 °C

with agitation at 250 rpm until OD600 reaches 0.8. Induce expression with 1 mL of IPTG 1,000× solution (*see* **Note 9**) and incubate with agitation at 250 rpm at 37 °C for 12 h.

2. For Max'SH production in rich medium or in M9 minimal medium, inoculate 1 L of 2YT medium or M9 medium containing chloramphenicol and ampicillin with *E. coli* BL21-pLysS (Invitrogen) transformed with the Max'SH-pET-3a construct (*see* **Note 10**) and grow at 37 °C with agitation at 250 rpm until OD600 reaches 0.9. Induce expression with 1 mL of IPTG 1,000× solution and incubate at 30 °C with agitation at 250 rpm for 4 h.
3. Centrifuge the bacterial culture at 15,000 × *g*, 4 °C, 5 min in a SLA-1500 rotor using 250 mL bottles. The culture is centrifuged sequentially to combine the equivalent of 1 L of culture per bottle (*see* **Note 11**).
4. In order to verify the successful protein expression, collect a 1 mL aliquot from the cultures and measure its OD600. Spin 800 μL of the aliquot at 16,060 × *g* for 1 min and discard the supernatant. Resuspend the pellet in a volume of IB Laemmli buffer equivalent to (0.1 × OD600 value) μL. This will ensure equal quantities of lysed cell extract in each sample and facilitate comparison between cells before and after protein expression. Vortex, sonicate for 5 s at maximum power, freeze in liquid nitrogen, and boil the samples for 2 min (repeat 3 times) prior to loading on a denaturing 16.5 acrylamide SDS-PAGE gel and proceed to electrophoresis followed by Coomassie blue staining or Western blot. The IB Laemmli buffer is optimized to allow solubilization of the inclusion bodies and to ensure adequate migration and protein separation to visualize the expression of c-Myc'SH.

3.2 Purification of the c-Myc b-HLH-LZ (c-Myc'SH)

1. Suspend the cell pellets in 3 mL of Lysis buffer per gram of pellet by vortexing. Add 150 μL of Triton solution per gram of culture pellet. Reduce the viscosity of the suspension by sonicating on ice, at power 15 for 6 × 15 s using an ultrasonic homogenizer.
2. Add the equivalent of 100 μL/g of culture pellet of bovine pancreatic DNase I and incubate 60 min at 37 °C with agitation at 50 rpm (*see* **Note 12**).
3. Centrifuge 20 min at 12,000 × *g* (13,000 rpm in a SS34 rotor in a Sorvall RC 5B Plus; use 35 mL centrifuge bottles) at 4 °C to pellet the inclusion bodies as well as high molecular weight complexes such as cell walls, ribosomes, and nondegraded genomic DNA. Lipids, soluble proteins, amino acids, sugars, and nucleic acids constitute the supernatant that is discarded after centrifugation.

4. Solubilize the pelleted inclusion bodies with 15 mL of Bull cracker buffer part A by vortexing. This acidic, high ionic strength and denaturant buffer enables complete solubilization of the c-Myc'SH construct from the inclusion bodies and will allow elimination of high molecular weight complexes and residual DNA by centrifugation (*see* **Note 13**).
5. Dilute the inclusion bodies solution 1:1 with Bull cracker part B.
6. Centrifuge 30 min at 30,000 ×*g* (19,000 rpm in a SS34 rotor in a Sorvall RC 5B Plus), 4 °C.
7. Purify the supernatant by cation exchange chromatography on 5, cationic exchange columns of 5 mL (HiTrap™ SP Sepharose HP columns from GE Healthcare or the equivalent) mounted in series on a FPLC system. Wash with 2 column volumes (CV) of FPLC buffer B at a flow rate of 5 mL/min. Equilibrate the columns with 5 CV (125 mL) of FPLC buffer A at a flow rate of 2.5 mL/min. Load the supernatant (equivalent to a maximum of 9 g of culture pellet per load) at 2.5 mL/min (*see* **Note 14**). Immediately after loading the supernatant, wash with 75 mL (3 CV) of U8 buffer, at a flow rate of 2.5 mL/min (*see* **Note 15**). Wash with 1 CV of buffer A. Wash with 3 CV of 10 % buffer B. Eluate with a gradient of buffer B of 10–35 % in 50 mL (0.5 %/mL) at 2.5 mL/min in 2.5 mL fractions. In these conditions, c-Myc'SH eluates at ~1.5 M NaCl (i.e., ~30 % v/v of the gradient) with a purity of 90 % (*see* **Note 16**).
8. Pool the fractions containing the pure construct and desalt on 5 consecutive HiTrap™ Desalting columns from GE Healthcare (or the equivalent) equilibrated with Desalting TFA buffer, at a 3.0 mL/min flow rate. Inject the pooled fractions (maximum volume of 7.5 mL per injection) and follow the elution with OD280 detector. The protein eluates within the first 15 mL after injection. Discard the following fractions showing low OD280 values and high conductivity. This method allows ~95 % recovery of 98–99 % desalted protein.
9. The purified protein can be concentrated either by centrifugation in Amicon®Ultra centrifugal filters Ultracel®-3K (Millipore™) or the equivalent or by lyophilization. For concentration using the Amicon®Ultra, follow manufacturer's indication (*see* **Note 17**). Alternatively, for lyophilization, add 50 % v/v acetonitrile to the desalted fractions, freeze in liquid nitrogen, and lyophilize. The foamy lyophilized protein obtained can be weighted and resolubilized to 100 μl/mg in 50 % v/v acetonitrile containing 0.05 % v/v TFA, aliquoted 1–10 mg per eppendorf, and lyophilized again.
10. Yields around 25 mg/L of culture in rich medium and 15 mg/L of culture in minimal M9 medium.

3.3 Purification of the Max b-HLH-LZ (Max'SH)

1. Suspend the cell pellets in 3 mL of Lysis buffer per gram of pellet by vortexing. Add 150 μL of Triton solution per gram of culture pellet. Reduce the viscosity of the suspension by sonicating, on ice, at power 15 for 6 × 15 s using an ultrasonic homogenizer.
2. Add the equivalent of 125 μL/g of culture pellet of bovine pancreatic DNase and incubate 30 min at 37 °C with agitation at 60 rpm (*see* **Note 12**).
3. Precipitate DNA by adding 90 μL per g of culture pellet of PEI solution, vortexing, and incubating on ice for 15 min (*see* **Note 18**).
4. Centrifuge 30 min at 30,000 × *g* (19,000 rpm in a SS34 rotor in a Sorvall RC 5B Plus), 4 °C. Unbroken cells, large cellular debris, and precipitated DNA/PEI complexes will be pelleted.
5. Dilute the supernatant 1:1 with a solution of 20 % v/v of FPLC buffer B/(A + B) to ensure optimal adhesion of the construct to the cationic column (*see* **Note 19**).
6. Purify the diluted supernatant by cation exchange chromatography on 5 HiTrap™ SP Sepharose HP columns (GE Healthcare) mounted in series on a FPLC system. Wash the columns with 2 CV of FPLC buffer B at a flow rate of 5 mL/min. Equilibrate with 5 CV of FPLC buffer A at a flow rate of 2.5 mL/min. Load the supernatant (equivalent to a maximum of 9 g of culture pellet per load) at 2.5 mL/min. Wash with 10 % v/v FPLC buffer B in buffer A. Eluate with a gradient of buffer B: 10–35 % in 50 mL (0.5 %/mL). Max'SH eluates at ~1.25 M NaCl (i.e., ~25 % of the gradient). The eluate contains >95 % pure protein (*see* **Note 20**).
7. Pool the fractions containing the pure construct and desalt (and optionally lyophilize) using the same procedures than for the c-Myc'SH construct (*see* **steps 8** and **9** in Subheading 3.2).
8. Yields around 75 mg/L of culture.

3.4 Production of Specific Heterodimeric Cross-linked Complexes

1. Dissolve 2 mg of each protein construct (i.e., c-Myc'SH and Max'SH) in 1 mL of buffer C. Add 50 mM DTT and incubate for 1 h at 37 °C to ensure complete reduction of the C-terminal cysteine residues (*see* **Note 21**).
2. Dilute 3,000× in buffer C and concentrate using Amicon®Ultra centrifugal filters Ultracel®-3K (Millipore™) or the equivalent to a final volume of 4 mL.
3. Add 1 mL of DMSO, mix by several inversions, and incubate at room temperature for up to 24 h (*see* **Note 22**).
4. Dilute 3,000× in H_2O_{dd} containing 0.1 % v/v TFA and purify by HPLC on a C18 reversed-phase column. Eluate with a gradient of 40–100 % v/v acetonitrile in 100 min. The specific heterodimers will elute around 56 % v/v acetonitrile.
5. Lyophilize the purified complexes.

3.5 Cationic PAGE, His-HEPES-KOH

1. For one resolving gel, mix 2 mL of acrylamide–urea solution, 2.5 mL of resolving gel buffer, 480 μL of H_2O_{dd}, 15 μL of 10 % APS, and 2 μL of TEMED in a 15 mL conical flask. Transfer the solution using a 10-mL glass pipet and cast approximately 4 mL of gel within a 0.75 mm × 7.25 cm × 10 cm gel cassette (leaving space for the stacking gel that will include the combs) (*see* **Note 23**). Gently overlay with 1 mL of isopropanol for polymerization (*see* **Note 24**). After polymerization (~20 min), remove the isopropanol.
2. Prepare the 4 % acrylamide stacking gel by mixing 130 μL of acrylamide–urea solution, 0.5 mL of Stacking gel buffer, 360 μL of H_2O_{dd}, 10 μL of 10 % APS, and 2 μL of TEMED. Pour above the stacking gel and place the comb immediately without introducing air bubbles. Let polymerize for up to 15 min.
3. Mount the electrophoresis system using the Anolyte and Catholyte buffers. Load the samples using the Cationic PAGE Loading buffer to a 1:1 ratio. Invert polarity and proceed to electrophoretic migration at 200 V and 18 mA for 120 min.

3.6 Electrophoretic Mobility Shift Assay with c-Myc'SH and Max'SH

1. Carefully wash the glass plates under hot water, rinse with 70 % ethanol, and dry completely.
2. Prepare a native 6 % acrylamide gel (*see* **Note 25**): mix in a 15 mL conical flask 1.45 mL of H_2O_{dd}, 2.5 mL of 2× TA buffer pH 8.0, 2.0 mL of 30 % acrylamide, 50 μL 10 % APS, 4 μL TEMED. Mix well and pour immediately ~5 mL of gel within a 0.75 mm × 7.25 cm × 10 cm gel casting cassette, place the combs quickly without introducing air bubbles (*see* **Note 26**).
3. Pre-run the gel 30 min at 100 V (*see* **Note 27**).
4. Prepare the samples by first adding the desired competitive DNA quantities to the Binding buffer, followed by addition of the protein(s) of interest and finally addition of the DNA probe. Where possible, minimize pipetting errors by using premixed solutions and distributing equal volumes in each reaction. Mix, quickspin using a microcentrifuge, and incubate the samples at room temperature for 15 min. As a control, always include a sample of labeled DNA without protein. The fluorescently labeled probe can be used at concentrations between 10 nM and 2 μM. Load the samples in the wells of the gel using the difference in refractive index as visual indicator (*see* **Note 28**).
5. Proceed to electrophoresis in 1× TA buffer at 100 V for 35 min (*see* **Note 29**).
6. Wearing gloves, disassemble the apparatus and remove one of the glass plates. If the gel remains attached to the spacer glass plate, carefully detach the sides of the gel using the plastic tool or a pipette tip. Gently rinse the gel and carefully pull one corner to lift the gel and place it onto the imaging system (*see* **Note 30**).

4 Notes

1. Utilization of higher-volume erlenmeyers allows thorough oxygenation of the medium during the culture. Baffled erlenmeyers will also increase oxygenation and hence bacterial growth and the maximal OD600 but are not mandatory.
2. The buffer is prepared without DTT and stored for up to 2 weeks at 4 °C and is used with addition of fresh DTT to a final concentration of 25 mM.
3. Acrylamide is a neurotoxic compound that is easily absorbed through skin. Wear gloves and a mask to avoid direct contact or inhalation when manipulating dry acrylamide. Acrylamide solutions are light sensitive and are stable for months if protected from light. Acrylamide and bis-acrylamide are slowly converted to acrylic and bisacrylic acid respectively when stored for longer periods.
4. Do not use protein solutions that are in a strong buffer with a pH value far from pH 7.7. Either adjust the pH with HCl or KOH or dialyze against 1× Sample buffer.
5. We used Alexa-Fluor™488 labeling because the excitation and emission wavelength allowed us to use these oligonucleotides in another set of fluorescence experiments in solution, but different fluorescent probes can be used. Although the sensitivity of fluorescent probes is lower than that of radioisotopes, this method avoids the cost and regulatory concerns associated with radioactivity. We used DNA probes of various lengths (from 18 to 80 base pairs). In order to obtain double stranded DNA, the two synthesized complementary single stranded oligonucleotides are mixed at a 1:1 ratio and are boiled for 5 min, followed by annealing by slowly cooling down to room temperature. This procedure works well for oligonucleotides that do not have a strong tendency to form secondary structure, and that do not bear long repeats. Complete annealing can be verified by running the DNA on a 6 % native gel along with a low molecular weight dsDNA marker and gel staining by incubation for 5 min in 30 mL of H_2O_{dd} containing either 0.001 % ethidium bromide or 0.01 % of fluorescent dye such as SYBRSafe™. **Caution**: Ethidium bromide is a powerful mutagenic agent and is also moderately toxic. Wearing 2 pairs of gloves is essential when working with this solution. Decontamination can be achieved using hypophosphorous acid or potassium permanganate [20].
6. Nonspecific DNA binding is prevented by addition of nonspecific competitor DNA such as poly dI-dC to the reaction buffer. As a general rule, the amount of poly dI-dC should be lower when working with purified proteins and DNA probes

than when working with crude extract [21]. This amount can be adjusted (lowered) to enable the detection of weaker complexes, and the poly dI-dC can also be replaced with competitor DNA probes of various specificities.

7. Sucrose can be substituted by glycerol, at the same final concentration.

8. In order to produce disulfide-linked and specific c-Myc'SH Max' heterodimers, a sequence coding for a C-terminal linker containing a cysteine residue (GSGC) was added to the c-Myc b-HLH-LZ construct after residue R^{435}. We also found that codon optimization of the c-Myc'SH construct for E. coli improved dramatically protein expression. The resulting optimized coding sequence for c-Myc'SH (V^{354} to R^{435}) is as follows: 5′-ATG GTG AAA CGC CGT ACC CAT AAT GTC TTG GAA CGC CAA CGC CGT AAC GAA CTG AAA CGC AGC TTC TTT GCG CTG CGT GAC CAG ATC CCG GAA CTG GAG AAC AAC GAG AAA GCA CCG AAA GTG GTT ATC TTG AAG AAA GCG ACG GCC TAT ATT CTG AGT GTT CAG GCC GAA GAG CAG AAA TTA ATT TCC GAA GAG GAT CTG CTC CGT AAA CGC CGT GAA CAA CTG AAG CAC AAA TTA GAG CAG CTG CGG GGC AGC GGC TGC TAA TGA-3′. This sequence is inserted between the *NdeI* and *BamHI* cloning sites in the pET-3a plasmid (Novagen), and the resulting construct is compatible with *E. coli* bacterial strains possessing an inducible system for T7 polymerase. Note: Addition of 1.0 mL of sterile 50 % glycerol can be used to help bacterial growth although it was not found to affect significantly (increase or decrease) the expression yield for this construct.

9. Although we found that constitutive expression occurs for c-Myc'SH in BL21-CodonPlus *E. coli* strain, utilization of IPTG ensures an optimal expression in the costly isotopically labeled M9 minimal medium.

10. The cDNA coding for the b-HLH-LZ domain of Max (residues A^{13} to E^{94}) was amplified from pVZ1-max. The resulting sequence was cloned in the pET-3a plasmid (Novagen) between the *NdeI* and *BamHI* cloning sites without adding a stop codon. This results in a construct with a C-terminal GSGC sequence after E^{94} by alternately making use of the stop codon following the *BamHI* restriction site of the pET-3a vector. In contrast to c-Myc'SH, the high expression and purification yields obtained for Max'SH did not require codon optimization.

11. The cells can be frozen at –20 °C in the Lysis buffer (resuspend cells in 3 mL of Lysis buffer per g of cells by vortexing) until further purification. This step was found to improve cell lysis and hence purification yields.

12. Upon cell lysis, the large amount of DNA released leads to highly viscous solutions. If after sonication and DNase I treatment, the solution is still viscous, repeat the sonication step and incubate in the presence of DNase I for up to 180 min. Extensive treatment with DNase I was not found to impair the protein integrity since the c-Myc'SH construct remains insoluble at these pH conditions.
13. We obtain a more complete solubilization when using Bull cracker A without DTT at room temperature. When the pellet is well fragmented, add DTT and first incubate for 30 min at 30 °C with mild agitation (60 rpm) and then overnight at 4 °C. The mixture is then vortexed and sonicated again until complete solubilization.
14. The Bull cracker buffer (part A+B) was optimized to allow optimal binding of the c-Myc'SH construct ($pI_{c\text{-}Myc'SH}$ ~10) to the cationic exchange resin while impairing the binding of contaminants with lower pIs.
15. The U8 buffer confers a neutral pH and lightly denaturing conditions that allow elimination of the contaminants bearing a similar mass/charge ratio than c-Myc'SH at pH 5.0 but that have a lower pI.
16. The fractions containing the c-Myc'SH construct can be pooled, diluted 1:1 with FPLC buffer A, reloaded on the column, and purified with the same buffer A/B gradient, to reach a purity of ~99 %. Alternatively, this polishing FPLC step can be performed in a sodium phosphate buffer at pH 2.8 to ensure maximum binding of the basic construct to the cationic exchange resin using an elution gradient with the same buffer containing 5 M NaCl.
17. Although high protein concentrations can be reached with this procedure, note that concentrations higher than 50 mg/mL can lead to the formation of a gel phase in the Amicon at 20 °C, depending on the buffer conditions and protein concentration.
18. Although the use of PEI helps in the precipitation of free DNA and the dissociation of the Max'SH construct from DNA present in solution, this step can be skipped without drastic loss in yield. Indeed, whereas this step optimizes the loading of the construct on the cationic exchange column, DNA precipitation is alternately achieved once the pH of the supernatant drops to pH 5.0 (**step 5** of the protocol). Note that if the PEI precipitation step is used, the use of DTT should be avoided since PEI tends to precipitate in the presence of DTT along with the protein of interest. However, DTT can be added to the purified fraction before the polishing run (*see* **Note 20**).
19. Depending on the efficiency of the DNase treatment and PEI precipitation, this step can result in the precipitation of numerous contaminants that can be removed with an additional

centrifugation step of 30 min at 30,000 × g (19,000 rpm in a SS34 rotor in a Sorvall RC 5B Plus), 4 °C.

20. A polishing FPLC run can be performed after pooling of the fractions containing Max'SH and reloading on the columns and repeating the elution gradient.
21. A different buffer C composition can be used (i.e., sodium phosphate- or Tris-based buffers). However, we found that using 100 mM KCl and pH values between 5 and 7 resulted in the best reduction and heterodimerization conditions.
22. The pH of the solution should be 5. After 45 min, the oxidation is already >99 % complete.
23. It is essential that the casting glasses are absolutely clean to ensure complete polymerization and to avoid any contaminants during migration. To this effect, thoroughly clean the glasses under warm water and rinse with 70 % ethanol. Finish up the drying using Kimwipes.
24. The difference between the densities of both liquids ensures that the resolving gel polymerizes without any wave at the top, allowing for a perfect migration and uniform separation of the samples during the subsequent electrophoretic migration.
25. The composition of the gel and of the binding buffer can be modified to optimize the electrophoretic conditions and separation of the complexes. For a complete discussion on this topic, please refer to [22].
26. We found that incubation of the gels at 4 °C overnight results in sharper bands after electrophoretic migration, probably due to a more complete acrylamide polymerization.
27. The pre-run should be performed at the temperature that will be used for the electrophoretic migration of the samples. For complexes formed only by Max'SH or c-Myc'SH and DNA, pre-run and electrophoretic migration can be performed at room temperature without affecting significantly the stability of the complexes.
28. We found that when working in the μM concentration range with the c-Myc'SH or Max'SH constructs, the order in which components are added affects the results of the experiments. Hence, to obtain heterodimeric complexes bound to DNA, it is best to add the components in the following order: c-Myc'SH, Max'SH, and labeled DNA probes.
29. Electrophoresis is performed at room temperature but can also be performed at 4 °C to study complexes of higher order. This can be achieved using a specifically designed cooling unit or by simply running the electrophoresis in a cold room.
30. Although a specific imaging setup using filters can be used to detect precisely certain emission wavelength, utilization of a common UV lamp detection setup works as well.

References

1. Adhikary S, Eilers M (2005) Transcriptional regulation and transformation by Myc proteins. Nat Rev Mol Cell Biol 6:635–645
2. Eilers M, Eisenman RN (2008) Myc's broad reach. Genes Dev 22:2755–2766
3. Grandori C et al (2000) The Myc/Max/Mad network and the transcriptional control of cell behavior. Annu Rev Cell Dev Biol 16:653–699
4. Whitfield JR, Soucek L (2012) Tumor microenvironment: becoming sick of Myc. Cell Mol Life Sci 69:931–934
5. Soucek L et al (2008) Modelling Myc inhibition as a cancer therapy. Nature 455:679–683
6. Bidwell GL et al (2009) Targeting a c-Myc inhibitory polypeptide to specific intracellular compartments using cell penetrating peptides. J Control Release 135:2–10
7. Mustata G et al (2009) Discovery of novel Myc-Max heterodimer disruptors with a three-dimensional pharmacophore model. J Med Chem 52:1247–1250
8. Prochownik EV, Vogt PK (2010) Therapeutic targeting of Myc. Genes Cancer 1:650–659
9. Beaulieu M-E, Mc Duff F-O, Frappier V, Montagne M, Naud J-F, Lavigne P (2012) J Mol Recognit 25:414–426. doi:10.1002/jmr.2203
10. Lebel R et al (2007) Direct visualization of the binding of c-Myc/Max heterodimeric b-HLH-LZ to E-box sequences on the hTERT promoter. Biochemistry 46:10279–10286
11. Mcduff F-O et al (2009) The Max homodimeric b-HLH-LZ significantly interferes with the specific heterodimerization between the c-Myc and Max b-HLH-LZ in absence of DNA: a quantitative analysis. J Mol Recognit 22:261–269
12. Montagne M et al (2012) The Max b-HLH-LZ can transduce into cells and inhibit c-Myc transcriptional activities. PLoS One 7:e32172
13. Naud J-F et al (2003) Improving the thermodynamic stability of the leucine zipper of max increases the stability of its b-HLH-LZ:E-box complex. J Mol Biol 326:1577–1595
14. Sauvé S et al (2007) The mechanism of discrimination between cognate and non-specific DNA by dimeric b/HLH/LZ transcription factors. J Mol Biol 365:1163–1175
15. Sauvé S et al (2004) The NMR solution structure of a mutant of the Max b/HLH/LZ free of DNA: insights into the specific and reversible DNA binding mechanism of dimeric transcription factors. J Mol Biol 342:813–832
16. Tam JPWCRLWZJW (1991) Disulfide bond formation in peptides by dimethyl sulfoxide. Scope and applications. J Am Chem Soc 113:6657–6662
17. Chrambach A, Jovin T (1983) Selected buffer systems for moving boundary electrophoresis on gels at various pH values, presented in a simplified manner. Electrophoresis 4:190–204
18. Garner MM, Revzin A (1981) A gel electrophoresis method for quantifying the binding of proteins to specific DNA regions: application to components of the Escherichia coli lactose operon regulatory system. Nucleic Acids Res 9:3047–3060
19. Fried M, Crothers DM (1981) Equilibria and kinetics of lac repressor-operator interactions by polyacrylamide gel electrophoresis. Nucleic Acids Res 9:6505–6525
20. Sambrook J, F. EF, Maniatis T (1989) Molecular cloning, a laboratory manual, 2nd edn. Cold Spring Harbor, New York
21. Larouche K et al (1996) Optimization of competitor poly(dI-dC).poly(dI-dC) levels is advised in DNA-protein interaction studies involving enriched nuclear proteins. Biotechniques 20: 439–444
22. Fried MG (1989) Measurement of protein-DNA interaction parameters by electrophoresis mobility shift assay. Electrophoresis 10: 366–376

Chapter 3

Analyzing Myc in Cell Transformation and Evolution

Markus Hartl and Klaus Bister

Abstract

The *myc* oncogene was originally identified as a transduced allele (v-*myc*) in the genome of a highly oncogenic avian retrovirus. The protein product (Myc) of the cellular c-*myc* proto-oncogene represents the key component of a transcription factor network controlling the expression of a large fraction of all human genes. Myc regulates fundamental cellular processes like growth, metabolism, proliferation, differentiation, and apoptosis. Mutational deregulation of c-*myc* leading to increased levels of the Myc protein is a frequent event in the etiology of human cancers. In this chapter, we describe cell systems and experimental strategies to monitor and quantify the oncogenic potential of *myc* alleles and to isolate and characterize transcriptional targets of Myc that are relevant for the cell transformation process. We also describe experimental procedures to study the evolutionary origin of *myc* and to analyze structure and function of the ancestral *myc* proto-oncogenes.

Key words Oncogene, Myc, Max, Transcription factor, Cell transformation, Cancer, Evolution, Protein–DNA interactions, Target gene

1 Introduction

The *myc* oncogene was originally identified as the transforming principle (v-*myc*) in the genome of avian acute leukemia virus MC29 encoding a single hybrid protein composed of partial structural (Gag) and Myc sequences [1, 2]. The highly oncogenic v-*myc* allele is derived from the chicken c-*myc* proto-oncogene by retroviral transduction [3, 4]. Subsequently, homologues of c-*myc* were identified in all vertebrate genomes. The protein product (Myc) of c-*myc* represents the key component of a transcriptional regulator network controlling the expression of a large fraction of all human genes [5, 6]. Myc is involved in the regulation of fundamental cellular processes like growth control, metabolism, proliferation, differentiation, and apoptosis. Deregulation of c-*myc* leading to increased levels of the Myc protein is a frequent mutational event in human tumorigenesis, occurring in about 30 % of all human cancers [4, 7]. Myc is a bHLH-Zip protein containing protein dimerization domains (helix-loop-helix and leucine zipper) and an adjacent

Laura Soucek and Nicole M. Sodir (eds.), *The Myc Gene: Methods and Protocols*, Methods in Molecular Biology, vol. 1012,
DOI 10.1007/978-1-62703-429-6_3, © Springer Science+Business Media, LLC 2013

DNA contact surface (basic region). Myc forms heterodimers with the bHLH-Zip protein Max, binds to specific DNA sequence elements (E-boxes) with the preferred consensus sequence 5′-CACGTG-3′, and is part of a transcription factor network including additional bHLH-Zip proteins [5, 6, 8]. Myc–Max heterodimers are involved in transcriptional activation of distinct target genes, but Myc has also been implicated in transcriptional repression. Several genes activated by Myc are related to processes of cell growth and metabolism, including protein synthesis, ribosomal biogenesis, glycolysis, mitochondrial function, and cell cycle progression [6, 9]. Most of the genes repressed by Myc are involved in cell cycle arrest, cell adhesion, and cell-to-cell communication [6]. Invertebrate orthologues of Myc and Max proteins have been identified in the triploblastic bilaterian organism *Drosophila melanogaster* [10]. *Drosophila* dMyc and dMax bind to a large number of E-boxes in the *Drosophila* genome and regulate the expression of many genes including key regulators of cell growth, cell size, and ribosome biogenesis [11]. Recently, ancestral forms of *myc* and *max* genes have been identified in the early diploblastic cnidarian *Hydra*, the most primitive metazoan organism employed so far for the structural, functional, and evolutionary analysis of these genes [12]. The principal design and the basic biochemical properties of the *Hydra* Myc and Max proteins are very similar to those of their vertebrate derivatives, suggesting that the principal functions of the Myc master regulator arose very early in metazoan evolution, at least 600 million years ago.

Here, we describe the essential methods to monitor and quantify the oncogenic potential of *myc* alleles in an avian fibroblast cell transformation system and to use that system to isolate and characterize target genes that are transcriptionally regulated by the Myc protein. Furthermore, we refer to methods to detect and analyze *myc* genes and their expression in evolutionary simple metazoan organisms.

2 Materials

All buffers, media, and solutions used for the procedures described in Subheading 3 (see below) are listed in alphabetical order.

1. 30 % (w/v) Acrylamide/0.8 % (w/v) bisacrylamide. Dissolve 30 g acrylamide and 0.8 g bisacrylamide in H_2O to 100 ml; filtrate (0.45 μm); store light-protected at 4 °C.
2. 1 % (w/v) Agarose/formaldehyde gel: 1 % (w/v) agarose, 6.57 % (v/v) formaldehyde, 1× RNA gel running buffer. Suspend 2 g agarose in 144 ml H_2O and boil in a microwave oven; cool to 65 °C; add 20 ml 10× RNA gel running buffer and 36 ml formaldehyde solution (36.5 %) warmed to 65 °C; mix and pour the gel (200 ml) in a chemical hood.

3. 10× Annealing buffer: 100 mM Tris–HCl pH 8.0, 50 mM $MgCl_2$, 650 mM KCl, 5 mM EDTA, 10 mM DTT. Mix 180 µl H_2O, 100 µl 1 M Tris–HCl pH 8.0, 50 µl 1 M $MgCl_2$, 650 µl 1 M KCl, 10 µl 0.5 M EDTA, and 10 µl 1 M DTT solution; store at −20 °C.
4. 1 mg/ml Aprotinin. Dissolve 5 mg aprotinin from bovine lung in 10 mM HEPES pH 7.9 to a final volume of 5 ml; store in 1-ml aliquots at −20 °C.
5. 10 % (w/v) Ammonium peroxydisulfate (APS). Dissolve 1 g APS in H_2O to 10 ml; store in 1-ml aliquots at −20 °C.
6. Avian cell culture medium: 1× Ham's F10 with glutamine, 0.295 % (w/v) TPB, 5 % (v/v) calf serum, 2 % (v/v) chicken serum, 0.188 % (w/v) $NaHCO_3$, 0.5 % (v/v) DMSO, 1× antibiotic-antimycotic. Mix 345 ml H_2O, 50 ml 10× Ham's F10 with glutamine, 50 ml 2.95 % (w/v) TPB, 25 ml calf serum, 10 ml chicken serum (heat-inactivated at 56 °C for 30 min), 12.5 ml 7.5 % (w/v) $NaHCO_3$, 2.5 ml DMSO (dimethyl sulfoxide), and 5 ml 100× antibiotic-antimycotic; store at 4 °C.
7. 1.8 % (w/v) Bacto agar. Suspend 1.8 g bacto agar in 100 ml H_2O; autoclave; store at 4 °C.
8. Boiling buffer: 10 mM sodium phosphate pH 7.2, 0.5 % (w/v) SDS (sodium dodecyl sulfate), 2 µg/ml aprotinin. Mix 9.65 ml H_2O, 100 µl 1 M sodium phosphate pH 7.2, and 250 µl 20 % (w/v) SDS; store at RT; prior to use add 1/500 vol 1 mg/ml aprotinin.
9. Buffered F10 medium: 1× Ham's F10 with glutamine, 0.188 % (w/v) $NaHCO_3$, 25 mM HEPES pH 7.3. Mix 425 ml H_2O, 50 ml 10x Ham's F10 with glutamine, 12.5 ml 7.5 % (w/v) $NaHCO_3$, and 12.5 ml 1 M HEPES pH 7.3; store at 4 °C.
10. 2.5 M calcium chloride ($CaCl_2$): Dissolve 18.38 g $CaCl_2 \times 2$ H_2O in H_2O to 50 ml; filter sterilize (0.22 µm); store at 4 °C.
11. 1× Cell culture lysis reagent (supplied by the Luciferase Assay System E1500 (Promega)): 25 mM Tris phosphate pH 7.8, 2 mM DTT, 2 mM DCTA (trans-1,2-diaminocyclohexane-N,*N*,*N*′,*N*′-tetraacetic acid), 10 % (v/v) glycerol, 1 % (v/v) Triton X-100 [4-(1,1,3,3-tetramethylbutyl)phenyl-polyethylene glycol]. Store at −20 °C.
12. ChIP-dilution buffer: 0.01 % (w/v) SDS, 1.1 % (v/v) Triton X-100, 1.2 mM EDTA, 16.7 mM Tris–HCl pH 8.0, 167 mM NaCl, 1 mM PMSF (phenylmethylsulfonyl fluoride), 2 µg/µl aprotinin, 1 µg/ml leupeptin, 1 µg/ml pepstatin A. Mix 167.4 ml H_2O, 100 µl 20 % (w/v) SDS, 22 ml 10 % (v/v) Triton X-100, 480 µl 0.5 M EDTA, 3.34 ml 1 M Tris–HCl pH 8.0, and 6.68 ml 5 M NaCl; filter sterilize (0.22 µm); store at 4 °C;

prior to use add 1/100 vol PMSF, 1/500 vol aprotinin, 1/1,000 vol leupeptin, 1/1,000 vol pepstatin A.

13. ChIP-protein A-bead slurry: 0.005 % (w/v) SDS, 0.55 % (v/v) Triton X-100, 0.6 mM EDTA, 8.35 mM Tris–HCl pH 8.0, 83.5 mM NaCl, 0.2 μg/μl salmon sperm DNA, 0.5 μg/μl BSA, 50 % (v/v) protein A sepharose CL-4B. Suspend 200 mg protein A sepharose CL-4B beads in 10 ml H_2O and incubate on ice for 30 min; wash 3× with each 10 ml H_2O (during the washing the beads should swell on ice for 30 min), then 1× with 2 ml of ChIP-dilution buffer; to the swollen pellet (~1 ml) add 1 ml of ChIP-dilution buffer, 40 μl of 10 mg/ml sonicated salmon sperm DNA, and 100 μl of 10 mg/ml BSA; store at 4 °C.
14. ChIP-elution buffer: 1 % (w/v) SDS, 0.1 M $NaHCO_3$. Mix 42.5 ml H_2O, 2.5 ml 20 % (w/v) SDS, and 5 ml 1 M $NaHCO_3$; filter sterilize (0.22 μm); store at RT.
15. ChIP-low salt washing buffer: 0.1 % (w/v) SDS, 1 % (v/v) Triton X-100, 2 mM EDTA, 20 mM Tris–HCl pH 8.0, 150 mM NaCl. Mix 168.2 ml H_2O, 1 ml 20 % (w/v) SDS, 20 ml 10 % (v/v) Triton X-100, 800 μl 0.5 M EDTA, 4 ml 1 M Tris–HCl pH 8.0, and 6 ml 5 M NaCl; filter sterilize (0.22 μm); store at 4 °C.
16. ChIP high-salt washing buffer: 0.1 % (w/v) SDS, 1 % (v/v) Triton X-100, 2 mM EDTA, 20 mM Tris–HCl pH 8.0, 500 mM NaCl. Mix 154.2 ml H_2O, 1 ml 20 % (w/v) SDS, 20 ml 10 % (v/v) Triton X-100, 800 μl 0.5 M EDTA, 4 ml Tris–HCl pH 8.0, and 20 ml 5 M NaCl; filter sterilize (0.22 μm); store at 4 °C.
17. ChIP-LiCl washing buffer: 0.25 M LiCl, 1 % (w/v) IGEPAL CA-630 (octylphenyl-polyethylene glycol), 1 % (w/v) sodium deoxycholate, 1 mM EDTA, 10 mM Tris–HCl pH 8.0. Mix 147.6 ml H_2O, 10 ml 5 M LiCl, 20 ml 10 % (v/v) IGEPAL CA-630, 20 ml 10 % (w/v) sodium deoxycholate, 400 μl 0.5 M EDTA, and 2 ml 1 M Tris–HCl pH 8.0; filter sterilize (0.22 μm); store at 4 °C.
18. Cloning bottom agarose: 1× Ham's F10 with glutamine, 0.295 % (w/v) Tryptose Phosphate Broth (TPB), 3.5 % (v/v) calf serum, 1.25 % (v/v) chicken serum, 0.188 % (w/v) $NaHCO_3$, 1.25 % (v/v) DMSO, 1× glutamine, 1× antibiotic-antimycotic, 0.625 % (w/v) sea plaque agarose. Heat 1 vol 1.5 % (w/v) sea plaque agarose in a microwave oven; cool to 45 °C in a water bath; mix with 45 °C warm 1.4 vol DC3 medium; keep at 45 °C in a water bath.
19. Cloning medium: 1.25× Ham's F10 with glutamine, 0.369 % (w/v) TPB, 12.5 % (v/v) calf serum, 5.75 % (v/v) chicken serum, 0.25 % (v/v) DMSO, 1.875× vitamins solution, 1.875× folic

acid solution, 1.25× glutamine, 1.25× antibiotic-antimycotic. Mix 201 ml H_2O, 50 ml 10× Ham's F10 with glutamine, 50 ml 2.95 % (w/v) TPB, 50 ml calf serum, 23 ml chicken serum (heat-inactivated at 56 °C for 30 min), 1 ml DMSO, 7.5 ml 100× vitamins solution, 7.5 ml 100× folic acid solution, 5 ml 100× L-glutamine (200 mM), and 5 ml 100× antibiotic-antimycotic; store at 4 °C.

20. Cloning top agarose: 1× Ham's F10 with glutamine, 0.295 % (w/v) TPB, 10 % (v/v) calf serum, 4.6 % (v/v) chicken serum, 0.2 % (v/v) DMSO, 1.5× vitamins solution, 1.5× folic acid solution, 1× glutamine, 1× antibiotic-antimycotic, 0.3 % (w/v) sea plaque agarose. Heat 1 vol 1.5 % (w/v) sea plaque agarose in a microwave oven; cool to 45 °C in a water bath; mix with 45 °C warm 4 vol cloning medium; keep at 45 °C in a water bath.
21. CsCl buffer ($d = 1.71$ g/cm^3): 5.7 M CsCl (cesium chloride), 25 mM sodium acetate pH 6.0. Dissolve 95.97 g CsCl in H_2O to 99.17 ml; add 0.83 ml 3 M sodium acetate pH 6.0; filter sterilize (0.22 μm); autoclave; store at RT; check the density (d) by determination of the refractory index (n) using a refractometer : $d = (10.8601 \times n) - 13.4974$.
22. DC3 medium: 1.714× Ham's F10 with glutamine, 0.506 % (w/v) TPB, 6.02 % (v/v) calf serum, 2.15 % (v/v) chicken serum, 0.323 % (v/v) $NaHCO_3$, 2.15 % (v/v) DMSO, 1.72× glutamine, 1.72× antibiotic-antimycotic. Mix 238.3 ml H_2O, 85.7 ml 10× Ham's F10 with glutamine, 85.7 ml 2.95 % (w/v) TPB, 30.1 ml calf serum, 10.75 ml chicken serum (heat-inactivated at 56 °C for 30 min), 21.5 ml 7.5 % (w/v) $NaHCO_3$, 10.75 ml DMSO, 8.6 ml 100× L-glutamine (200 mM), and 8.6 ml 100× antibiotic-antimycotic; store at 4 °C.
23. 50× Denhardt's reagent: 1 % (w/v) BSA, 1 % (w/v) PVP, 1 % (w/v) Ficoll 400. Dissolve 5 g BSA (bovine serum albumin), 5 g PVP (polyvinylpyrrolidone), and 5 g Ficoll 400 in H_2O to 500 ml; store in 50-ml aliquots at −20 °C.
24. Dilution buffer: 10 mM sodium phosphate pH 7.2, 187.5 mM NaCl, 1.25 % (v/v) IGEPAL CA-630, 1.25 % (w/v) sodium deoxycholate, 2 μg/ml aprotinin. Mix 81.5 ml H_2O, 1 ml 1 M sodium phosphate pH 7.2, 3.75 ml 5 M NaCl, 1.25 ml IGEPAL CA-630, and 12.5 ml 10 % (w/v) sodium deoxycholate; store at RT; prior to use add 1/500 vol 1 mg/ml aprotinin.
25. DMEM medium (−Met): DMEM (1×, w/o: sodium pyruvate, L-Gln, L-Met, L-Cys), 5 % (v/v) fetal bovine serum, 1× glutamine, 1× sodium pyruvate, 1× cysteine. Mix 46 ml DMEM (1× w/o: sodium pyruvate, L-Gln, L-Met, L-Cys), 2.5 ml dialyzed fetal bovine serum, 0.5 ml 100× glutamine (200 mM), 0.5 ml 100× sodium pyruvate (100 mM), and 0.5 ml 100× cysteine (50 mM); store at 4 °C.

26. 1 M DTT solution: 1 M DTT, 10 mM sodium acetate pH 5.2. Dissolve 1.545 g DTT (dithiothreitol) in H_2O to 9.97 ml, add 33 μl 3 M sodium acetate pH 5.2; filter sterilize (0.22 μm); store in 1-ml aliquots at −20 °C.
27. 0.5 M EDTA. Dissolve 93.06 g EDTA (ethylenediaminetetraacetic acid) ×2 H_2O disodium salt (Titriplex III) in 400 ml H_2O; adjust pH to 8.0 with ~10 g NaOH; add H_2O to 500 ml; autoclave; store at RT.
28. 5× EMSA buffer: 50 mM Tris–HCl pH 7.5, 2.5 mM EDTA, 325 mM KCl, 25 mM $MgCl_2$, 5 mM DTT, 0.5 μg/μl BSA, 50 % (v/v) glycerol. Mix 40 μl H_2O, 50 μl 1 M Tris–HCl pH 7.5, 5 μl 0.5 M EDTA, 325 μl 1 M KCl, 25 μl 1 M $MgCl_2$, 5 μl 1 M DTT solution, 50 μl 10 mg/ml BSA, and 500 μl glycerol; store at −20 °C.
29. 1 mg/ml Ethidium bromide. Dissolve 10 mg ethidium bromide in H_2O to 10 ml; store light protected at RT.
30. Focus agar: 1× Ham's F10 with glutamine, 0.295 % (w/v) TPB, 3.5 % (v/v) calf serum, 1.25 % (v/v) chicken serum, 0.188 % (v/v) $NaHCO_3$, 1.25 % (v/v) DMSO, 1× glutamine, 1× antibiotic-antimycotic, 0.75 % (w/v) bacto agar. Heat 1 vol 1.8 % (w/v) bacto agar in a microwave oven; cool to 45 °C in a water bath; mix with 45 °C warm 1.4 vol DC3 medium; keep at 45 °C in a water bath.
31. 100× Folic acid solution: 1.81 mM folic acid, 1 M $NaHCO_3$. Dissolve 8.4 g $NaHCO_3$ in 50 ml H_2O; use this solution to dissolve 80 mg folic acid; add H_2O to 100 ml; filter sterilize (0.22 μm); store in 10-ml aliquots at −20 °C.
32. Giemsa staining solution. Mix 1 vol Giemsa solution (azure-eosin-methylene blue) and 9 vol H_2O heated to 80 °C; store at RT.
33. GITC buffer: 4 M guanidine thiocyanate, 25 mM sodium acetate pH 6.0, 0.835 % (v/v) 2-mercaptoethanol. Dissolve 94.53 g guanidine thiocyanate in H_2O to 196.66 ml; add 1.67 ml 3 M sodium acetate pH 6.0; store at RT; prior to use add 1.67 ml 2-mercaptoethanol.
34. 1.25 M Glycine. Dissolve 4.69 g glycine in H_2O to 50 ml; filter sterilize (0.22 μm); store at 4 °C.
35. Growth medium: 1× Ham's F10 with glutamine, 10 % (v/v) calf serum, 0.188 % (w/v) $NaHCO_3$, 25 mM HEPES pH 7.3, 1× antibiotic-antimycotic. Mix 370 ml H_2O, 50 ml 10× Ham's F10 with glutamine, 50 ml calf serum, 12.5 ml 7.5 % (w/v) $NaHCO_3$, 12.5 ml 1 M HEPES pH 7.3, and 5 ml 100× antibiotic-antimycotic; store at 4 °C.
36. 2× HBS (HEPES buffered saline): 280 mM NaCl, 10 mM KCl, 1.5 mM Na_2HPO_4, 12 mM glucose, 50 mM HEPES

pH 7.05. Dissolve 6.55 g NaCl, 0.296 g KCl, 0.106 g $Na_2HPO_4 \times 2\ H_2O$, 1.19 g glucose $\times H_2O$, and 4.77 g HEPES (*N*-2-hydroxyethylpiperazine-*N'*-2-ethanesulfonic acid) in 360 ml H_2O; adjust to pH 7.05 with 0.5 M NaOH; add H_2O to 400 ml; filter sterilize (0.22 μm); store in 50-ml aliquots at −20 °C.

37. 1 M HEPES pH 7.3 or 7.9. Dissolve 23.83 g HEPES (*N*-2-hydroxyethylpiperazine-*N'*-2-ethanesulfonic acid) in 90 ml H_2O; adjust to pH 7.3 or 7.9 with 10 M NaOH; add H_2O to 100 ml; filter sterilize (0.22 μm); store at -20 °C.
38. Hybridization solution: 50 % (v/v) formamide, 6× SSC, 5× Denhardt's reagent, 5 mM EDTA, 0.2 % (w/v) SDS, 100 μg/ml salmon sperm DNA. Mix 35 ml H_2O, 250 ml formamide, 150 ml 20× SSC, 50 ml 50× Denhardt's reagent, 5 ml 0.5 M EDTA, and 5 ml 20 % (w/v) SDS; adjust to pH 7.4 with concentrated HCl; store light-protected at RT; prior to use boil 1/100 vol salmon sperm DNA (10 mg/ml) for 5 min; chill on ice; add to the hybridization solution.
39. 1 M KCl. Dissolve 18.64 g KCl (potassium chloride) in H_2O to 250 ml; autoclave; store at RT.
40. 1 mg/ml Leupeptin. Dissolve 5 mg leupeptin trifluoroacetate salt in H_2O to 5 ml; store in 1-ml aliquots at −20 °C.
41. 5 M LiCl. Dissolve 21.2 g LiCl (lithium chloride) in H_2O to 100 ml; autoclave; store at RT.
42. Luciferase Assay Reagent. Contains beetle luciferin and is supplied by the Luciferase Assay System E1500 (Promega); store in 1-ml aliquots at −80 °C.
43. 1 M $MgCl_2$. Dissolve 20.33 g $MgCl_2 \times 6\ H_2O$ in H_2O to 100 ml; autoclave; store at RT.
44. 5 M NaCl. Dissolve 146.1 g NaCl in H_2O to 500 ml; autoclave; store at RT.
45. 1 M $NaHCO_3$. Dissolve 4.20 g $NaHCO_3$ in H_2O to 50 ml; store at RT.
46. 1 M NaH_2PO_4. Dissolve 27.6 g $NaH_2PO_4 \times H_2O$ in H_2O to 200 ml; autoclave; store at RT.
47. 1 M Na_2HPO_4. Dissolve 35.6 g $Na_2HPO_4 \times 2\ H_2O$ in H_2O to 200 ml; autoclave; store at RT.
48. 2 M NaOH. Dissolve 8 g NaOH pellets in H_2O to 100 ml; store at RT.
49. 0.1 M NaOH, 5 mM EDTA. Mix 49 ml H_2O, 0.5 ml 10 M NaOH, and 0.5 ml 0.5 M EDTA; store at RT.
50. PBS (phosphate-buffered saline): 10 mM sodium phosphate pH 7.2, 150 mM NaCl. Mix 960 ml H_2O, 10 ml 1 M sodium phosphate pH 7.2, and 30 ml 5 M NaCl; autoclave; store at 4 °C.

51. PBS-PI: 10 mM sodium phosphate pH 7.2, 150 mM NaCl, 1 mM PMSF, 2 μg/μl aprotinin, 1 μg/μl leupeptin, 1 μg/ml pepstatin A. To 100 ml PBS add 1 ml 0.1 M PMSF, 200 μl aprotinin 1 mg/ml, 100 μl leupeptin 1 mg/ml, and 100 μl pepstatin A; store at 4 °C.
52. 1 mg/ml Pepstatin. Dissolve 5 mg pepstatin A in DMSO to 5 ml; store in 1-ml aliquots at −20 °C.
53. PGM (primary growth medium): 1× Ham's F10 with glutamine, 0.295 % (w/v) TPB, 8 % (v/v) calf serum, 2 % (v/v) chicken serum, 0.188 % (w/v) $NaHCO_3$, 1× antibiotic-antimycotic. Mix 332.5 ml H_2O, 50 ml 10× Ham's F10 with glutamine, 50 ml 2.95 % (w/v) TPB, 40 ml calf serum, 10 ml chicken serum (heat-inactivated at 56 °C for 30 min), 12.5 ml 7.5 % (w/v) $NaHCO_3$, and 5 ml 100× antibiotic-antimycotic; store at 4 °C.
54. 0.1 M PMSF. Dissolve 174 mg PMSF in isopropanol to 10 ml; store in 1-ml aliquots at −20 °C.
55. Proteinase K solution: 10 mg/ml proteinase K. Dissolve 100 mg proteinase K (10 mg/ml) in H_2O to 10 ml; store in 1-ml aliquots at −20 °C.
56. RIPA buffer: 10 mM sodium phosphate pH 7.2, 150 mM NaCl, 1 % (v/v) IGEPAL CA-630, 1 % (w/v) sodium deoxycholate, 0.1 % (w/v) SDS, 2 μg/ml aprotinin. Mix 422.5 ml H_2O, 5 ml 1 M sodium phosphate pH 7.2, 15 ml 5 M NaCl, 5 ml IGEPAL CA-630, 50 ml 10 % (w/v) sodium deoxycholate, and 2.5 ml 20 % (w/v) SDS; filter sterilize (0.22 μm); store at 4 °C; prior to use add 1/500 vol aprotinin 1 mg/ml.
57. RIPA-BSA: 10 mM sodium phosphate pH 7.2, 150 mM NaCl, 1 % (v/v) IGEPAL CA-630, 1 % (w/v) sodium deoxycholate, 0.1 % (w/v) SDS, 2 μg/ml aprotinin, 2 mg/ml BSA. Dissolve 100 mg BSA in RIPA buffer to 50 ml; store at 4 °C; prior to use add 1/500 vol aprotinin 1 mg/ml.
58. RIPA-protein A-bead slurry: 10 mM sodium phosphate pH 7.2, 150 mM NaCl, 1 % (v/v) IGEPAL CA-630, 1 % (w/v) sodium deoxycholate, 0.1 % (w/v) SDS, 10 % (v/v) protein A sepharose CL-4B. Wash 100 mg protein A sepharose CL-4B beads 3× with each 20 ml H_2O (during the washing the beads should swell on ice for 30 min); determine the volume of the swollen beads; add 9 vol of RIPA buffer; store at 4 °C.
59. RIPA-sucrose: 10 mM sodium phosphate pH 7.2, 150 mM NaCl, 1 % (v/v) IGEPAL CA-630, 1 % (w/v) sodium deoxycholate, 0.1 % (w/v) SDS, 2 μg/ml aprotinin, 10 % (w/v) sucrose. Dissolve 10 g sucrose in RIPA buffer to 100 ml; filter sterilize (0.22 μm); store at 4 °C; prior to use add 1/500 vol aprotinin 1 mg/ml.

60. RNA buffer A: 1 M NaCl, 40 mM Tris–HCl pH 7.5, 1 mM EDTA, 0.1 % (w/v) SDS. Mix 150.6 ml H_2O, 40 ml 5 M NaCl, 8 ml 1 M Tris–HCl pH 7.5, and 400 μl 0.5 M EDTA; autoclave; add 1 ml 20 % (w/v) SDS; store at RT.
61. RNA buffer B: 0.1 M NaCl, 20 mM Tris–HCl pH 7.5, 1 mM EDTA, 0.1 % (w/v) SDS. Mix 190.6 ml H_2O, 4 ml 5 M NaCl, 4 ml 1 M Tris–HCl pH 7.5, and 400 μl 0.5 M EDTA; autoclave; add 1 ml 20 % (w/v) SDS; store at RT.
62. RNA buffer C: 10 mM Tris–HCl pH 7.5, 1 mM EDTA, 0.05 % (w/v) SDS. Mix 197.1 H_2O, 2 ml 1 M Tris–HCl pH 7.5, and 400 μl 0.5 M EDTA; autoclave; add 0.5 ml 20 % (w/v) SDS; store at RT.
63. 10× RNA gel running buffer: 200 mM MOPS [3-(*N*-morpholino)-propanesulfonic acid], 80 mM sodium acetate, 10 mM EDTA. Dissolve 41.2 g MOPS in ~800 ml H_2O; add 26.6 ml 3 M sodium acetate pH 5.2, and 20 ml 0.5 M EDTA; adjust to pH 7.0 with 2 M NaOH; add H_2O to 1,000 ml; store light-protected at 4 °C.
64. RNA sample buffer: 6 % (v/v) formaldehyde, 70 % (v/v) formamide, 0.03 % (w/v) bromophenol blue, 1× RNA gel running buffer, 1 % (w/v) Ficoll 400. Dissolve 0.1 g Ficoll 400 in 1 ml 10× RNA gel running buffer, 0.3 ml 1 % (w/v) bromophenol blue, 1.65 ml formaldehyde solution (36.5 %), and 7 ml formamide; store in 1-ml aliquots at −20 °C.
65. 20 % (w/v) SDS (sodium dodecyl sulfate). Dissolve 20 g SDS in H_2O to 100 ml; store at RT.
66. 1× SDS gel-loading buffer: 60 mM Tris–HCl pH 6.8, 3 % (w/v) SDS, 10 % (v/v) glycerol, 0.005 % (w/v) bromophenol blue, 5 % (v/v) 2-mercaptoethanol. Mix 6.35 ml H_2O, 0.6 ml 1 M Tris–HCl pH 6.8, 1.5 ml 20 % (w/v) SDS, 1 ml glycerol, and 50 μl 1 % (w/v) bromophenol blue; store at RT; prior to use add 1/20 vol 2-mercaptoethanol.
67. SDS lysis buffer: 1 % (w/v) SDS, 10 mM EDTA, 50 mM Tris–HCl pH 8.0, 1 mM PMSF, 2 μg/μl aprotinin, 1 μg/ml leupeptin, 1 μg/ml pepstatin A. Mix 44 ml H_2O, 2.5 ml 20 % (w/v) SDS, 1 ml 0.5 M EDTA, and 2.5 ml 1 M Tris–HCl pH 8.0; filter sterilize (0.22 μm); store at RT; prior to use add 1/100 vol 0.1 M PMSF, 1/500 vol aprotinin 1 mg/ml, 1/1,000 vol leupeptin 1 mg/ml, and 1/1,000 vol pepstatin A 1 mg/ml.
68. 1.5 % (w/v) Sea plaque agarose. Suspend 1.5 g sea plaque agarose in 100 ml H_2O; autoclave; store at 4 °C.
69. Shock solution: 15 % (v/v) glycerol in buffered F10 medium. Mix 85 ml buffered F10 medium and 15 ml glycerol; prepare fresh; prior to use warm to 37 °C.

70. 3 M Sodium acetate pH 5.2 or 6.0. Dissolve 24.61 g sodium acetate in 80 ml H_2O; adjust to pH 5.2 or 6.0 with acetic acid; add H_2O to 100 ml; autoclave; store at RT.
71. 1 M Sodium phosphate pH 7.2. Mix 136.8 ml 1 M Na_2HPO_4 and 63.2 ml 1 M NaH_2PO_4; check pH; autoclave; store at RT.
72. 20× SSC: 3 M NaCl, 0.3 M sodium citrate. Dissolve 175 g NaCl and 88 g sodium citrate in H_2O to 1,000 ml; autoclave; store at RT.
73. Stripping solution: 5 mM Tris–HCl pH 8.0, 0.1× Denhardt's reagent, 2 mM EDTA. Mix 989 ml H_2O, 5 ml 1 M Tris–HCl pH 8.0, 2 ml 50× Denhardt's reagent, and 4 ml 0.5 M EDTA; store at 4 °C.
74. 10× T4 PNK buffer: 700 mM Tris–HCl pH 7.5, 100 mM $MgCl_2$, 50 mM DTT. Mix 150 μl H_2O, 700 μl 1 M Tris–HCl pH 7.5, 100 μl 1 M $MgCl_2$, and 50 μl 1 M DTT solution; store at −20 °C.
75. 10× TBE (Tris/Borate/EDTA) buffer: 890 mM Tris, 890 mM boric acid, 20 mM EDTA. Dissolve 108 g Tris base and 55 g boric acid in H_2O to 960 ml; add 40 ml 0.5 M EDTA; autoclave; store at RT.
76. TE buffer: 10 mM Tris–HCl pH 8.0, 1 mM EDTA. Mix 988 ml H_2O, 10 ml 1 M Tris–HCl pH 8.0, and 2 ml 0.5 M EDTA; autoclave; store at RT.
77. 2.95 % (w/v) TPB (Tryptose Phosphate Broth): Dissolve 29.5 g TPB in H_2O to 1,000 ml; autoclave; store at 4 °C.
78. Transfer buffer: 10 % (v/v) calf serum in TS (Tris saline) buffer. Mix 10 ml calf serum and 90 ml TS buffer; store at 4 °C.
79. 1 M Tris–HCl pH 6.8 or 7.5 or 8.0. Dissolve 60.57 g Tris base in 400 ml H_2O; adjust to pH 6.8 or 7.5 or 8.0 with concentrated HCl; add H_2O to 500 ml; autoclave; store at RT.
80. Trypsin–EDTA solution: 1× trypsin–EDTA (0.5 % trypsin, 5.3 mM EDTA) in TS buffer. Mix 5 ml 10× trypsin–EDTA and 45 ml TS buffer; store at 4 °C.
81. TR solution: 5× trypsin–EDTA in TS buffer. Mix 7.5 ml 10× trypsin–EDTA and 7.5 ml TS buffer; store at 4 °C.
82. TS (Tris saline) buffer: 150 mM NaCl, 5 mM KCl, 25 mM Tris–HCl pH 7.5, 5 mM glucose. Dissolve 0.99 g glucose × H_2O in ~200 ml H_2O, add 30 ml 5 M NaCl, 5 ml 1 M KCl, 25 ml 1 M Tris–HCl 7.5, and H_2O to 1,000 ml; autoclave; store at 4 °C.
83. Washing solution 2× SSC: 2× SSC, 0.1 % (w/v) SDS, 1 mM EDTA. Mix 893 ml H_2O, 100 ml 20× SSC, 5 ml 20 % (w/v) SDS, and 2 ml 0.5 M EDTA; store at RT.
84. Washing solution 0.2× SSC: 0.2× SSC, 0.1 % (w/v) SDS, 1 mM EDTA. Mix 983 ml H_2O, 10 ml 20× SSC, 5 ml 20 % (w/v) SDS, and 2 ml 0.5 M EDTA; store at RT.

3 Methods

3.1 Preparation and Cultivation of Primary Quail Embryo Fibroblasts (QEF)

1. Incubate fertilized quail eggs for 9 days at 37.8 °C in an egg incubator with a water-saturated atmosphere (>60 % humidity) and automatic turning device.
2. Clean the egg with ethanol and open at the shallow site using a spatula.
3. Remove the broken shell with a sterile forceps, then the membrane under the egg shell with another sterile forceps.
4. Gently pull out the embryo with closed forceps under its neck.
5. Put the embryo onto a petri dish. Remove head, arms, and legs and transfer the body into a 50 ml tube.
6. Wash the body with 10 ml *TS buffer* (37 °C), shake briefly, decant the supernatant, and thoroughly homogenize the body with the round end of a spatula for 5 min.
7. Add 10 ml *TR solution* (37 °C), shake gently, incubate for 5 min at room temperature (RT), and then carefully pipette 5× up and down using a 10-ml wide-mouth pipette.
8. Let clumps settle down for 5 min and then decant the supernatant into a prepared 50-ml tube containing 10 ml ice-cold *PGM* (*see* **Note 1**).
9. Repeat this washing step with following solutions kept at 37 °C: 5 ml *TS buffer*, 5 ml *TS buffer*, 5 ml *TR solution*, 5 ml *TS buffer*, 5 ml *TS buffer*.
10. Finally, mix cell suspension by inverting the 50-ml tube with the cells and centrifuge at 150 × g for 20 min at 4 °C.
11. Decant the supernatant, add 20 ml *PGM* to the pellet, and pipette 25× up and down with a wide-mouth pipette.
12. Let the clumps settle down for 5 min and transfer the supernatant into a fresh 50-ml tube.
13. From this supernatant, remove 0.5 ml and dilute to 5 ml with *TS buffer* for determination of the cell number.
14. Rinse a hemocytometer and its cover slip with ethanol, then put the cover slip onto the chamber.
15. Pipette one drop of the diluted cell suspension at the edge of the cover slip and put the device onto a microscope.
16. Count the cells except the dark and smooth blood cells in the four fields encompassing 16 squares (0.1 mm^3) each and determine the average cell number of one field [calculation of the cell number/ml: *N* (average number of cells in one field) × 10 (dilution factor) × 10^4]. The total number of cells should be in the range of 10^8 for a 9 day-old embryo.
17. Plate 6×10^6 cells per 100-mm dish containing 10 ml *PGM*.

18. Incubate cells at 37 °C in a 5 % CO_2 humidified atmosphere and replace the medium after 3 days with 12 ml *PGM* per dish.
19. Culture the cells at 37 °C (5 % CO_2) in a water-saturated atmosphere.
20. For transfer, wash the cells of one dish with 5 ml *TS buffer*, aspirate, add 2 ml *trypsin–EDTA solution*, and aspirate.
21. Incubate at 37 °C for 5–10 min (*see* **Note 2**), then suspend the cells in 2 ml of *transfer buffer* and transfer into a conical 15-ml tube.
22. Wash the dish with 3 ml of culture medium and add to the cell suspension.
23. From the final volume of 5 ml, dilute 0.1 ml with 9.9 ml of ISOTON™ II Diluent.
24. Count the cells using a Coulter counter; determine the number of cells/ml and the total cell number of the trypsinized dish.
25. Calculate the volume of cell suspension ($3–6 \times 10^6$) to be seeded into 100-mm dishes containing 10 ml of *avian cell culture medium*.
26. Change the medium after 2–3 days.

3.2 Infection of QEF with myc-Carrying Retroviruses

1. Calculate the amount of QEFs needed and remove medium from an appropriate number of dishes.
2. Wash the cells of each dish with 5 ml *TS buffer*, aspirate, add 2 ml *trypsin–EDTA solution*, and aspirate.
3. Incubate at 37 °C for 5–10 min, then suspend the cells in 2 ml *transfer buffer* per dish, and transfer into a conical tube.
4. Wash each dish with 3 ml of *growth medium* and add to the cell suspension.
5. Dilute 0.1 ml of the cell suspension with 9.9 ml of ISOTON® II Diluent.
6. Count the cells using a Coulter counter and determine the number of cells/ml.
7. Dilute the cell suspension with *growth medium* to a concentration of 3×10^6 cells/ml.
8. Seed 1 ml of this suspension into a 100-mm dish containing 10 ml of *growth medium* and incubate at 37 °C (5 % CO_2) for 30 min.
9. Meanwhile, quickly thaw a viral MC29 stock in a water bath at 37 °C (*see* **Note 3**).
10. Add 1 ml of the viral stock to the cells and incubate over night at 37 °C (5 % CO_2).
11. The next day aspirate the culture medium from the cells and replace with 10 ml of *avian cell culture medium*.

12. Cultivate the cells for 2 weeks to allow a complete spread of the virus.
13. To collect MC29 virus, remove the 2–3 days-old culture medium from confluent grown cells and freeze at −80 °C.

3.3 DNA Transfection of myc-Carrying Retroviral Vector Plasmids

1. Prepare a cell suspension from avian fibroblasts as described above.
2. Dilute the cell suspension with *growth medium* to a concentration of 1.25×10^6 cells/ml and seed each 1 ml of the cell suspension into 60-mm dishes containing 4 ml of *growth medium* (*see* **Note 4**).
3. Incubate over night at 37 °C (5 % CO_2). The cells should be almost confluent the next day.
4. Before transfection replace the medium with fresh *growth medium* and incubate at 37 °C for 1 h (5 % CO_2).
5. To prepare the transfection cocktail, add 50 μl of *2.5 M* $CaCl_2$ to 450 μl of H_2O containing 5 μg of cesium chloride-purified plasmid DNA. Drop this mixture into a conical tube containing 0.5 ml of *2× HBS*. Vortex at medium speed for 2 s.
6. After the solution has become turbid due to calcium phosphate precipitation, slowly drop the transfection mixture into the medium of the seeded cells.
7. Gently swirl the dishes and incubate at 37 °C for 4 h.
8. Remove the medium and add 2 ml of 37 °C warm *shock solution* (*see* **Note 5**).
9. Incubate for 2 min at RT then add 5 ml of *buffered F10 medium*.
10. Aspirate, then add 5 ml of *buffered F10 medium*.
11. Aspirate, then add 5 ml of *avian cell culture medium* and incubate the cells at 37 °C (5 % CO_2).

3.4 Focus Assay of myc-Transformed Cells

1. Transfect cells seeded on 60-mm dishes with varying amounts of plasmid DNA (0.5–5.0 μg) using the calcium phosphate method, or infect cells grown on 60-mm dishes with 300 μl of serial dilutions (10^0–10^{-5}) from a viral MC29 stock (*see* **Note 6**).
2. After transfection or infection cultivate the cells in 5 ml of *growth medium* at 37 °C (5 % CO_2).
3. The next day remove the medium from the confluent cells and add 4 ml of liquid *focus agar* (*see* **Note 7**).
4. Let the agar harden for 1 h at RT, then incubate the cells at 37 °C (5 % CO_2).
5. Overlay cells every 2–3 days with 2 ml of *focus agar*. Foci of transformed cells are detectable after 1–2 weeks.

6. To pick single foci, prepare a flexible plastic tube attached to a Pasteur pipette and MP-24 wells filled with each 1 ml of *avian cell culture medium* (*see* **Note 8**).
7. Under the microscope detach the focus and carefully suck until the focus is visible in the pipette tip (*see* **Note 9**).
8. Transfer the focus into the MP-24 well and rinse the pipette with culture medium.
9. Incubate at 37 °C (5 % CO_2) for 1 day until cells have been attached to the dish. Replace medium.
10. After cells have grown to confluence, transfer cells from the entire well into an MP-6 well, then to a 60-mm dish, and finally to a 100-mm dish to propagate cells derived from a single focus.
11. For scoring the number of foci, remove the agar block from the cells and wash twice with *TS buffer*.
12. Add 5 ml of absolute ethanol to fix the cells. Incubate for 2 min at RT.
13. Aspirate ethanol, cover the cells with 2 ml of *Giemsa staining solution*, and incubate for 30 min at RT (do not rock the dishes).
14. Remove the staining solution and wash 3× with water. Air-dry the dishes and score foci by counting and photography (*see* Fig. 1).
15. Determine the number of foci per microgram of transfected DNA to quantify the transformation efficiency.

3.5 Colony Assay of myc-Transformed Cells

1. Transfect or infect cells plated on 60-mm dishes as described above.
2. 2–3 days after transfection or infection, transfer the cells into a 100-mm dish.
3. Cultivate the cells by transferring them every 4–7 days until they have adopted a transformed morphology (3–6 passages).
4. Prepare MP-6 dishes filled with each 1 ml of liquid *cloning bottom agarose* per well. Let the agar harden for 30 min (*see* **Note 7**).
5. Prepare an adequate amount of *cloning top agarose* and keep at 45 °C in a water bath.
6. Transfer trypsinized cells suspended in 5 ml of *transfer buffer* into a conical tube.
7. Determine the cell number and transfer each 10,000–100,000 cells into 3-ml tubes.
8. Add each 2 ml of liquid *cloning top agarose* (45 °C) into the tubes and mix with a 2-ml wide-mouth pipette.

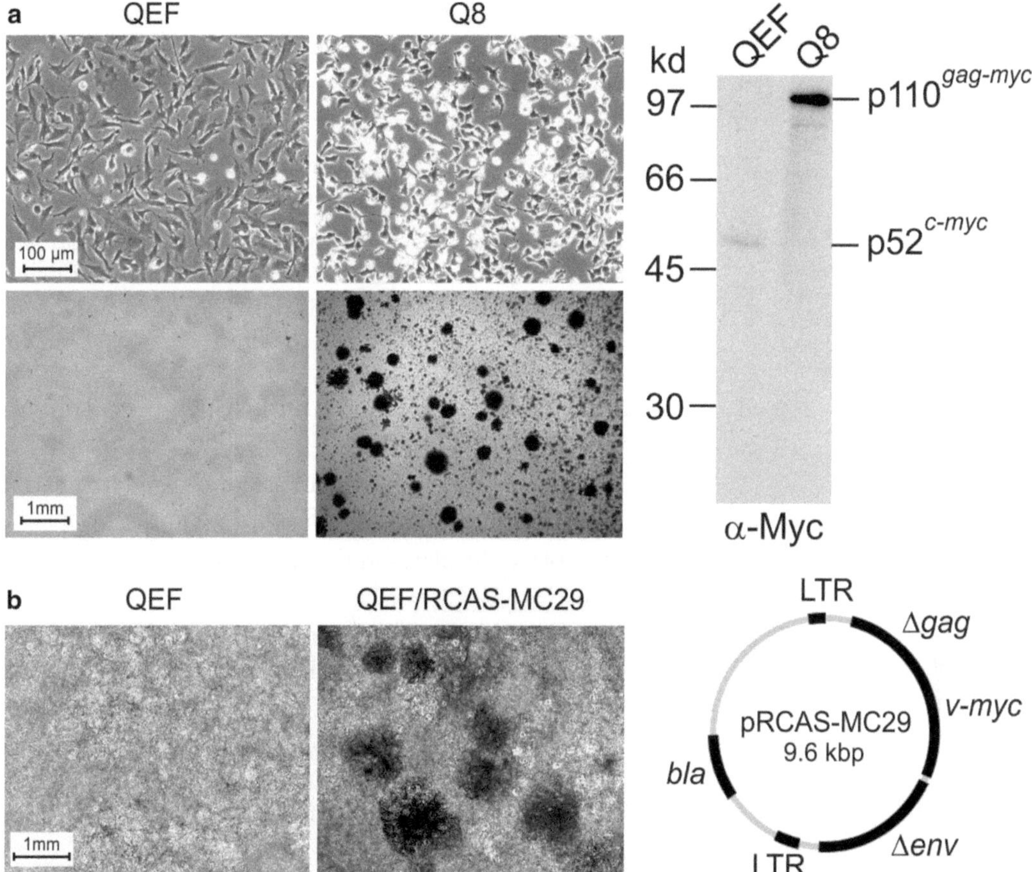

Fig. 1 Transformation of quail embryo fibroblasts by the v-*myc* oncogene. (**a**) *Top left panels*: phase-contrast micrographs of normal quail embryo fibroblasts (QEF) and of the MC29-transformed quail cell line Q8 [2]. *Bottom left panels*: colony formation by Q8 cells in soft agar, in comparison to control QEF. *Right panel*: immunoprecipitation and sodium dodecyl sulfate-polyacrylamide gel electrophoresis (SDS-PAGE) of radiolabeled v-Myc and endogenous c-Myc proteins from Q8 or QEF cell extracts, respectively. (**b**) *Left panels*: focus formation of QEF transfected with a plasmid construct containing the MC29 provirus (pRCAS-MC29), in comparison to control QEF. Bright-field micrographs were taken 2 weeks after DNA transfection. *Right panel*: map of pRCAS-MC29. Modified and adapted from [13, 14]

9. Pour the mixtures into the prepared MP-6 wells. Let the agar harden for 2 h at RT, then incubate at 37 °C (5 % CO_2).
10. Overlay the cells every 2–3 days with 1 ml of *cloning top agarose* per well (*see* **Note 10**). Colonies consisting of transformed cells are detectable after 1–2 weeks (*see* Fig. 1).
11. Determine the number of colonies formed by 1,000 cells to quantify the transformation efficiency.
12. To generate cell clones derived from single colonies, prepare a flexible plastic tube attached to a Pasteur pipette and MP-24 wells filled with each 1 ml of *avian cell culture medium*.

13. Localize the colony under the microscope and carefully suck until the colony is visible in the pipette.
14. Hold the pipette into the MP-24 well and rinse with culture medium until the colony has been transferred into the well.
15. Incubate at 37 °C (5 % CO_2) for 1 d until cells have been attached to the dish. Replace medium.
16. After cells have grown to confluence, transfer cells from the entire well to an MP-6 well, then to a 60-mm dish, and finally to a 100-mm dish to propagate cells derived from a single agar colony.

3.6 Northern Analysis of c-myc or v-myc RNAs

1. Grow the cells to near confluence and change the medium 1 day prior to RNA isolation.
2. Aspirate the culture medium with a Pasteur pipette attached to a water pump and wash cells with 5 ml of *TS buffer* per plate.
3. Aspirate and place the plates in tilted position to remove residual liquid.
4. Spread 2–4 ml *GITC buffer* per plate onto the cells and agitate plates on a rocking platform to distribute the buffer and to detach the lysed cells.
5. Put the plates in a tilted position and transfer the lysate into a plastic tube.
6. Draw the lysate three times through a 20-G needle to shear the chromosomal DNA.
7. Transfer 4 ml of *CsCl buffer* into an ultracentrifuge polyallomer tube.
8. Carefully layer 7.5–8 ml of cell lysate onto the CsCl-cushion without mixing both phases (consider the maximal loading capacity per ultracentrifuge tube which is lysate from 10^8 cells) and fill up the tube to 2 mm from the top. Balance the tubes with *GITC buffer*.
9. Insert the tubes into the rotor buckets [6] and centrifuge samples at 107,000 × *g* (SW41 rotor) at 21 °C for 16–24 h.
10. After the run has been completed, remove both phases by pipetting until approximately 1 ml liquid remains at the bottom of the tube (*see* **Note 11**).
11. Quickly invert the tube and let drain all liquid onto a paper towel.
12. With a sterile scalpel, cut off the bottom of the tube.
13. Dissolve the RNA pellet in 2 × 200 μl *0.3 M sodium acetate pH 6.0* for 3 min each and transfer into a microcentrifuge tube (*see* **Note 12**).
14. Add 1 ml of absolute ethanol, store at −80 °C for 30 min, and centrifuge at 13,000 × *g* for 15 min at 4 °C.

15. Wash the pellet in 70 % ethanol and dry in a SpeedVac (do not dry completely!).
16. Dissolve the RNA in 100 μl H_2O per 10^8 cells. Store the RNA solution at −80 °C.
17. Determine the optical densities (OD) at 260 and 280 nm; for pure RNA the OD_{260}:OD_{280} ratio is 2.0.
18. Calculate the RNA concentration by using the correlation 1 OD_{260} = 40 μg/ml. Yields vary between 0.5 mg and 2 mg of total RNA per 10^8 cells.
19. For poly(A)$^+$-RNA selection, fill 100–200 mg oligo-dT powder (for 0.5–5 mg of total RNA) into a 10-ml tube and suspend by inverting several times in 5 ml of *RNA buffer B*.
20. After 10-min equilibration, fill the suspension into a sterile Econo-Column chromatography column (0.7 × 5 cm; 2 ml).
21. Rinse the column with 4 ml of *0.1 M NaOH, 5 mM EDTA*.
22. Wash the column with H_2O until the pH becomes neutral (check with pH paper).
23. Equilibrate the column with 5 ml of *RNA buffer A*.
24. Meanwhile, heat the total RNA in a volume of 500 μl at 65 °C for 5 min.
25. Add 500 μl *RNA buffer A* (65 °C) to the RNA, mix, and leave at RT for 2 min.
26. Load the RNA solution onto the column, collect the eluate, and heat again at 65 °C for 5 min.
27. Reload the RNA as before.
28. Wash the column 5 times with 1 ml of *RNA buffer B*.
29. Elute the poly(A)$^+$ RNA with 2 ml of *RNA buffer C*.
30. For a second round of poly(A)$^+$ RNA selection, use a column with 50 mg of oligo-dT powder and repeat the procedure as described above.
31. Elute the poly(A)$^+$ selected RNA with 4 × 450 μl *buffer C* into four microcentrifuge tubes.
32. To each tube add 50 μl *3 M sodium acetate pH 6.0* and 1 ml absolute ethanol, mix, and store at −80 °C for 30 min.
33. Spin at 13,000 × *g* for 20 min at 4 °C.
34. Wash the pellets with 70 % ethanol and centrifuge as above.
35. Dry the pellets in a SpeedVac and dissolve the poly(A)$^+$ RNA in the appropriate amount of H_2O to achieve a concentration of ~1 μg/μl. Store the RNA solution at -80 °C.
36. Determine the RNA concentration. The yield after the first selection should be 5–10 % from the total RNA (10–20-fold enrichment) and 1–2 % after the second selection (50–100-fold enrichment).

37. For gel electrophoresis of isolated RNAs, pour a *1 % (w/v) agarose/formaldehyde* gel and let the gel harden for at least 1 h.
38. To 4–8 μl of RNA (30 μg total RNA or 1–5 μg poly(A)$^+$-selected RNA dissolved in 4–8 μl H_2O), add 16 μl of *RNA sample buffer*.
39. Heat samples to 65 °C for 10 min, centrifuge briefly, and then add 1 μl of *1 mg/ml ethidium bromide*.
40. Load the samples onto the gel and run at 30–40 V const. overnight (500 Vh) in *1× RNA gel running buffer* until the bromophenol blue dye has migrated ~10 cm.
41. Photograph the gel with an aligned ruler next to the size marker and cut the gel to an appropriate size (e.g. 15 cm × 12 cm).
42. Rinse the gel in deionized water for 5 min to remove the formaldehyde.
43. Equilibrate the gel in *10× SSC* for 30 min.
44. Prepare a nylon membrane sheet of the appropriate size and equilibrate first in sterile water for 10 min, then in *10× SSC* for 30 min.
45. With a tray and a board, prepare a blotting device and fill the tray with *10× SSC*.
46. Cut two sheets of Whatman™ paper (20 cm × 30 cm), soak them in *10× SSC*, and create a wick to support the gel.
47. Put the inverted gel onto the device and insulate the area around the gel with Parafilm™.
48. Layer the nylon membrane onto the gel and remove air bubbles with a pipette.
49. Layer three pieces of Whatman™ paper (cut to the gel size and soaked in *10× SSC*) onto the membrane and remove air bubbles.
50. Layer paper towels onto the Whatman™ papers and put a 0.5–1 kg-weight onto the top.
51. Transfer for 16–24 h.
52. After the transfer, mark positions of the slots with a ball pen and put the membrane into 500 ml of *2× SSC* for 10 min to remove rests of agarose.
53. Bake the membrane wrapped between two sheets of Whatman™ paper in a vacuum oven for 2 h at 80 °C.
54. After baking, wrap the blot in aluminum foil and store dark and dry.
55. For detection of transferred RNAs by nucleic acid hybridization, pre-hybridize the filter in 100 ml *hybridization solution* in a plastic tray overnight at 37 °C.
56. Boil a radioactive-labeled DNA probe (pierce a hole into the tube cap) for 5 min in a water bath for denaturation and then chill on ice.

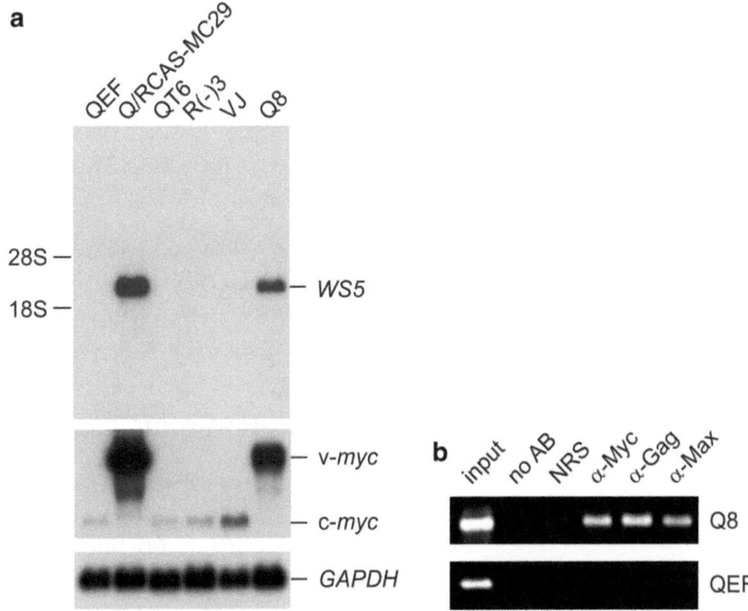

Fig. 2 Transcriptional activation of the *WS5* target gene by the v-Myc oncoprotein. (**a**) Northern analysis using poly(A)$^+$-selected RNAs from normal QEF, QEF transfected by pRCAS-MC29 (Q/RCAS-MC29), quail cell lines transformed by the oncogenes v-*src* [R(−)3], v-*jun* (VJ), or chemically transformed by methylcholanthrene (QT6), and the v-*myc* (MC29) transformed cell line Q8. Hybridization was performed using cDNA probes specific for *WS5*, v/c-*myc*, and the glyceraldehyde 3-phosphate dehydrogenase (*GAPDH*) gene. (**b**) Chromatin immunoprecipitation (ChIP) analysis of Q8 and QEF cells using no antibody (no AB), normal rabbit serum (NRS), and polyclonal antisera directed against v/c-Myc, Gag, or Max. Polymerase chain reaction (PCR) analysis was performed to amplify a *WS5* promoter specific region containing a cluster of four Myc binding sites (E-boxes). Modified and adapted from [15]

57. After pre-hybridization transfer the filter into a new tray containing 40 ml *hybridization solution* and the denatured radiolabeled probe.
58. Hybridize the filter overnight at 37 °C.
59. The next day wash the filter three times for 10 min each in *washing solution* 2× *SSC* at 37 °C.
60. Then wash the filter three times for 20 min each in *washing solution 0.2× SSC* at 60 °C.
61. Let the filter dry on a Whatman™ paper. Then wrap into a plastic foil and autoradiograph at −80 °C using an intensifier screen (*see* Fig. 2).
62. To remove the hybridized probe from the blot, incubate the filter sequentially for 1 h at 65 °C and 1 h at 70 °C in *stripping solution*.

3.7 Immunoprecipitation of Myc Proteins

1. Seed 2.5×10^6 cells onto 60-mm dishes and incubate in *avian cell culture medium* for 1 day at 37 °C and 5 % CO_2.
2. Wash the cells twice with 2.5 ml PBS and once with 1 ml DMEM.
3. Incubate the cells for 1 h at 37 °C (5 % CO_2) in 2.5 ml *DMEM medium (–Met)*.
4. Remove the medium by aspiration and replace it with 1 ml *DMEM medium (-Met)* containing 50 μCi ^{35}S-methionine (>1,000 Ci/mmol) per dish.
5. Incubate the dishes for 2–3 h at 37 °C (5 % CO_2) with occasional rocking.
6. Wash the cells twice with 2.5 ml ice-cold *PBS*, aspirate, and drain the last traces.
7. Place the dishes onto a surface chilled at 0 °C.
8. Scrape the cells of one 60-mm dish into 250 μl *boiling buffer* and transfer into a 1.5-ml reaction tube (*see* **Note 13**).
9. Pierce a hole into the cap and heat for 2 min at 100 °C in a boiling water bath, then cool on ice for 15 min.
10. Dilute the lysate with 1 ml *dilution buffer* and pipette up and down several times to shear chromosomal DNA.
11. Centrifuge at 20,000 × *g* for 60 min at 4 °C.
12. Transfer the clarified lysates into cooled fresh tubes and determine the activity from 6.25 μl (1/200 vol) dissolved in 10 ml ReadyMix™ (Beckmann) using a scintillation counter.
13. Cell extracts may be stored at this stage at –80 °C.
14. For each precipitate, prepare 1.5-ml reaction tubes containing 1–5 μl of antiserum or pre-immune serum diluted in 200 μl *RIPA-BSA*.
15. Dilute clarified lysate corresponding to $1–2 \times 10^7$ cpm to a volume of 500 μl with *RIPA buffer* and mix with the diluted antiserum.
16. Incubate on ice for 1–2 h.
17. Add 100 μl of *RIPA-protein A-bead slurry* and incubate for 60 min at 4 °C on a rocking platform.
18. Pellet the immune complexes by centrifugation at 20,000 × *g* for 20 s at 4 °C.
19. Discard the supernatant and resuspend the pellets in 650 μl *RIPA buffer*.
20. Transfer this suspension onto a cushion of 650 μl *RIPA-sucrose*.
21. Centrifuge at 13,000 × *g* for 15 min at 4 °C.
22. Discard the supernatant and wash the pellets 3× with 1 ml *RIPA buffer* and once with 1 ml *PBS*.

23. After the last wash suspend the pellet in 40 μl *1× SDS gel-loading buffer* and incubate at 95 °C for 5 min.
24. Centrifuge at 15,000 × *g* for 5 min at RT.
25. Load the supernatant (*see* **Note 14**) directly onto a SDS-polyacrylamide gel (*see* Fig. 1).

3.8 Analysis of Myc-DNA Interactions by Electrophoretic Mobility Shift Assay (EMSA)

1. To prepare a radiolabeled double-stranded DNA probe, mix 100 pmol each of two single-stranded complementary oligonucleotides (18–22 mers), 5 μl *10× annealing buffer*, and H_2O to 50 μl (DNA concentration: ~25 ng/μl).
2. Incubate at 95 °C –100 °C in a boiling water bath for 3 min and then switch off the heating plate.
3. Leave until the reaction mix has cooled down to RT (~3 h). Store at −20 °C.
4. To 75 ng of the annealed DNA, add 1 μl *10× T4 PNK buffer*, 3 μl ^{32}P-γ-ATP (3,000 Ci/mmol), H_2O to 9 μl, and 1 μl of T4 DNA polynucleotide kinase.
5. Incubate at 37 °C for 1 h.
6. Add 30 μl H_2O, 10 μl *5 M LiCl*, 200 μl absolute ethanol, mix, and store at −80 °C for 1 h.
7. Collect the precipitated oligonucleotide by centrifugation at 15,000 × *g* for 30 min at 4 °C.
8. Remove the supernatant and wash the pellet with 1 ml of 70 % (v/v) ethanol. Recover the DNA by centrifugation as above.
9. Remove the supernatant and dry the pellet in a SpeedVac for 5 min at RT.
10. Dissolve the pellet in 50 μl *1× annealing buffer* and determine the Čerenkov activity from 1/100 vol. An activity of 0.5–1.0×10^8 cpm/μg DNA should be obtained. Store the DNA probe at −20 °C.
11. For the EMSA reaction, dilute the DNA probe to 10,000 cpm/μl with *1× EMSA buffer* resulting into a DNA concentration of 7.5 nM (~0.1 ng/μl).
12. Determine the protein concentration and the molarity of the recombinant protein (e.g., a 1 μg/μl solution of a protein with a M_r of 20,000 is 50 μM).
13. Dilute the recombinant protein solution with *1× EMSA buffer* to a concentration of 1 μM (20 ng/μl for a protein with M_r of 20,000).
14. To 5 μl protein solution, add 17.5 μl *1× EMSA* buffer and 2.5 μl DNA probe (final concentrations: protein, 200 nM; DNA, 0.75 nM).
15. To titer the optimal amount of protein which is required to consume all DNA, set up the same binding reaction using

serial 1:2 dilutions of the 1-μM protein stock solution resulting in final concentrations of 100, 50, 25 nM, etc.

16. Incubate the reactions (final volume: 25 μl) at 25 °C for 45 min.
17. To detect protein–DNA interactions, perform gel electrophoresis as described below.
18. The protein concentration [P] which is necessary to bind to 50 % of the applied DNA probe is equivalent to the dissociation constant (K_d) according to $K_d = [P] \times [DNA]/[P\text{–}DNA]$.
19. To specify the binding protein, an appropriate antibody (0.5–1 μl) may be included in the reaction (*see* **Note 15**).
20. For detection of DNA binding by gel electrophoresis, pour a 5 % (w/v) polyacrylamide gel, mix 31.3 ml H_2O with 6.7 ml *30 % (w/v) acrylamide/0.8 % (w/v) bisacrylamide*, 2 ml *10× TBE*, 250 μl *10 % (w/v) APS*, and 35 μl TEMED (*N,N,N′,N′*-tetramethylethylenediamine).
21. Cast a vertical gel using 1.5-mm spacers and let polymerize for 1 h.
22. Remove the comb; rinse the slots with *0.5× TBE* and insert the glass plates with the gel into an appropriate gel chamber. As running buffer use *0.5× TBE*.
23. Load the samples using a Hamilton syringe. Into an extra slot, fill 25 μl *1× EMSA buffer* containing 0.1 % (w/v) bromophenol blue.
24. Run at constant 20 mA for approximately 2 h until the bromophenol blue dye has passed 2/3 of the gel.
25. Rock the gel for 15 min in 10 % (v/v) acetic acid at RT (*see* **Note 16**).
26. Put the inverted gel onto a glass plate and drain excess fluid.
27. Layer two sheets of Whatman™ 3 MM CHR paper onto the gel and dry for 75 min at 65 °C on a gel dryer.
28. Wrap the gel into a plastic foil and autoradiograph at −80 °C using an intensifier screen for 2–16 h (*see* Fig. 3). For quantification of radioactive signals, use a phosphorimager.

3.9 Analysis of Myc-Induced Transcriptional Activation by Reporter Gene Assay

1. Co-transfect cells in triplicate seeded on 60-mm dishes with appropriate expression and reporter plasmids (total 5 μg).
2. After 1–2 days wash cells with 2.5 ml *PBS buffer* per dish and aspirate.
3. Add 250 μl *1× cell culture lysis reagent* per dish (*see* **Note 17**).
4. Place the dishes onto an ice-cooled sheet of aluminum foil and rock for 3 min.

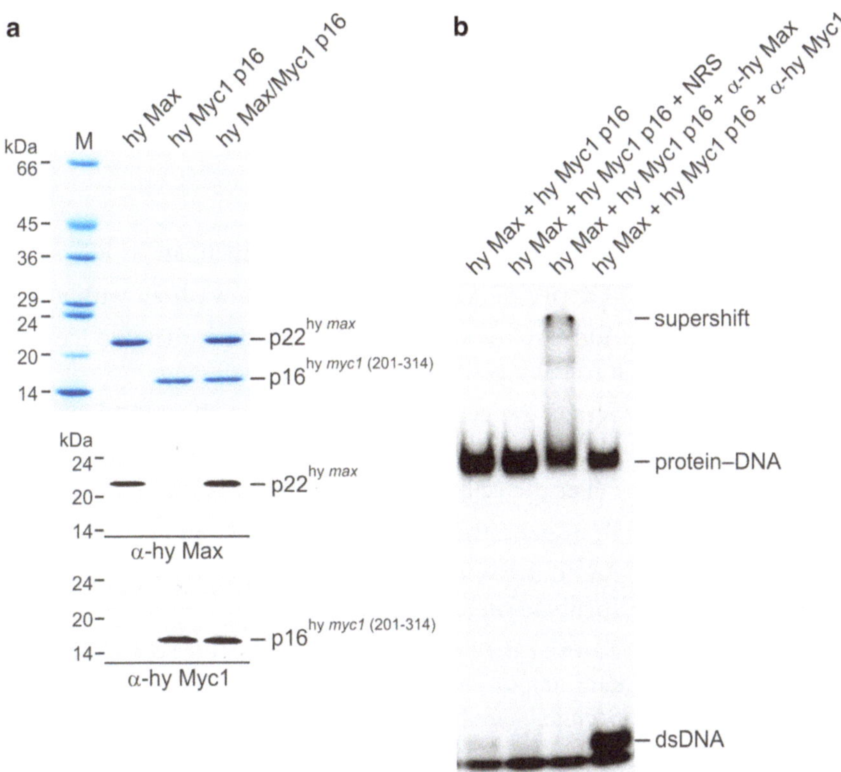

Fig. 3 Biochemical properties of recombinant *Hydra* Myc and Max proteins. (**a**) SDS-PAGE of 2-μg (Coomassie Brilliant Blue staining) or 50-ng (immunoblotting) aliquots of purified recombinant *Hydra* Max p22, *Hydra* Myc1 p16, and a 1:1 mixture of both proteins. (**b**) EMSA using the recombinant proteins shown in (**a**) and a radiolabeled double-stranded oligodeoxynucleotide containing the Myc/Max-binding motif 5′-CACGTG-3′. Antibodies were added to the binding reactions as indicated. Modified and adapted from [12]

5. Scrape the cells using a rubber policeman and transfer into a 1.5-ml reaction tube.
6. Vortex the tube for 15 s, then centrifuge at 4 °C for 10 min at 15,000 × *g*.
7. Transfer the supernatant into a fresh tube and keep on ice. Cell extracts may be stored at this stage at −80 °C.
8. Determine the protein concentration (e.g., by Bradford assay).
9. Thaw the *Luciferase Assay Reagent* and keep at 25 °C (*see* **Note 18**).
10. Program a luminometer (e.g., LUMAT LB 9507, Berthold Technologies) to perform a 10-s measurement.
11. Dispense 5 μl of the cell lysate into a luminometer tube and add 40 μl *Luciferase Assay Reagent*.
12. Put the tube into the luminometer and initiate reading.

3.10 Analysis of Myc Binding to Target Promoters by ChIP Assay

1. Use cells from three 100-mm dishes grown to subconfluency (~10^7 cells/dish). Determine the cell number from an extra dish.
2. Remove the culture medium and add 9.725 ml of fresh culture medium and 275 μl of 36.5 % (v/v) formaldehyde to a final concentration of 1 % (v/v).
3. Incubate at RT for 15 min.
4. Add 1 ml of *1.25 M glycine* and incubate at RT for 5 min.
5. Remove the culture medium and wash 2× with ice-cold *PBS*.
6. Add 2 ml of ice-cold *PBS-PI* and incubate on ice for 3 min on a rocking platform.
7. Scrape the cells and transfer the cell suspension into a conical tube. Wash the dish with each 4 ml of *PBS-PI*.
8. Pellet the cells by centrifugation at 720 × *g* for 10 min at 4 °C.
9. Resuspend the cells in 6 ml *PBS-PI* and centrifuge as above.
10. Add 0.33 ml of *SDS-lysis buffer* per 10^7 cells and homogenize. Incubate on ice for 10 min.
11. Transfer the lysate (~1 ml) into a 10-ml reaction tube and sonicate at constant 40 W with 25 % power for 10 s (Branson Sonifier) (*see* **Note 19**).
12. Repeat the sonication 3–5× keeping the samples and the Sonifier tip cooled. The sheared chromatin should have DNA sizes between 0.2 and 1 kbp (see below) (*see* **Note 20**).
13. Centrifuge the extract at 15,000 × *g* for 10 min at 4 °C to pellet cell debris.
14. Save the supernatant (undiluted extract) which can be stored at −80 °C.
15. To monitor sonication, mix 4 μl of *5 M NaCl* with 100 μl of undiluted extract and incubate at 65 °C for 4 h. Recover the liberated DNA by phenol/chloroform extraction and analyze 10, 5, and 2 μl aliquots from the aqueous phase by agarose gel electrophoresis.
16. For immunoprecipitation dilute the supernatant (~0.9 ml) in a tenfold volume of *ChIP-dilution buffer*.
17. To 40 μl of the diluted extract, add 410 μl *ChIP-elution buffer* and store at −20 °C (input sample).
18. To preclear the lysate, add 75 μl of *ChIP-protein A-bead slurry* to 1 ml of the sheared and diluted chromatin extract. Incubate at 4 °C for 1 h using a rotary shaker.
19. Centrifuge at 1,500 × *g* for 5 min at 4 °C.
20. Transfer the precleared supernatant into a fresh 1.5-ml reaction tube and add 2 μl of specific antiserum. Include each reaction with pre-immune serum and without antiserum as a control.
21. Incubate overnight at 4 °C using a rotary shaker.

22. Add 75 µl of *ChIP-protein A-bead slurry* and incubate at 4 °C for 1 h using a rotary shaker.
23. Centrifuge at 1,500 × *g* for 5 min at 4 °C.
24. Discard the supernatant and add 1 ml of *ChIP-low salt wash buffer* to the pellet.
25. Incubate at 4 °C for 5 min using a rotary shaker and centrifuge as above.
26. Wash as above with 1 ml of *ChIP high-salt wash buffer.*
27. Wash as above with 1 ml of *ChIP-LiCl wash buffer.*
28. Wash 2× with 1 ml of *TE buffer.*
29. To the pellet add 225 µl of *ChIP-elution buffer* and incubate for 15 min at RT using a rotary shaker.
30. Centrifuge at 1,500 × *g* for 5 min at 4 °C and recover the supernatant containing the cross-linked chromatin complexes.
31. Repeat the elution and then pool the supernatants resulting in a final volume of 450 µl.
32. Dissociate the cross-linked material (including the Input sample) by adding 18 µl of *5 M NaCl* to a final concentration of 192 mM and incubate at 65 °C for 4 h. Samples can be stored at −20 °C at this time.
33. For DNA purification and subsequent polymerase chain reaction (PCR) analysis, add 9 µl of *0.5 M EDTA*, 18 µl *1 M Tris–HCl pH 6.5*, and 1.8 µl of a *10 mg/ml proteinase K* solution to each sample.
34. Incubate for 1 h at 45 °C.
35. Add 450 µl Roti-phenol/chloroform/isoamyl alcohol (25:24:1) and vortex each sample for 1 min (*see* **Note 21**).
36. Centrifuge the samples at 13,000 × *g* for 5 min at RT.
37. Transfer the supernatant to a fresh tube containing 450 µl chloroform/isoamyl alcohol (24:1) and repeat the extraction step (*see* **Note 21**).
38. Transfer the supernatant to a fresh 1.5-µl reaction tube and add 50 µl of *3 M sodium acetate pH 5.2*, 1 µl of 20 µg/µl glycogen as an inert carrier, and 1 ml of absolute ethanol.
39. Mix and incubate at −80 °C for 90 min.
40. Centrifuge the samples at 15,000 × *g* for 30 min.
41. Discard the supernatant carefully and wash the pellet with 1 ml of 70 % (v/v) ethanol.
42. Repeat the centrifugation for 15 min, discard the supernatant, and dry the pellet in a SpeedVac.
43. Resuspend the pellet in 50 µl of sterile H_2O and store at −20 °C. Use 5–10 µl for the PCR (25–35 cycles) reaction.

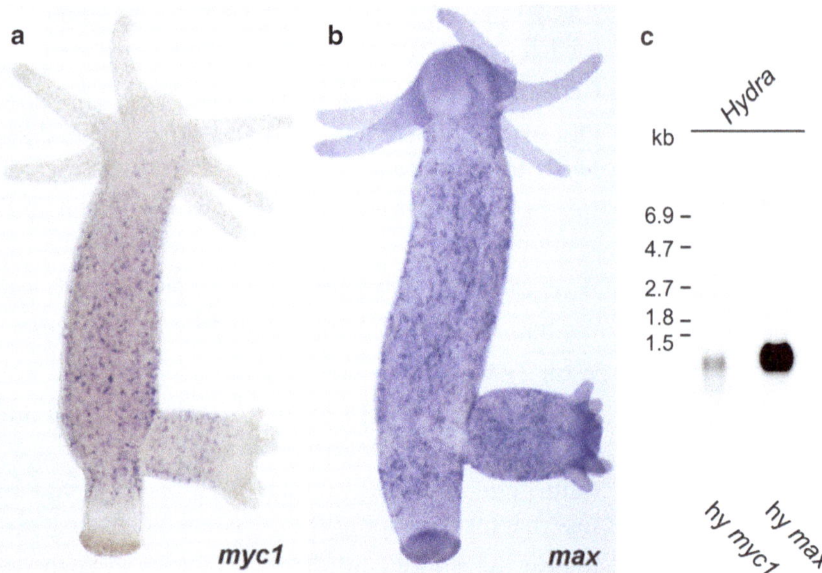

Fig. 4 Expression of *Hydra myc1* and *max*. (**a**, **b**) In situ hybridization showing that *myc1* and *max* are activated in cells belonging to the interstitial stem cell lineage in the gastric region of intact, budding polyps. In addition, *max* is expressed at a lower level in the epithelium throughout the entire body column. (**c**) Northern analysis using poly(A)$^+$-selected RNAs from whole *Hydra* animals and *myc1* or *max* specific cDNA probes. Modified and adapted from [12]

Appropriate PCR conditions have to be determined depending on the melting temperatures of primers and template.

44. Visualize PCR products by agarose gel electrophoresis and quantify ethidium bromide stained bands (*see* Fig. 2). Alternatively, PCR products can be quantified by real time PCR.

3.11 Analysis of myc Expression by In Situ Hybridization

1. Cultivate the strains *Hydra vulgaris* or *Hydra magnipapillata* at 18 °C and feed daily with freshly hatched *Artemia nauplii*. For usage in experiments, collect animals 24 h after the last feeding [12].
2. For whole mount in situ hybridization with digoxigenin-labeled RNA probes, proceed as described in [12, 16] (*see* Fig. 4).
3. For in situ hybridizations in single cell preparations, macerate *Hydra* as described in [17]. Then fix macerates in paraformaldehyde and hybridize as described in [12, 18].

4 Notes

1. The first decantation of the cell suspension may be difficult because of high viscosity. If necessary, do not decant the entire supernatant in order to avoid transferring cell clumps. During the next extraction steps, the viscosity of the cell suspension is significantly lower.

2. Check complete detachment of trypsinized cells under the microscope.
3. Do not leave the viral stock longer than necessary at 37 °C. Immediately after thawing, add the virus to the cells. Also, avoid repeated freeze–thaw cycles of viral stocks. Ideally, they are stored in aliquots and used only once.
4. After seeding, carefully move the dishes in horizontal and vertical directions to distribute the cells evenly on the surface. Do not swivel the dishes circularly! Repeat the rocking step after 30 min and after 60 min to avoid that most of the cells settle in the middle of the dish. That would result in an uneven cell distribution, which is detrimental for foci scoring.
5. Use wide-mouth pipettes for slowly dropping these solutions onto the cells. Distribute the single drops over the entire cell surface to minimize any stress caused by splashing liquid onto the cells, which are very sensitive during this transfection step. After glycerol shock and washing, check cell viability under the microscope. Cells may look slightly different after the shocking procedure but they will recover soon.
6. Prepare serial dilutions using *transfer buffer* as a diluent.
7. Boiling of agar or agarose preparations in a microwave oven requires special care. Wear safety goggles. Check that the agar has been completely dissolved and put the hot glass vessel into the 45 °C-water bath. Agar clumps in the overlay disturb the transparency and make scoring of foci or colonies difficult.
8. Pasteur pipettes which have a ~30° kink approximately 1 cm away from the tip are recommended to detach foci from the dish. To produce such a device, hold a pipette on top of a Bunsen burner flame and wait until the glass melts locally. Then, immediately remove the pipette from the flame, let it cool, and sterilize.
9. Use a flexible polyethylene tube attached with one end to the narrow site of a blue tip. Fill the tip with a cotton plug and use this inverted tip as a mouthpiece. The other end of the tube is attached to the kinked Pasteur pipette.
10. Let the agar harden for at least 2 h at RT not exceeding 25 °C.
11. Change the tip when you enter the CsCl phase. Try to remove any suspended precipitates of DNA or proteins before inverting the tube.
12. Pipette up and down several times to dissolve the RNA pellet completely.
13. Use a sterile rubber policeman to scrape the lysate from the dish. Afterwards, decontaminate the scraping device by submerging it in a 1 % (w/v)-SDS solution for several days.

14. Take care that no material from the sepharose pellet becomes loaded onto the SDS-polyacrylamide gel.
15. Addition of specific antibodies directed against the DNA binding protein may lead either to a ternary antibody-protein–DNA complex with lower gel electrophoretic mobility or to inhibition of DNA binding. The latter can be observed when the epitope of the antibody is localized in the DNA binding domain of the protein.
16. Remove the upper glass plate and leave the gel on the lower one. Submerge plate with attached gel in acetic acid solution. After fixing, use the upper glass plate to invert the gel.
17. Rock the dish immediately after addition of lysis buffer to quickly cover all cells.
18. Use aliquots of Luciferase Reagent stored at −80 °C. Once the reagent is thawed, keep it protected from light.
19. The tip of the Sonifier should be immersed as close as possible towards the bottom of the tube but not touch it. Otherwise, foam could be generated by the sonication and make subsequent pipetting difficult.
20. The optimal number of sonication cycles depends on the cellular material and can be determined in a pilot experiment with subsequent analysis of the chromatin fragment sizes obtained after various cycles of sonication.
21. It is important to remove the upper aqueous phases as accurately as possible for quantitative recovery of the released DNA fragments.

Acknowledgment

This work was supported by Austrian Science Fund (FWF) grant P23652 to K.B.

References

1. Duesberg PH, Bister K, Vogt PK (1977) The RNA of avian acute leukemia virus MC29. Proc Natl Acad Sci U S A 74:4320–4324
2. Bister K, Hayman MJ, Vogt PK (1977) Defectiveness of avian myelocytomatosis virus MC29: Isolation of long-term nonproducer cultures and analysis of virus-specific polypeptide synthesis. Virology 82:431–448
3. Bister K, Jansen HW (1986) Oncogenes in retroviruses and cells: biochemistry and molecular genetics. Adv Cancer Res 47:99–188
4. Vogt PK (2012) Retroviral oncogenes: a historical primer. Nat Rev Cancer 12:639–648
5. Eisenman RN (2001) Deconstructing myc. Genes Dev 15:2023–2030
6. Eilers M, Eisenman RN (2008) Myc's broad reach. Genes Dev 22:2755–2766
7. Nesbit CE, Tersak JM, Prochownik EV (1999) MYC oncogenes and human neoplastic disease. Oncogene 18:3004–3016
8. Blackwood EM, Eisenman RN (1991) Max: a helix-loop-helix zipper protein that forms a

sequence-specific DNA-binding complex with Myc. Science 251:1211–1217

9. Dang CV (1999) c-Myc target genes involved in cell growth, apoptosis, and metabolism. Mol Cell Biol 19:1–11
10. Gallant P, Shiio Y, Cheng PF, Parkhurst SM, Eisenman RN (1996) Myc and Max homologs in *Drosophila*. Science 274:1523–1527
11. Orian A, van Steensel B, Delrow J et al (2003) Genomic binding by the *Drosophila* Myc, Max, Mad/Mnt transcription factor network. Genes Dev 17:1101–1114
12. Hartl M, Mitterstiller AM, Valovka T, Breuker K, Hobmayer B, Bister K (2010) Stem cell-specific activation of an ancestral *myc* protooncogene with conserved basic functions in the early metazoan *Hydra*. Proc Natl Acad Sci U S A 107:4051–4056
13. Hartl M, Karagiannidis AI, Bister K (2006) Cooperative cell transformation by Myc/Mil(Raf) involves induction of AP-1 and activation of genes implicated in cell motility and metastasis. Oncogene 25:4043–4055
14. Hartl M, Nist A, Khan MI, Valovka T, Bister K (2009) Inhibition of Myc-induced cell transformation by brain acid-soluble protein 1 (BASP1). Proc Natl Acad Sci U S A 106:5604–5609
15. Reiter F, Hartl M, Karagiannidis AI, Bister K (2007) *WS5*, a direct target of oncogenic transcription factor Myc, is related to human melanoma glycoprotein genes and has oncogenic potential. Oncogene 26:1769–1779
16. Grens A, Gee L, Fisher DA, Bode HR (1996) *CnNK-2*, an NK-2 homeobox gene, has a role in patterning the basal end of the axis in *Hydra*. Dev Biol 180:473–488
17. David CN, MacWilliams H (1978) Regulation of the self-renewal probability in *Hydra* stem cell clones. Proc Natl Acad Sci U S A 75:886–890
18. Kurz EM, Holstein TW, Petri BM, Engel J, David CN (1991) Mini-collagens in *Hydra* nematocytes. J Cell Biol 115:1159–1169

Chapter 4

Identifying Myc Interactors

Romina Ponzielli, William B. Tu, Igor Jurisica, and Linda Z. Penn

Abstract

In this chapter, we discuss in detail two essential methods used to evaluate the interaction of Myc with another protein of interest: co-immunoprecipitation (Co-IP) and in vitro pull-down assays. Co-IP is a method that, by immunoaffinity, allows the identification of protein–protein interactions within cells. We provide methods to conduct Co-IPs from whole-cell extracts as well as cytoplasmic and nuclear-enriched fractions. By contrast, the pull-down assay evaluates whether a bait protein that is bound to a solid support can specifically interact with a prey protein that is in solution. We provide methods to conduct in vitro pull-downs and further detail how to use this assay to distinguish whether a protein–protein interaction is direct or indirect. We also discuss methods used to screen for Myc interactors and provide an in silico strategy to help prioritize hits for further validation using the described Co-IP and in vitro pull-down assays.

Key words Myc, Binding proteins, Interactors, Interactome, Protein–protein interaction, Co-immunoprecipitation, In vitro pull-down, Chromatin immunoprecipitation

1 Introduction

Identifying the molecular mechanisms of proto-oncogenic and oncogenic Myc function will provide insights critical for the development of novel anticancer therapeutics targeting deregulated Myc in tumor cells. It is well appreciated that protein–protein interactions (PPI) play a major role in Myc function, but surprisingly, a relatively small number of Myc-binding proteins has been identified and validated to date. Moreover, the region(s) of Myc important for the majority of these interactions also remain(s) poorly characterized. The Myc protein is divided into an N-terminal domain (NTD) involved in transactivation and transrepression of gene transcription and a C-terminal domain (CTD) essential for DNA binding.

Identification of the Myc interactome started with the discovery of Max, a critical partner for Myc target gene activation and repression. Max was identified by screening a cDNA expression library with a radiolabeled fusion protein containing the Myc CTD [1].

Laura Soucek and Nicole M. Sodir (eds.), *The Myc Gene: Methods and Protocols*, Methods in Molecular Biology, vol. 1012, DOI 10.1007/978-1-62703-429-6_4, © Springer Science+Business Media, LLC 2013

In the past 15 years, additional Myc interactors have been identified using a variety of methods. For example, traditional yeast two-hybrid (Y2H) library screens were used to identify Myc-interacting proteins. The Y2H assay is a tool that facilitates the study of protein–protein interactions by measuring transcription of a reporter gene. Basically, in the two-hybrid assay, two fusion proteins are created: the protein of interest (bait), which is fused to a DNA-binding domain, and its potential binding partner (prey), which is fused to an activation domain. If the two proteins interact, they will form a transcriptional activator that will activate a reporter gene whose product can be quantified. Using Myc fragments as bait has led to the identification of several binding proteins, such as Miz-1, Bin1, BRCA1, and INI1 [2–5]. However, the entire Myc NTD could not be used in traditional Y2H approaches due to the inherent activation conferred by the Myc NTD when used as bait. To overcome this issue, we established the repressed transactivator assay (RTA) to identify novel Myc NTD interactors [6]. With this modified Y2H system, successful capture of a Myc NTD interactor turns the reporter genes from "ON" to "OFF," which is opposite to the usual "OFF" to "ON" reporters of traditional Y2H assays. Briefly, the Myc NTD bait is expressed as a fusion protein with the yeast Gal4-DNA-binding domain (Gal4-Myc NTD). When Gal4-Myc NTD is recruited to the promoter, reporter gene expression is turned "ON." With the RTA, the prey proteins are fused to the repression domain of the yeast Tup1 protein. When the Tup1-fusion prey protein interacts with the Gal4-Myc NTD, the reporter is turned "OFF." Using the RTA and Gal4-Myc NTD as bait, we screened a prey library generated from a medulloblastoma-derived cell line and identified a novel Myc interactor, JPO2 [7]. In this study we demonstrated that expression of the novel transcription factor, JPO2, potentiates Myc anchorage-independent growth, and its expression is associated with metastasis in medulloblastomas. Moreover, the region of Myc essential for JPO2 interaction was mapped to residues 1–43 of the Myc N-terminus [7]. The JPO2 repressor (R1/RAM2/CDC7L/JPO2) has recently been shown to have functionality beyond medulloblastoma [8].

In the past 5 years, advances in proteomics using high mass accuracy and high-resolution mass spectrometry (MS) techniques have enabled a more global approach to identify the Myc interactome (as reviewed in ref. 9). In our recent review article, we compared three independent studies that had been conducted to identify Myc binding proteins using MS. Surprisingly only a limited number of protein hits was overlapping between the three studies, which may be a function of different cells and methodology used by each group [9]. Thus, when studying the Myc interactome, it is essential to take into consideration the cellular context as well as the method used for the study. Clearly, more Myc interactome studies are required to further understand the mechanism of Myc action.

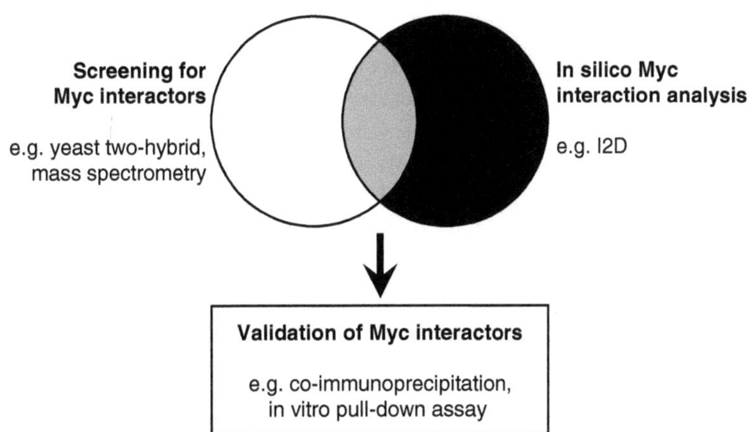

Fig. 1 Identifying Myc interactors. Conducting a protein–protein interactor screen can identify known and novel Myc-binding proteins (*white circle*). Screening approaches can include yeast two-hybrid or mass spectrometry-based techniques. From such screens a list of putative interactors will be identified (*white circle*). To help prioritize which hits to validate as likely interactors, the list of potential Myc-binding proteins generated from the screen can be compared to in silico Myc interactor databases, such as I2D (*black circle*). The overlap (*grey*) may have a higher probability of being true positives and will likely also include known Myc interactors. The hits that overlap (*grey*) can serve as positive controls to evaluate the reliability of the screen and for downstream validation. Novel interactors will reside in both the overlap (*grey*) and nonoverlapping regions (*white*) with presumably higher and lower probability, respectively. The downstream validation will then proceed using assays such as co-immunoprecipitation and in vitro pull-downs, as discussed in the text (*white rectangle*)

As outlined in the above paragraphs, there are many approaches to identify novel Myc interactors; however, one of the most challenging steps takes place when the screen is complete and one has to prioritize hits for further validation. One strategy to accomplish this goal is to overlap the list of identified Myc interactors obtained from a screening method of choice (Fig. 1, white circle) with an independent in silico analysis of the Myc interactome (Fig. 1, black circle).

Several PPI databases may be queried directly to obtain a Myc-specific PPI network. These include the Biomolecular Interaction Network Database (BIND) [10], the Biological General Repository for Interaction Datasets (BioGRID) [11], the Database of Interacting Proteins (DIP) [12], Human Protein Reference Database (HPRD) [13], the molecular interaction database (IntAct) [14], and the molecular interaction database (MINT) [15]. One of the first challenges to query such resources is to translate a list of names or IDs of putative Myc interactors into a suitable format that can be used to query such resources. An alternative and complementary approach is to query comprehensive integrated servers, such as the open source software for collecting, storing, and querying biological pathways (cPath) [16], Interologous Interaction Database (I2D) (http://ophid.utoronto.ca/i2d) [17], the reference index for protein interaction data (iRefIndex) [18], or Proteomics Standards Initiative common query interface (PSICQUIC) [19, 20]. Several of these also support diverse identifiers from gene and protein names, including Genbank IDs, Swiss-Prot, and Affymetrix probe IDs. Local data may be loaded from a specific file format, such as ASCII TAB-delimited or Proteomics Standards Initiative-Molecular Interaction (PSI-MI) [14] file format. For further enrichment analysis, it may be useful to generate multiple random graphs to form background distributions and then identify enrichment of Gene Ontology (GO) terms, disease biomarkers, or pathways. Alternatively, overrepresented network properties may be identified, such as frequency of motifs and graphlets [21–23].

For example, Myc has many known and predicted interactions in I2D ver. 2, which contains 17,229 human proteins and 222,917 physical human protein interactions. Querying I2D ver. 2 with P01106 Swiss-Prot ID for MYC results in a network of 861 proteins and 7,355 interactions. Using source information, one can simply "zoom in" only on interactomes generated from Y2H- or MS-based data [24]. Alternatively, one can only use direct interactions from PSICQUIC web server to identify 823 proteins. Detangling this MYC interactome "hairball" can be done by considering additional parameters (such as co-localization and co-expression of binding partners and presence of compatible interaction domains) or considering other data that may prioritize certain proteins or pathways within a certain biological context. This can be done interactively in network visualization tools, such as Network Analysis, Visualization, and Graphing TORonto (NAViGaTOR) (http://ophid.utoronto.ca/navigator/) or Cytoscape (http://www.cytoscape.org/).

Annotating the resulting protein interaction network can be done on multiple levels. Tools such as the Database for Annotation, Visualization, and Integrated Discovery (DAVID) [25] help identify enriched pathways, diseases, GO terms, etc. Identifying other connected pathway members from Reactome [26], the Kyoto Encyclopedia of Genes and Genomes (KEGG) [27], Pathway Commons (http://www.pathwaycommons.org/), WikiPathways [28], or other sources may be needed, as enrichment analysis will usually give different results depending on which pathway database is used.

Once prioritization of the putative Myc-binding proteins is accomplished, the next step is to validate the individual candidates. The essential methods used to identify and determine an interaction are Co-IPs and pull-downs. In this chapter, we will discuss and describe these validation methods (Fig. 1, white rectangle) that we use in our lab, which are essentially consensus protocols used throughout the Myc field.

Co-IP assays: Co-IP is a method that, by immunoaffinity, allows the identification of protein–protein interactions at physiological or pathological levels within cells. An antibody against a specific protein will form an immune complex that can be extracted from a cell lysate through precipitation using Protein A or G beads. Any protein interacting with the immunoprecipitated protein will be bound to the immune complex and precipitated as well. After a series of washes, the immune complexes are eluted from the beads, resolved by SDS-PAGE, and then visualized by immunoblot to interrogate the presence of Myc (positive control) and the identity of the interacting proteins. For a detailed review on Co-IPs and their implementation in interactome studies, please refer to Markham et al. [29]. As Myc is a transcription factor, it is mainly found in the nuclear fraction of the cell. However, since it has also been reported to be located in the cytoplasm [30], we include in this chapter a protocol to Co-IP Myc using whole-cell extracts (WCE) as well as a protocol using cell fractionation, in which the Myc Co-IP can be conducted from the cytoplasmic or nuclear fraction.

In vitro pull-down assays: This assay evaluates whether a bait protein that is bound to a solid support can specifically interact with a prey protein that is in solution. Briefly, the immobilized bait protein is incubated with the prey protein, and after several wash steps, an interaction is scored by resolving the eluate and detecting the bait and prey proteins using SDS-PAGE and immunoblotting, respectively. The solid support used to immobilize the bait is often agarose beads. The bait may be directly or, more commonly, indirectly bound to the bead. As such, the pull-down assay consists of a bait protein that is tagged and retained on an immobilized affinity ligand that is specific for the tag. For example, the protein tag may be glutathione S-transferase (GST), polyhistidine (polyHis), or streptavidin, which is retained by glutathione-, metal chelate (cobalt or nickel)-, or biotin-coated agarose beads, respectively. Once the tagged protein is immobilized by the affinity ligand, a complex is formed. This immobilized bait protein is then incubated with the prey protein in vitro. The prey protein is often generated by overexpression in mammalian cells, reticulocytes, bacteria, or insect cells. An extract is then prepared and incubated with the bait.

In vitro pull-down assays are often used to evaluate direct PPIs that are predicted by previous methods such as Co-IP. The main limitation of Co-IP is that proteins that interact in a complex may not necessarily do so directly, because the interaction may be mediated by other molecules in the complex. To determine whether a given PPI is due to a direct interaction, the in vitro pull-down assay with bait and prey proteins produced in bacteria can be used.

A direct interaction cannot be evaluated through the use of prey protein extracts from mammalian cells or reticulocyte lysates. In this chapter we will emphasize the use of the pull-down assay as a method to validate direct PPIs.

There are a number of complementary assays that can be performed to determine if an identified interactor is a transcriptional regulator in complex with Myc, including an eletrophoretic mobility shift assay (EMSA) and chromatin immunoprecipitation (ChIP). EMSA is an in vitro assay employed to characterize protein–DNA or protein–RNA interactions, in which a protein or a mixture of protein interactors can be evaluated for its binding to the same DNA or RNA fragment. By contrast, ChIP allows the identification of chromatin regions bound by proteins of interest *in vivo*. In a ChIP reaction, the chromatin-cross-linked protein complexes can be immunoprecipitated in one reaction with an antibody against Myc and in another reaction with an antibody against a potential interactor. These reactions are then assayed by quantitative real-time PCR (Q-PCR) to determine whether the binding of each protein of interest occurs at the same region of the genome. We have optimized the experimental parameters for high-throughput ChIP experiments, in which detailed protocols for Myc ChIP as well as Q-PCR are described [31]. To further interrogate the potential for two proteins to bind to the same region of DNA, re-ChIP can be performed, in which the cross-linked samples are first immunoprecipitated using an antibody against one of the proteins, followed by a second sequential immunoprecipitation using an antibody specific for a second protein of interest, followed by Q-PCR. To identify protein–DNA interactions as well as potential protein complexes interacting on a broader genome-wide scale, two similar techniques are used: ChIP followed by microarray hybridization (ChIP-chip) or high-throughput sequencing (ChIP-seq). The ChIP-seq technology can interrogate the entire human genome and offers higher resolution, less noise, and a better coverage than ChIP-chip and in recent years has become the technique of choice for genome-wide protein-DNA interaction studies. All these additional techniques are presented in other chapters of this book.

An advantage for the Myc field is that the Encyclopedia of the DNA Elements (ENCODE) project at University of California, Santa Cruz (UCSC), has included Myc in its publicly available studies. Thus, the Myc-binding sites across a small panel of cell lines can be readily visualized. Detailed information on how to use this useful tool can be accessed at the Genome UCSC webpage (http://genome.ucsc.edu/ENCODE/aboutScaleup.html).

Finally, once the regions of interaction between Myc and its direct binding partner are known (*see* **Notes 1** and **2**), further analysis at the level of nuclear magnetic resonance (NMR) or crystallography analysis is worth considering. With these methodologies, it is possible to achieve atomic resolution and better understanding of the structural basis of the interaction, as was performed for Myc-Max and Myc-Bin1 interactions [32–34]. This knowledge can be instrumental to the development of inhibitors that disrupt the interaction, which may ultimately lead to novel therapeutic agents that block Myc activities.

2 Materials

Prepare all solutions using ultrapure water (prepared by purifying deionized water to attain a sensitivity of 18 MΩ cm at 25 °C). Carefully follow all disposal regulations when disposing waste materials.

2.1 Components of Co-IP Using Whole-Cell Extract

1. 1× phosphate-buffered saline (PBS) at 4 °C.
2. 500× protease inhibitor cocktail: 5 mg/mL antipain hydrochloride (Sigma-Aldrich), dissolved in ethanol; 10 mg/mL leupeptin A (Sigma-Aldrich), dissolved in H_2O; 10 mg/mL aprotinin (Sigma-Aldrich), dissolved in H_2O; and 10 mg/mL pepstatin A (Sigma-Aldrich), dissolved in 10 % acid acetic in methanol. The latter will take at least 3 h to dissolve. Once the four components are dissolved, prepare the cocktail as follows: 1 mL of each component, 16 mL of H_2O. Aliquot into small volumes and keep at −80 °C. Freeze-thaw only twice.
3. IP WCE Buffer: 25 mM Tris pH 7.6, 0.5 % Nonidet P40 (NP-40) (IGEPAL, Sigma-Aldrich), 250 mM NaCl, 3 mM ethylenediamine tetraacetic acid (EDTA), 3 mM ethylene glycol tetraacetic acid (EGTA). Store at 4 °C. Prior to use add 1× protease inhibitor cocktail, 1 mM phenylmethylsulfonyl fluoride (PMSF), and 0.5 mM dithiothreitol (DTT).
4. IP WCE Wash Buffer: 25 mM Tris pH 7.6, 1 % NP-40 (IGEPAL, Sigma-Aldrich), 200 mM NaCl, 3 mM EDTA, and 3 mM EGTA. Store at 4 °C. For more stringent washes, prepare the wash buffer at higher concentration of NaCl (500–750 mM).
5. Use Bradford assay to quantify proteins.
6. Bovine serum albumin (BSA) standard curve to quantify proteins by Bradford assay.
7. Protein G sepharose (50 % slurry; GE Healthcare).
8. Sodium dodecyl sulfate (SDS) buffer: 1 % SDS [v/v], 10 % glycerol [v/v], 0.08 M Tris pH 6.8, 1 mM β-mercaptoethanol [v/v], and 1 % bromophenol blue [w/v].
9. Boiling block.
10. Reagents for SDS-PAGE and immunoblotting. SDS-PAGE [SDS-polyacrylamide gel (7–12 %), Electrophoresis buffer

(25 mM Tris-Base, 192 mM Glycine, 0.1 % SDS)]. Immunoblotting [0.45 μM pore PVDF membrane (Immobilon-P, Millipore Corporation, Bedford, MA, USA), Immunoblotting buffer (25 mM Tris-Base, 192 mM Glycine, 20 % methanol), blocking buffer (0.1 % PBS-Tween 20 (Sigma-Aldrich)), 5 % nonfat milk].

11. Antibodies for Co-IP: We use a homemade N262-rabbit antibody (Ab) to IP Myc and the pre-bleed from the same rabbit as a control. If using commercial N262-rabbit Ab (e.g., sc-764, Santa Cruz biotechnology), we recommend starting with 1 μg of Ab. In this case use 1 μg of normal rabbit IgG as a control. For the reciprocal Co-IP, use the Ab against the protein of interest for which the concentration should be addressed empirically.
12. Antibodies for immunoblotting: To detect Myc we use 9E10 homemade monoclonal antibody, which can be purchased from commercial vendors (e.g., sc-40, Santa Cruz biotechnology). To detect the interactor use an Ab against the protein of interest for which the concentration should be addressed empirically.

2.2 Components of Co-IP Using Nuclear or Cytoplasmic Extracts

1. 1× PBS at 4 °C.
2. 500× protease inhibitor cocktail: same as above Subheading 2.1, **item 2**.
3. Buffer A: 10 mM HEPES pH 7.9, 10 mM KCl, 0.1 mM EDTA, and 0.1 mM EGTA. Store at 4 °C. Prior to use add 1× protease inhibitor cocktail, 1 mM PMSF, and 1 mM DTT.
4. NP-40 (IGEPAL, Sigma-Aldrich).
5. Buffer B: 20 mM HEPES pH 7.9, 400 mM NaCl, 1 mM EDTA, and 1 mM EGTA. Store at 4 °C. Prior to use add 1× protease inhibitor cocktail, 1 mM PMSF, and 1 mM DTT.
6. Buffer C: 20 mM HEPES pH 7.9, 1 mM EDTA, and 1 mM EGTA. Prior to use add 1× protease inhibitor cocktail and 1 mM PMSF.
7. Protein G sepharose (50 % slurry; GE Healthcare).
8. IP Wash Buffer: Buffers C and D in equal volumes, 0.1 % NP-40 (IGEPAL, Sigma-Aldrich).
9. SDS buffer: 1 % SDS [v/v], 10 % glycerol [v/v], 0.08 M Tris pH 6.8, 1 mM β-mercaptoethanol [v/v], and 1 % bromophenol blue [w/v].
10. Boiling block.
11. Reagents for SDS-PAGE and immunoblotting: *See* Subheading 2.1, **item 10**.

12. Antibodies for Co-IP: same as above Subheading 2.1, **item 11.**
13. Antibodies for immunoblotting: same as above Subheading 2.1, **item 12.**

2.3 Components of GST Pull-Downs

1. Appropriate sources of protein include lysates from protein expression systems for both GST-tagged protein (bait) and free protein (prey) (i.e., extract from E. coli with ectopic expression of prey protein or baculovirus containing prey protein-infected insect cells), lysates derived from in vitro transcription/translation reactions, and previously purified proteins from a variety of cell systems.
2. Glutathione agarose beads (GE Healthcare).
3. Wash buffer: 20 mM Tris pH 8.0, 500 mM NaCl, 10 mM β-mercaptoethanol, 10 % glycerol [v/v], 1 mM PMSF, and 2 mM benzamidine.
4. High-salt wash buffer: 20 mM Tris pH 8.0, 500 mM NaCl, 10 mM β-mercaptoethanol, 10 % glycerol [v/v], 1 mM PMSF, and 2 mM benzamidine.
5. Low-salt wash buffer: 20 mM Tris pH 8.0, 150 mM NaCl, 10 mM β-mercaptoethanol, 10 % glycerol [v/v], 1 mM PMSF, and 2 mM benzamidine.
6. Elution Buffer: 10 mM reduced glutathione, 50 mM Tris pH 8.0, and 500 mM NaCl.
7. Reagents to perform SDS-PAGE, Coomassie staining, and immunoblotting. SDS-PAGE [SDS-polyacrylamide gel (7–12 %), Electrophoresis buffer (25 mM Tris-Base, 192 mM Glycine, 0.1 % SDS)]. Coomassie staining (1 g Coomassie R250, 100 mL glacial acetic acid, 400 mL methanol, 500 mL ddH_2O). Immunoblotting [0.45 μM pore PVDF membrane (Immobilon-P, Millipore Corporation), Immunoblotting buffer (25 mM Tris-Base, 192 mM Glycine, 20 % methanol), blocking buffer (0.1 % PBS-Tween 20 (Sigma-Aldrich)), 5 % nonfat milk].
8. Antibodies for immunoblotting: same as above Subheading 2.1, **item 12.**

3 Methods

3.1 Co-IP Using WCE (See Notes 1–5)

1. Cells are seeded 1 day in advance, such that 70–80 % confluency is achieved the following day.
2. Wash the cells 2× in cold PBS at 4 °C.
3. Manually scrape the cells in cold PBS and transfer to a 15 mL polypropylene tube.

4. Spin-down the cells for 3 min at 259 × *g* at 4 °C.
5. Resuspend cell pellet in IP WCE Buffer (500 μL per 10 cm tissue culture plate) to lyse the cells.
6. Vortex tubes gently and then incubate on ice for 10 min.
7. Spin-down lysates for 3 min at 4,500 × *g* at 4 °C.
8. Transfer supernatant containing cell lysate to a fresh tube.
9. Perform protein quantification by the Bradford protocol and extrapolate from a standard curve generated by a dilution series of BSA.
10. In 1.5 mL tubes, incubate 500 μg of total WCE per IP for 3 h at 4 °C with 10 μL of homemade rabbit polyclonal c-Myc antibody (N262) or 10 μL of pre-bleed. If commercial N262 antibody is used, incubate with 1 μg of N262-rabbit or 1 μg of normal rabbit IgG. The final volume is 500 μL. Dilute proteins with IP WCE Buffer, if necessary.
11. Keep 50 μg of total WCE as positive control for protein expression to be analyzed in **step 16** with the immunoprecipitated samples.
12. Add 50 μL of protein G sepharose to each tube and incubate samples overnight at 4 °C.
13. Wash the antibody–protein complexes with IP WCE wash buffer. Precipitate the complexes with the beads by centrifuging for 30 s at 956 × *g* at 4 °C. Discard supernatant and repeat three more times.
14. Recover the complexes by adding 25 μL of SDS Lysis Buffer.
15. Boil for 5 min. Harvest supernatant. These immunoprecipitated samples can be stored at −20 °C.
16. After boiling analyze immunoprecipitated samples by SDS-PAGE and immunoblotting.
17. To detect c-Myc by immunoblotting, incubate membrane with antihuman c-Myc monoclonal antibody (9E10; 1:1,000).
18. Proteins can be visualized by chemiluminescence reagents. When quantification is important, then the detection method of choice is LI-COR (LI-COR Biosciences) as precision and linearity of fluorescence detection allows an accurate quantification.

3.2 Co-IP Using Nuclear or Cytoplasmic Extracts (See Notes 1–5)

1. Cells are seeded 1 day in advance, such that 70–80 % confluency is achieved the following day.
2. Wash the cells 2× in cold PBS at 4 °C.
3. Manually scrape the cells in cold PBS and transfer to a 15 mL polypropylene tube.
4. Spin-down the cells for 3 min. at 259 × *g* at 4 °C.

5. Resuspend cell pellet in Buffer A (500 μL per 10 cm tissue culture plate) to lyse the cells.
6. Incubate on ice for 15 min.
7. Add NP-40 (IGEPAL) to a final concentration of 0.5 %.
8. Vortex for 2 s at maximum.
9. Spin-down at 353 × *g* for 5 min at 4 °C.
10. Keep supernatant on ice if doing cytoplasmic Co-IP and continue on **step 15**.
11. For nuclear extraction resuspend pellet in Buffer B (500 μL per 10 cm tissue culture plate).
12. Vortex for 2 s.
13. Incubate on ice for 10 min.
14. Spin-down at 4,500 × *g* for 10 min at 4 °C.
15. Dilute supernatant containing the nuclear fraction with Buffer C (500 μL per 10 cm tissue culture plate). At this point, if studying cytoplasmic fractions, dilute the cytoplasmic fraction with Buffer C.
16. Incubate each lysate for 3 h at 4 °C in 1.5 mL tubes with 10 μL of homemade rabbit polyclonal c-Myc antibody (N262) or 10 μL of pre-bleed. If commercial N262 antibody is used, incubate with 1 μg of N262-rabbit or 1 μg of normal rabbit IgG.
17. Keep 10 % of each fraction to be analyzed in **step 22** with the immunoprecipitated samples and used as a control for protein expression.
18. Add 50 μL of protein G sepharose to each tube and incubate for 1 h at 4 °C.
19. Wash the antibody–protein complexes with IP Wash Buffer. Precipitate the complexes with protein G sepharose beads by centrifuging for 30 s at 956 × *g* at 4 °C. Discard supernatant and repeat three more times.
20. Recover the complexes by adding 25 μL of SDS Lysis Buffer.
21. Boil for 5 min. Supernatant containing immunoprecipitated samples can be stored at −20 °C.
22. After boiling analyze immunoprecipitated samples by SDS-PAGE and immunoblotting.
23. To detect c-Myc by immunoblotting, incubate membrane with antihuman c-Myc monoclonal antibody (9E10; 1:1,000).
24. Proteins can be visualized by chemiluminescence reagents. When quantification is important, then the detection method of choice is LI-COR (LI-COR Biosciences) as precision and linearity of fluorescence detection allows an accurate quantification.

3.3 GST Pull-Downs (See Notes 1 and 6)

1. Incubate 100 μg of the GST control and the GST-tagged proteins with 50 μL of 50 % slurry of glutathione agarose beads for 1 h at 4 °C to allow for binding.
2. Wash 2× with 1 mL of Wash Buffer to remove unbound proteins.
3. Incubate the GST-bead complexes with 100 μg of prey protein for 1 h at 4 °C to allow binding between the two proteins.
4. Wash 1× with high-salt wash buffer.
5. Wash 1× with low-salt wash buffer.
6. Elute the bound complexes with Elution Buffer.
7. Resolve by SDS-PAGE.
8. Visualize by Coomassie staining and/or immunoblotting to detect with antibodies against Myc and the interactor.

4 Notes

1. To further characterize the interaction regions of the two proteins, differential binding assays are performed. In this type of assays, deletions, point mutations, small fragments, or larger domains are cloned into mammalian expression vectors to perform Co-IP with cells expressing these constructs. This allows the regions of interaction to be mapped. Differential binding assays have been successfully used to identify interactors that are dependent on a specific region of Myc for binding (e.g., Myc box (MB) II region and TRRAP). In addition, these assays are conducted for high-throughput screenings of protein–protein interactions.
2. Ideally, reciprocal Co-IPs are performed. However, in some cases, the reciprocal Co-IP may not show the expected interaction. This Co-IP assay relies on the accessibility of the antibody to bind the recognized epitope on the protein which could potentially be masked by the studied interactor or other proteins within the complex.
3. Low affinity or transient protein–protein interactions can sometimes be difficult to assess by Co-IP unless the interaction can be stabilized. This can be achieved by employing methods such as cross-linking the sample proteins prior to Co-IP. The sensitivity of this method will depend on the spatial distribution of the chemical groups of the proteins involved in the cross-link and will often require empirical optimization. Another approach is to treat the cells with proteasomal inhibitor, such as MG132, for 4–6 h to allow the protein–protein interaction to be temporarily stabilized.

4. Many protein interactions remain intact using non-denaturing buffers, as described above for Co-IP methods; however, using buffers with low ionic strength, such as ≤120 mM NaCl, can help retain very weak interactions. Again, this would need to be empirically tested.
5. Background or nonspecific binding when performing a Co-IP can be addressed by adjusting the salt concentration of the Co-IP buffer (ranges from 100 to 1,000 mM NaCl), decreasing the amount of primary antibody, or pre-clearing the lysates. Lysates can be pre-cleared by incubating the samples in the presence of the Protein A or G sepharose beads and/or nonspecific antibody from the same species as the primary antibody.
6. As in **Note 1** above, further characterization of the interaction regions of two proteins can also be achieved by conducting in vitro pull-downs with the regions or fragments of interest.

Acknowledgments

We would like to acknowledge the members of the Penn lab, Manpreet Kalkat and Lindsay Lustig, for helpful discussions and critical review of this chapter. This work was supported by Canadian Institutes of Health Research (CIHR #MOP123289, L.Z.P), Ontario Research Fund (GL2-01-030, I. J., L.Z.P.), Canada Foundation for Innovation (CFI #12301 and CFI #203373 I. J.), and Canada Research Chairs Program (CRC #203373 L.Z.P. and CRC #225404, I.J.). Additional support was provided by the Ontario Ministry of Health and Long-Term Care. The views expressed do not necessarily reflect those of the OMOHLTC.

References

1. Blackwood EM, Eisenman RN (1991) Max: a helix-loop-helix zipper protein that forms a sequence-specific DNA-binding complex with Myc. Science 251:1211–1217
2. Cheng SW, Davies KP, Yung E et al (1999) c-MYC interacts with INI1/hSNF5 and requires the SWI/SNF complex for transactivation function. Nat Genet 22:102–105
3. Peukert K, Staller P, Schneider A et al (1997) An alternative pathway for gene regulation by Myc. EMBO J 16:5672–5686
4. Sakamuro D, Elliott KJ, Wechsler-Reya R et al (1996) BIN1 is a novel MYC-interacting protein with features of a tumour suppressor. Nat Genet 14:69–77
5. Wang Q, Zhang H, Kajino K et al (1998) BRCA1 binds c-Myc and inhibits its transcriptional and transforming activity in cells. Oncogene 17:1939–1948
6. Hirst M, Ho C, Sabourin L et al (2001) A two-hybrid system for transactivator bait proteins. Proc Natl Acad Sci U S A 98:8726–8731
7. Huang A, Ho CS, Ponzielli R et al (2005) Identification of a novel c-Myc protein interactor, JPO2, with transforming activity in medulloblastoma cells. Cancer Res 65:5607–5619
8. Johnson S, Stockmeier CA, Meyer JH et al (2011) The reduction of R1, a novel repressor protein for monoamine oxidase A, in major depressive disorder. Neuropsychopharmacology 36:2139–2148
9. Kalkat M, Wasylishen AR, Kim SS et al (2011) More than MAX: discovering the Myc interactome. Cell Cycle 10:374–375

10. Bader GD, Donaldson I, Wolting C et al (2001) BIND–the biomolecular interaction network database. Nucleic Acids Res 29:242–245
11. Reguly T, Breitkreutz A, Boucher L et al (2006) Comprehensive curation and analysis of global interaction networks in Saccharomyces cerevisiae. J Biol 5:11
12. Xenarios I, Rice DW, Salwinski L et al (2000) DIP: the database of interacting proteins. Nucleic Acids Res 28:289–291
13. Eskin E, Sharan R, Halperin E (2006) A note on phasing long genomic regions using local haplotype predictions. J Bioinform Comput Biol 4:639–647
14. Hermjakob H, Montecchi-Palazzi L, Lewington C et al (2004) IntAct: an open source molecular interaction database. Nucleic Acids Res 32:D452–D455
15. Zanzoni A, Montecchi-Palazzi L, Quondam M et al (2002) MINT: a molecular INTeraction database. FEBS Lett 513:135–140
16. Cerami EG, Bader GD, Gross BE et al (2006) cPath: open source software for collecting, storing, and querying biological pathways. BMC Bioinformatics 7:497
17. Brown KR, Jurisica I (2007) Unequal evolutionary conservation of human protein interactions in interologous networks. Genome Biol 8:R95
18. Razick S, Magklaras G, Donaldson IM (2008) iRefIndex: a consolidated protein interaction database with provenance. BMC Bioinformatics 9:405
19. Orchard S, Kerrien S, Abbani S et al (2012) Protein interaction data curation: the International Molecular Exchange (IMEx) consortium. Nat Methods 9:345–350
20. Orchard S, Kerrien S, Jones P et al (2007) Submit your interaction data the IMEx way: a step by step guide to trouble-free deposition. Proteomics 7(Suppl 1):28–34
21. Geraci J, Liu G, Jurisica I (2012) Algorithms for systematic identification of small subgraphs. Methods Mol Biol 804:219–244
22. Przulj N, Corneil DG, Jurisica I (2004) Modeling interactome: scale-free or geometric? Bioinformatics 20:3508–3515
23. Przulj N, Corneil DG, Jurisica I (2006) Efficient estimation of graphlet frequency distributions in protein–protein interaction networks. Bioinformatics 22:974–980
24. Miller AK, Marsh J, Reeve A et al (2010) An overview of the CellML API and its implementation. BMC Bioinformatics 11:178
25. da Huang W, Sherman BT, Tan Q et al (2007) DAVID bioinformatics resources: expanded annotation database and novel algorithms to better extract biology from large gene lists. Nucleic Acids Res 35:W169–W175
26. Vastrik I, D'Eustachio P, Schmidt E et al (2007) Reactome: a knowledge base of biologic pathways and processes. Genome Biol 8:R39
27. Kanehisa M, Goto S, Kawashima S et al (2002) The KEGG databases at GenomeNet. Nucleic Acids Res 30:42–46
28. Pico AR, Kelder T, van Iersel MP et al (2008) WikiPathways: pathway editing for the people. PLoS Biol 6:e184
29. Markham K, Bai Y, Schmitt-Ulms G (2007) Co-immunoprecipitations revisited: an update on experimental concepts and their implementation for sensitive interactome investigations of endogenous proteins. Anal Bioanal Chem 389:461–473
30. Conacci-Sorrell M, Ngouenet C, Eisenman RN (2010) Myc-nick: a cytoplasmic cleavage product of Myc that promotes alpha-tubulin acetylation and cell differentiation. Cell 142:480–493
31. Ponzielli R, Boutros PC, Katz S et al (2008) Optimization of experimental design parameters for high-throughput chromatin immunoprecipitation studies. Nucleic Acids Res 36:e144
32. Andresen C, Helander S, Lemak A et al (2012) Transient structure and dynamics in the disordered c-Myc transactivation domain affect Bin1 binding. Nucleic Acids Res 40:6353–6366
33. Beaulieu ME, McDuff FO, Frappier V et al (2012) New structural determinants for c-Myc specific heterodimerization with Max and development of a novel homodimeric c-Myc b-HLH-LZ. J Mol Recognit 25:414–426
34. Nair SK, Burley SK (2003) X-ray structures of Myc-Max and Mad-Max recognizing DNA. Molecular bases of regulation by proto-oncogenic transcription factors. Cell 112:193–205

Chapter 5

Detection of c-Myc Protein–Protein Interactions and Phosphorylation Status by Immunoprecipitation

Colin J. Daniel, Xiaoli Zhang, and Rosalie C. Sears

Abstract

Co-immunoprecipitation is an invaluable technique in evaluating native protein–protein interactions in vitro and in vivo. However, it can be difficult to detect interactions of a very transient nature, particularly interactions with phosphatases and kinases. The evaluation of the phosphorylation status of c-Myc can also be challenging with the current commercially available phosphorylation sensitive antibodies. Here, we describe two protocols: one for the co-immunoprecipitation of endogenous c-Myc to detect protein–protein interactions and second, for the immunoprecipitation of endogenous c-Myc to probe for phosphorylation status.

Key words c-Myc, Immunoprecipitation, Phosphorylation, GSK3β, B56α, PP2A, Axin1, HBP1

1 Introduction

The c-Myc proto-oncoprotein is a tightly regulated transcription factor that is involved in the regulation of many cellular processes including proliferation, growth, apoptosis, and differentiation [1]. The phosphorylation status of c-Myc has been shown to play a pivotal role in c-Myc activity, stability, and degradation [2–7]. Two key phosphorylation sites, conserved across species and Myc family proteins, threonine 58 (T58) and serine 62 (S62) in the N-terminal region, can affect c-Myc protein stability and its ubiquitinylation by the SCF^{FBW7} E3 ubiquitin ligase and degradation by the 26S proteosome [2, 8–11]. Previous work has shown that c-Myc protein turnover via this mechanism involves GSK3β (the kinase that phosphorylates T58) and PP2A (the phosphatase that dephosphorylates S62), as well as the multi-domain scaffold protein Axin1 [12, 13]. c-Myc has also been shown to interact with chromatin-associated factors, including, recently, the tumor suppressor protein HBP1 [14].

Co-immunoprecipitation (Co-IP) of c-Myc and its associated binding proteins can be difficult due to the relatively low levels of

Laura Soucek and Nicole M. Sodir (eds.), *The Myc Gene: Methods and Protocols*, Methods in Molecular Biology, vol. 1012, DOI 10.1007/978-1-62703-429-6_5,

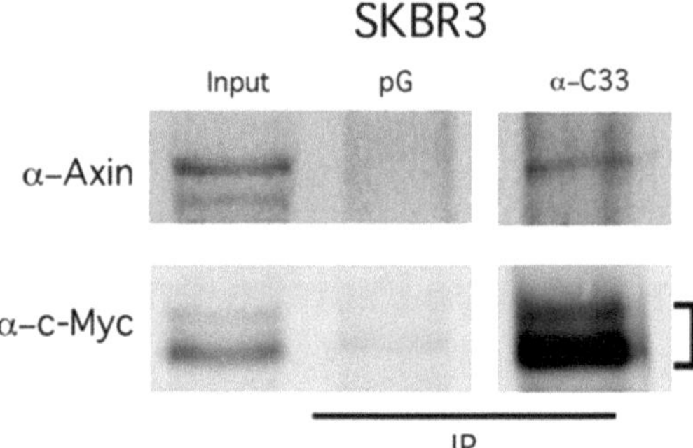

Fig. 1 Co-immunoprecipitation of endogenous c-Myc in the breast cancer cell line SKBR3. Endogenous c-Myc was immunoprecipitated from SKBR3 cell lysates using α-C33 c-Myc antibody and immunoblotted for endogenous Axin1 and total c-Myc (α-Y69). Protein G beads alone were used as an IP control

endogenous c-Myc and its localization in the nucleus, often associated with chromatin, where extraction of c-Myc without disrupting protein–protein interactions can be challenging. Here, we describe a protocol to co-immunoprecipitate endogenous c-Myc in order to look at its association with a variety of proteins, including GSK3β, PP2A-B56α, Pin1, HBP1, and the scaffold protein Axin1 [12, 13]. An example is shown of an endogenous c-Myc Co-IP with Axin1 in the breast cancer cell line SKBR3 (Fig. 1). Cells are harvested in a specific Co-IP buffer that is stringent enough to extract c-Myc from the nucleus while maintaining protein–protein interactions. The harvesting and extraction process includes several sonication steps to help with the extraction of c-Myc from the nucleus and must be carried out carefully to preserve protein associations.

The second protocol presented here is to detect the phosphorylation status of c-Myc by immunoprecipitation (IP). Detection of the phospho-S62 form of c-Myc can be particularly difficult, and it has been recently reported that the current commercially available phospho-S62 antibody reacts with a protein in fetal bovine serum used in tissue culture growth media [7]. To overcome this problem, we have found that immunoprecipitation of c-Myc prior to western blot analysis is the best method to eliminate this background band and increase detection of pS62-Myc. The IP buffer used is more stringent than the Co-IP buffer, and sonication steps are still required to extract soluble c-Myc out of the nucleus.

If antibodies of the same species are used for both the immunoprecipitation and western blot analysis, conjugation of the antibody for the IP step to Sepharose beads is recommended to minimize the IgG heavy-chain background that migrates close to c-Myc. This step will provide for a much cleaner and easier interpretation of results.

Standard SDS-PAGE using a 10 % bis-acrylamide resolving and a 4 % stacking gel was used to separate samples. Protein is then transferred to a fluorescence-specific PVDF transfer membrane using standard transfer protocols, and immune blotting is carried out and detected using the LI-COR Odyssey Infrared Imager system. All of our western blots are visualized and quantified using this system because it has a much broader linear range compared to chemiluminescence (ECL) and film. We find that ECL and film will usually exaggerate differences between band intensities. In addition, the LI-COR scanner allows two color detection such that analysis of total c-Myc and phosphorylated c-Myc can be simultaneously detected, allowing band overlay and a more precise calculation of the ratio between total and phosphoprotein.

We will finish this report by providing additional insight in to the validation and interpretation of some commercially available phospho-specific antibodies in the western blot analysis of c-Myc.

2 Materials

Prepare solutions using ultrapure Millipore water (mH_2O) and use reagents of analytical grade where possible and of electrophoresis grade for western blotting.

2.1 Tissue Culture

1. HEK 293 cells (ATCC).
2. SKBR3 cells (a gift from Dr. Gail Clinton's lab, OHSU, Portland, OR).
3. MOLM 14 cells (a gift from Dr. Brian Druker's lab, OHSU, Portland, OR).
4. Growth media: 10 % fetal bovine serum (FBS) (Thermo Scientific), Dulbecco's Modified Eagle Medium (DMEM), 2 mM L-glutamine, 1 mM sodium pyruvate, 0.1 mM nonessential amino acids, 100 units/mL penicillin, and 100 μg/mL streptomycin (Invitrogen). Combine components under sterile conditions to the indicated concentrations and store at 4 °C.
5. Growth media for MOLM 14 s: 10 % defined FBS (Thermo Scientific), Roswell Park Memorial Institute (RPMI) 1640 medium, 2 mM L-glutamine, 100 units/mL penicillin, and 100 μg/mL streptomycin (Invitrogen). Combine components under sterile conditions to the indicated concentrations and store at 4 °C.
6. Incubator: 37 °C and 5 % CO_2.
7. P100 or P150 tissue culture plates.
8. T25 tissue culture flasks.

2.2 Antibody-Bead Conjugation

1. Protein G-Sepharose Fast Flow beads (GE Healthcare).
2. Protein A IPA300 (Repligen).
3. 1 % acetylated BSA (molecular biology grade, Sigma).
4. 0.2 M Na borate pH 9: Add 3.815 g Na borate to about 40 mL mH_2O. Heat slightly and stir to dissolve. Use HCl to pH to 9.0 and bring up volume to 50 mL (*see* **Note 1**).
5. 0.2 M ethanolamine pH 8: Make fresh; add 500 μL ethanolamine to about 30 mL of mH_2O. Use HCl to pH to 8.0 and bring up volume to 40 mL (*see* **Note 2**).
6. 20 mM DMP in 0.2 M Na borate pH 9: Add 51.8 mg dimethylpimelimidate (DMP) to 10 mL 0.2 M Na borate pH 9 (*see* **Note 3**).
7. 1× Sterile phosphate-buffered saline (PBS).
8. 10 % Na azide solution.
9. 1.5 mL microfuge tube rotator (VWR).

2.3 Immunoprecipitation

1. Co-IP buffer: 20 mM Tris pH 7.5, 12.5 % glycerol, 0.2 % NP40 (Ipegal), 200 mM NaCl, 1 mM EDTA, 1 mM EGTA, and 1 mM DTT.
2. IP buffer: 20 mM Tris pH 7.5, 50 mM NaCl, 0.5 % Triton X-100, 0.5 % deoxycholic acid (DOC), 0.5 % SDS, and 1 mM EDTA.
3. Phosphatase inhibitors: 10 mM NaF, 0.1 mM NaVan, and 3 mg/mL β-glycerol phosphate.
4. Protease inhibitors: 1.6 mg/mL IAA, 0.2 mg/mL AEBSF, 1 μg/mL aprotinin, 0.5 μg/mL leupeptin, and 0.7 μg/mL pepstatin.
5. 5× SDS sample buffer (50 mL): 48 % glycerol, 0.3 M Tris pH 6.8, 10 % SDS (w/v), 3.4 M β-mercaptoethanol, and add bromophenol blue for color.
6. Cell scraper (VWR, Radnor, PA).
7. Sonicator (Branson Sonifier 450).
8. Micro-centrifuge (VWR).

2.4 Western-Blot Analysis

1. Immobilon-FL PVDF membrane (Millipore).
2. Odyssey Blocking Buffer (LI-COR Biosciences).
3. AquaBlock (Eastcoast Bio).
4. LI-COR Odyssey Infrared Imager (LI-COR Biosciences).
5. PBS-T buffer: Phosphate-buffered saline pH 7.5 and 0.1 % Tween 20.
6. Rocking platform (VWR).

2.5 Primary and Secondary Antibodies

1. Total c-Myc: N262 (rabbit polyclonal), C-33 (mouse monoclonal), and C-19 (rabbit polyclonal) (Santa Cruz Biotechnology Inc); Y69 (rabbit monoclonal) (Abcam, Cambridge, MA).
2. Phospho c-Myc: pS62 c-Myc 33A12E10 (mouse monoclonal) (Abcam);pT58 c-Myc (rabbit polyclonal) (Applied Biological Materials).
3. α-V5 (mouse monoclonal) (Invitrogen).
4. α-HA (mouse monoclonal) (Applied Biological Materials).
5. α-Axin1 (C76H11, rabbit monoclonal) (Cell Signaling).
6. Goat anti-mouse or -rabbit Alexa Fluor 680 (Molecular Probes).
7. Goat anti-mouse or -rabbit IRDye800 (Rockland).

3 Methods

3.1 Antibody to Bead Conjugation

1. Incubate 500 μL 50 % bead slurry with 1 mL of 1 % acetylated BSA for 1 h rotating at 4 °C.
2. Wash three times with 1 mL cold PBS by spinning at 2,000 × *g* for 1 min and aspirating.
3. Resuspend bead pellet to a 50 % bead slurry and add Na azide to 0.1 % to preserve (*see* **Note 4**).
4. Combine 150–400 μL 50 % bead slurry with 15–40 μL of antibody (*see* **Note 5**).
5. Incubate for 2 h to overnight at 4 °C on a rotator.
6. Wash twice with 10× bead-bed volume with 0.2 M Na borate pH 9 (0.75–2 mL).
7. Resuspend beads in 10× bead-bed volume with 20 mM DMP in 0.2 M Na borate pH 9 (0.75–2 mL).
8. Incubate for 30 min at room temperature on rotator. Do not incubate for longer than 30 min.
9. Wash once with 10× bead-bed volume with 0.2 M ethanolamine pH 8 (0.75–2 mL).
10. Resuspend bead in 10× bead-bed volume with 0.2 M ethanolamine pH 8 (0.75–2 mL).
11. Incubate for 2 h at room temperature on rotator.
12. Wash three times with 10× bead-bed volume with sterile PBS (0.75–2 mL).
13. Resuspend beads in 1–1.5× bead-bed volume with sterile PBS and add 10 % Na azide to 0.1 % to preserve. Store at 4 °C for up to a week.

3.2 Co-immunoprecipitation

1. Prepare the Co-IP buffer, add all inhibitors, and place on ice (*see* **Note 6**).
2. Aspirate media from cells. Wash once with 5 mL of cold PBS and aspirate dry. Repeat if using larger P150 tissue culture plates.
3. Add 1 mL or 2.5 mL of cold Co-IP buffer with inhibitors to a P100 or P150 tissue culture plate, respectively.
4. Gently rock the plate side to side to coat all of the plate with buffer and incubate on ice or at 4 °C for 5 min to lyse.
5. Pipette cells up and down five times to break up cell clumps.
6. Carefully scrape and collect cells in a 1.5 mL microfuge tube for a P100 or a 5 mL tube for a P150. Place on ice.
7. Sonicate for 10 pulses at 1 output and 15 % duty (*see* **Note 7**).
8. Incubate on ice for 20 min.
9. Centrifuge at 16,000 × *g* for 10 min at 4 °C.
10. Transfer cleared lysate to new tubes on ice.
11. Transfer 5 % volume of cleared lysate to a new tube as an input sample and place on ice.
12. Add about 15 μL of antibody-conjugated beads to the lysate and rotate tubes at 4 °C for 2–4 h (*see* **Note 8**).
13. Wash beads by pelleting beads at 2,000 × *g* for 1 min at 4 °C, aspirating off supernatant and adding 400 μL of Co-IP buffer. Gently flick tubes, invert several times, and respin (*see* **Note 9**).
14. Repeat for a total of four times (*see* **Note 10**).
15. Resuspend washed beads in 40 μL of 1.5× SDS sample buffer and gently flick tubes.
16. Add 5× SDS sample buffer to input samples to a final concentration of 1.5× (*see* **Note 11**) and gently flick tubes.
17. Boil samples for 10 min prior to SDS-PAGE (*see* **Note 12**).
18. Perform a standard SDS-PAGE (*see* **Note 13**).
19. Transfer protein to an Immobilon-FL PVDF membrane (*see* **Note 14**).
20. Block the membrane in Odyssey Blocking Buffer or AquaBlock for 45 min to an hour with gentle rocking.
21. Prepare primary antibody by diluting 1:500–1:1,000 in 1:1 AquaBlock and PBS-T (*see* **Note 15**).
22. Incubate membrane with primary antibody overnight at 4 °C with gentle rocking.
23. Wash membrane three times with PBS-T.
24. Prepare secondary antibody by diluting 1:10,000 in 1:1 AquaBlock and PBS-T (*see* **Note 16**).

25. Incubate membrane for 1 h at room temperature with gentle rocking.
26. Wash membrane twice with PBS-T and once with PBS (*see* **Note 17**).
27. Image blot on LI-COR Odyssey Infrared Imager (*see* Fig. 1).

3.3 Immunoprecipitation for Detection of c-Myc Serine 62 Phosphorylation

1. Prepare the IP buffer, add all inhibitors, and place on ice.
2. Add 1 or 2.5 mL of cold IP buffer with inhibitors to a P100 or P150 tissue culture plate, respectively.
3. Gently rock the plate side to side to coat all of the plate with buffer and incubate on ice or at 4 °C for 5 min to lyse.
4. Pipette cells up and down five times to break up cell clumps (*see* **Note 18**).
5. Carefully scrape and collect cells in a 1.5 mL microfuge tube for a P100 or a 5 mL tube for a P150. Place on ice.
6. Sonicate for 3 × 10 pulses at 1 output and 15 % duty (*see* **Note 19**).
7. Incubate on ice for 20 min.
8. Centrifuge at 16,000 × *g* for 10 min at 4 °C.
9. Transfer cleared lysate to new tubes on ice (*see* **Note 20**).
10. Transfer 5 % volume of cleared lysate to a new tube as an input sample and place on ice.
11. Add 1–2 μg of antibody and incubate overnight at 4 °C with tubes rotating, then add 15 μL of beads for 1 h. (*see* **Note 21**), then add 15 μL of beads for 1 h.
12. Wash beads by pelleting beads at 2,000 × *g* for 1 min at 4 °C, aspirating off supernatant, and adding 400 μL of IP buffer. Gently flick tubes, invert several times, and respin.
13. Repeat for a total of four times.
14. Resuspend washed beads in 40 μL of 1.5× SDS sample buffer and gently flick tubes.
15. Add 5× SDS sample buffer to input samples to a final concentration of 1.5× and gently flick tubes.
16. Perform **steps 17** through **27** as described above in Subheading 3.2.

3.4 Detection of c-Myc Phosphorylation by Commercially Available Phospho-Specific Antibodies

We have been able to identify T58 phosphorylated c-Myc using the phospho-T58 antibody (α-pT58) from ABM (*see* Fig. 2a, b) [4, 6]. The specificity of this antibody has been shown in [6, 7] and is demonstrated here by immunoprecipitation of ectopically expressed, HA-tagged wild-type or T58A mutant c-Myc in HEK 293 cells and immunoblotting with α-pT58 and α-HA (*see* Fig. 2a). In this result, no band is detected in the T58A mutant c-Myc lane when α-pT58 is used for detection in the western blot. Phosphorylation at T58 is

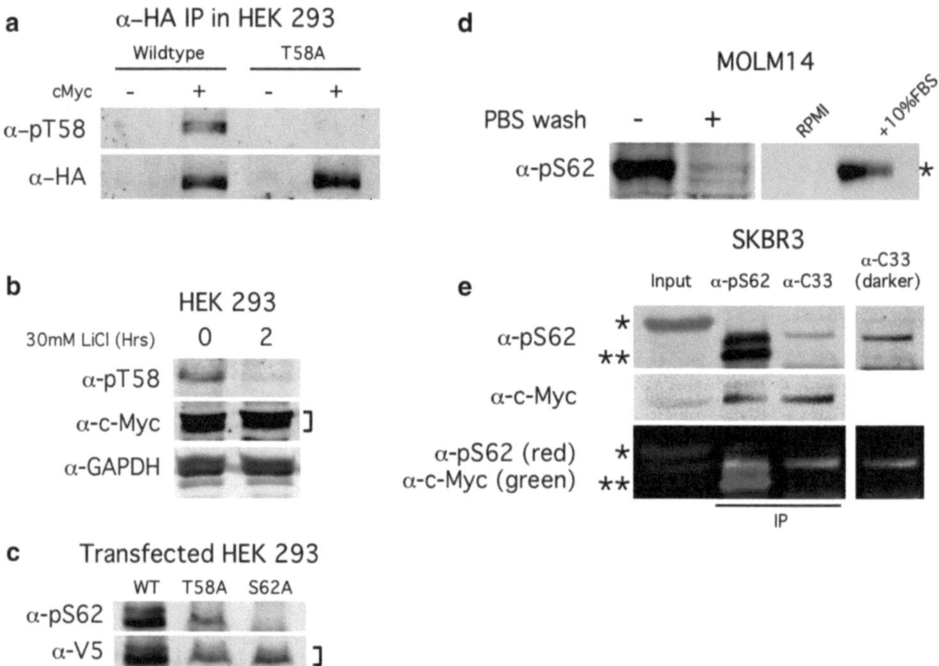

Fig. 2 Detection of phosphorylated c-Myc. (**a**) pT58 antibody is reactive with wild-type c-Myc but not the T58A mutant. HA-tagged c-Myc-WT and c-Myc-T58A were expressed in HEK 293 cells (+) and immunoprecipitated with α-HA antibody. Immunoprecipitates were detected by western blotting with α-HA (ABM) and α-pT58 (ABM). (**b**) Inhibition of the GSK3β kinase by LiCl reduces a band reactive to the pT58 antibody. Endogenous c-Myc was detected by the N262 α-c-Myc antibody; α-pT58 (ABM) was used to detect pT58-Myc, and α-GAPDH was used as a loading control. (**c**) Ectopic V5-tagged c-Myc-WT, c-Myc-T58A, and c-Myc-S62A were expressed in HEK 293 cells and detected by α-V5 and α-pS62 antibodies. S62-phosphorylated c-Myc was only detected in the WT and T58A lanes. It should be noted that expression of endogenous c-Myc inhibits expression of ectopic c-Myc in this system. (**d**) The 33A12E10 phospho-S62 antibody cross-reacts with a protein in fetal bovine serum that migrates to a similar region as c-Myc in SDS-PAGE (indicated by *asterisk*). Washing cells with PBS prior to harvesting significantly reduces this nonspecific band. (**e**) Endogenous c-Myc was immunoprecipitated with α-pS62 and α-C33 antibodies from SKBR3 cell lysates and detected using both the α-pS62 and total c-Myc antibody Y69 from Abcam. The nonspecific background band (indicated by *asterisk*) is eliminated once the IP is performed to reveal a c-Myc-specific band recognized by both antibodies (*double asterisk* indicates IgG heavy chain). The panels on the far right are darker scans from the α-C33 immunoprecipitation. Dual color detection of pS62 and total c-Myc demonstrates an overlapping (*yellow*) IP'd c-Myc band (color figure online)

catalyzed by the GSK3β kinase, and treatment with LiCl can inhibit this kinase [8]. To further demonstrate the specificity of this antibody here, HEK 293 cells were treated with 30 mM LiCl for 2 h resulting in the reduction of T58 phosphorylated endogenous c-Myc (*see* Fig. 2b). It is important to note that we have also reported detection of pT58 c-Myc with an antibody raised against a dually phosphorylated (pT58/pS62) peptide (Cell Signaling) [8]. In this case, membrane blocking and primary antibody detection must be performed in 5 % and 2.5 % non-fat milk/PBS-T, respectively, to inhibit pS62 detection.

We have verified the specificity of the phospho-S62 antibody 33A12E10 from Abcam in [6, 7]. This is demonstrated here by expressing V5-tagged forms of wild-type (WT), T58A, and S62A c-Myc with transient transfection in HEK 293 cells. Cell lysates were probed with α-V5 and α-pS62 antibodies. S62-phosphorylated c-Myc was detected in the c-Myc-WT and c-Myc-T58A transfected cells, but not the c-Myc-S62A transfected cells with the 33A12E10, pS62 antibody (*see* Fig. 2c). As previously mentioned, this commercially available phospho-S62 antibody reacts with a protein found in fetal bovine serum used in tissue culture media. This background band runs within the size range for c-Myc at around 67 kDa. It is possible to reduce the amount of cross-reactivity by washing cells with PBS prior to lysis as we have recently reported [7]. This is illustrated here in a lysate prepared from the acute monocytic leukemia cell line MOLM14 (*see* Fig. 2d). However, due to variations in cell type and density, simple washing may not be reliable or consistent especially when using cells that express low levels of endogenous c-Myc. As described in this chapter, we have found that immunoprecipitation prior to western blotting provides for clearer results (*see* Fig. 2e). Here we used the commercial α-pS62 and α-c-Myc (C33) antibody to immunoprecipitate endogenous c-Myc in the breast cancer cell line SKBR3, which expresses high levels of S62-phosphorylated c-Myc [12]. In the input lane, c-Myc is detected with a total c-Myc antibody, but the α-pS62 antibody primarily detects the background serum band (indicated by asterisk). However, after immunoprecipitation, the prominent enriched band is reactive to both pS62 and total c-Myc antibodies (IgG heavy chain is indicated by double asterisk). Overlap between the pS62 and total c-Myc antibody staining appears yellow in the merged red and green colored blot (color version online).

4 Notes

1. This solution should be made fresh but can be aliquoted into 5 × 10 mL portions and kept frozen for future use. We typically do not use aliquots older than 2 weeks.
2. This solution should also be made fresh but can be aliquoted into 4 × 10 mL portions and kept frozen for future use. We typically do not use aliquots older than 2 weeks. Titrating to the right pH for this solution can be difficult. Initial pH is around 12 but will quickly drop below 8 if using 12 M HCl. It is recommended to use diluted 6 M HCl and add dropwise.
3. This reagent should be made fresh on the day of bead conjugation.
4. Blocking beads with acetylated BSA can reduce the amount of nonspecific binding. This step can be done prior to actual antibody-bead conjugation and stored at 4 °C for several weeks.

5. Typical antibody-bead ratio range from 1:10 to 1:50. Make sure beads used are isotype compatible. We find that pA beads are much cleaner than pG alone. A combination of pA/G can be used to minimize background.
6. The buffer is typically made fresh on the day of the experiment. Inhibitors are added prior to lysing of cells.
7. Sonication setting can vary depending on cell type. Typically, the lowest setting is favored. Sonication is important to help the release of endogenous c-Myc from the nucleus.
8. Add the appropriate amount of antibody-conjugated beads to achieve about 1–2 μg of antibody per IP. If the antibody is not conjugated to beads, go ahead and add the equivalent amount of antibody (1–2 μg) to the IP and incubate for the indicated times followed by the addition of 15 μL of beads for an additional 1 h. Antibody incubation times vary depending on antibody used. Typically, incubations do not go overnight as unstable complexes may be lost after such extended periods.
9. Supernatant is aspirated using a flathead pipette tip with a 0.2 mm bore. If the bead pellet is too small to visualize, you can add about 15 μL of pre-blocked beads to make a larger more visible pellet.
10. Bead washing conditions can be varied based on cell type and co-immunoprecipitated protein of interest. Typically, if there is a high amount of background binding of nonspecific proteins to the beads, washing buffer volume and/or repetitions can be increased. In some instances, if the background is still high, the amount of antibody-conjugated beads used can be decreased. Another way of minimizing background is to pre-clear the lysate with 50 μL of pre-blocked beads for an hour at 4 °C before performing the Co-IP.
11. We use a 1.5× final concentration of SDS sample buffer to account for additional dilution by the remaining volume of Co-IP buffer in the bead pellet.
12. Heating the sample at 37 °C for 30 min instead of boiling can sometimes be used to prevent aggregation of some larger proteins of interest.
13. c-Myc typically runs between 55 and 60 kDa. We normally use a 10 % bis-acrylamide resolving gel with a 4 % stacker.
14. Our lab currently uses the LI-COR Odyssey Infrared Imager system for all of our western blotting. Standard chemiluminescent western blot analysis should work just the same without much lost in signal.
15. Phosphorylation-specific antibodies are used at 1:500, and the total c-Myc antibodies can be used at 1:1,000. When looking at c-Myc phosphorylation, it is preferred to do sequential probing

with the phosphorylation-specific antibody first followed by a total c-Myc antibody the following day.

16. We have found that using a secondary antibody conjugated to the 680 nm dye yields much cleaner results than the dye in the 800 nm channel. We also typically reuse our secondary antibody dilutions several times before discarding.
17. We find that removing detergent in the last wash step improves signal on the Odyssey.
18. At this point, the lysate will appear rather "snotty" depending on the confluency of cells on the plate. If the lysate is too thick to even pipette up and down, additional IP buffer can be added.
19. Sonication settings can vary depending on cell type. Typically the lowest setting is favored. Unlike the Co-IP buffer, the IP buffer requires more sonication passes to effectively break up cellular debris. Be sure to keep the lysate from getting warm by placing back on ice every 10 pulses.
20. Most of the time, there may not be a visible pellet after the spin. However, be aware if the lysate is not completely homogenous and is still "snotty." If so, additional passes using the sonicator are required.
21. If using antibody-conjugated beads, add the appropriate amount to achieve about 1–2 μg of antibody per IP.

Acknowledgments

The authors would like to thank Julienne R. Escamilla-Powers, Deanne Tibbits, and Hugh K. Arnold for a helpful discussion. This work was supported by Department of Defense Breast Cancer Research Program Award BC061306, Susan G. Komen Breast Cancer Foundation Awards BCTR0201697 and BCTR0706821, and National Institutes of Health Awards 1 R01 CA129040 and CA100855 to R.C.S.

References

1. Meyer N, Penn LZ (2008) Reflecting on 25 years with MYC. Nat Rev Cancer 8:976–990
2. Sears RC (2004) The life cycle of C-myc: from synthesis to degradation. Cell Cycle 3: 1133–1137
3. Yeh E, Cunningham M, Arnold H, Chasse D, Monteith T, Ivaldi G, Hahn WC, Stukenberg PT, Shenolikar S, Uchida T et al (2004) A signalling pathway controlling c-Myc degradation that impacts oncogenic transformation of human cells. Nat Cell Biol 6:308–318
4. Wang X, Cunningham M, Zhang X, Tokarz S, Laraway B, Troxell M, Sears RC (2011) Phosphorylation regulates c-Myc's oncogenic activity in the mammary gland. Cancer Res 71: 925–936
5. Hemann MT, Bric A, Teruya-Feldstein J, Herbst A, Nilsson JA, Cordon-Cardo C, Cleveland JL, Tansey WP, Lowe SW (2005) Evasion of the p53 tumour surveillance network by tumour-derived MYC mutants. Nature 436:807–811

6. Zhang X, Farrell AS, Daniel CJ, Arnold H, Scanlan C, Laraway BJ, Janghorban M, Lum L, Chen D, Troxell M et al (2012) Mechanistic insight into Myc stabilization in breast cancer involving aberrant Axin1 expression. Proc Natl Acad Sci U S A 109:2790–2795
7. Tibbitts DC, Escamilla-Powers JR, Zhang X, Sears RC (2012) Studying c-Myc serine 62 phosphorylation in leukemia cells: concern over antibody cross-reactivity. Blood 119: 5334–5335
8. Sears R, Nuckolls F, Haura E, Taya Y, Tamai K, Nevins JR (2000) Multiple Ras-dependent phosphorylation pathways regulate Myc protein stability. Genes Dev 14:2501–2514
9. Sears R, Leone G, DeGregori J, Nevins JR (1999) Ras enhances Myc protein stability. Mol Cell 3:169–179
10. Welcker M, Orian A, Jin J, Grim JE, Harper JW, Eisenman RN, Clurman BE (2004) The Fbw7 tumor suppressor regulates glycogen synthase kinase 3 phosphorylation-dependent c-Myc protein degradation. Proc Natl Acad Sci U S A 101:9085–9090
11. Yada M, Hatakeyama S, Kamura T, Nishiyama M, Tsunematsu R, Imaki H, Ishida N, Okumura F, Nakayama K, Nakayama KI (2004) Phosphorylation-dependent degradation of c-Myc is mediated by the F-box protein Fbw7. EMBO J 23:2116–2125
12. Arnold HK, Zhang X, Daniel CJ, Tibbitts D, Escamilla-Powers J, Farrell A, Tokarz S, Morgan C, Sears RC (2009) The Axin1 scaffold protein promotes formation of a degradation complex for c-Myc. EMBO J 28:500–512
13. Arnold HK, Sears RC (2006) Protein phosphatase 2A regulatory subunit B56alpha associates with c-myc and negatively regulates c-myc accumulation. Mol Cell Biol 26:2832–2844
14. Escamilla-Powers JR, Daniel CJ, Farrell A, Taylor K, Zhang X, Byers S, Sears R (2010) The tumor suppressor protein HBP1 is a novel c-myc-binding protein that negatively regulates c-myc transcriptional activity. J Biol Chem 285:4847–4858

Chapter 6

An Efficient Way of Studying Protein–Protein Interactions Involving HIF-α, c-Myc, and Sp1

Kenneth K.-W. To and L. Eric Huang

Abstract

Protein–protein interaction is an essential biochemical event that mediates various cellular processes including gene expression, intracellular signaling, and intercellular interaction. Understanding such interaction is key to the elucidation of mechanisms of cellular processes in biology and diseases. The hypoxia-inducible transcription factor HIF-1α possesses a non-transcriptional activity that competes with c-Myc for Sp1 binding, whereas its isoform HIF-2α lacks Sp1-binding activity due to phosphorylation. Here, we describe the use of in vitro translation to effectively investigate the dynamics of protein–protein interactions among HIF-1α, c-Myc, and Sp1 and to demonstrate protein phosphorylation as a molecular determinant that functionally distinguishes HIF-2α from HIF-1α.

Key words c-Myc, HIF-1α, HIF-2α, In vitro coupled transcription/translation, Protein–protein interaction, Sp1, Rabbit reticulocyte lysate, Wheat germ extract

1 Introduction

Numerous studies have provided compelling evidence that both HIF-1α and HIF-2α play an important role in cellular adaptation to hypoxia as a typical transcription factor [1, 2]. Some recent studies, however, indicate a non-transcriptional role for HIF-1α and HIF-2α to exert their functions by engaging in protein–protein interactions in hitherto unrecognized aspects of the hypoxic response [3, 4]. At the biochemical level, there is a dynamic relationship between HIF-1α and c-Myc competing for Sp1 binding [5, 6], while HIF-2α lacks Sp1 binding activity due to phosphorylation [7]. This biochemical nuance accounts for the functional difference between HIF-1α and HIF-2α in the HIF-1α–c-Myc pathway [4].

In vitro translation is very useful in a variety of applications to study proteins of interest. This chapter describes the use of this tool to reveal the dynamics of protein–protein interactions among HIF-1α, c-Myc, and Sp1 and to demonstrate phosphorylation as

Laura Soucek and Nicole M. Sodir (eds.), *The Myc Gene: Methods and Protocols*, Methods in Molecular Biology, vol. 1012, DOI 10.1007/978-1-62703-429-6_6, © Springer Science+Business Media, LLC 2013

a molecular restraint to prevent HIF-2α from binding to Sp1. In a typical experiment, HIF-1α is produced from the in vitro coupled transcription/translation system, as described below, and radiolabeled with the addition of [^{35}S]methionine. HIF-1α interaction with the endogenous Sp1 from the in vitro translation system is captured then by immunoprecipitation using Sp1 antibody. The protein complex is subjected to sodium dodecyl sulfate polyacrylamide gel electrophoresis (SDS-PAGE) and visualized by autoradiography. While the experimental illustration is specific to the investigation of HIF-1α, the general strategy should be applicable to the study of protein–protein interactions in general.

2 Materials

2.1 In Vitro Coupled Transcription/Translation

1. TNT rabbit reticulocyte lysate (Promega).
2. TNT wheat germ extract (Promega).
3. TNT T7 RNA polymerase (Promega).
4. TNT reaction buffer (Promega).
5. Amino acid mixture, minus methionine, 1 mM (Promega).
6. 10 mCi/mL [^{35}S]methionine (1,000 Ci/mmol) (Amersham Pharmacia Biotech; *see* **Note 1**).
7. pcDNA3-based eukaryotic expression plasmids under the control of T7 promoter for HIF-1α, HIF-1α fragment, HIF-2α fragment, and c-Myc.
8. Nuclease-free water.
9. λ protein phosphatase (New England Biolabs).
10. Resveratrol (Sigma).

2.2 Immunoprecipitation

1. Anti-Sp1 antibody (Santa Cruz Biotech).
2. Protein A/G agarose.
3. NETN buffer: 100 mM NaCl, 1 mM EDTA, 20 mM Tris (pH 8.0), and 0.5 % Nonidet P-40.
4. RIPA buffer: 50 mM Tris, 150 mM NaCl, 0.1 % SDS, 0.5 % sodium deoxycholate, and 1 % Nonidet P-40.
5. Precast 12 % or 4–12 % polyacrylamide gels.
6. Phosphate-buffered saline (PBS).

3 Methods

The following experimental procedures involve the use of radioactive isotope ^{35}S. General handling precautions for radiation safety should be taken.

3.1 In Vitro Coupled Transcription/Translation

There are a number of in vitro coupled transcription/translation kits available commercially, which provide the convenience of combining in vitro transcription and translation in a single reaction. The procedures described below refer to the TNT Quick Coupled Transcription/Translation System from Promega (Madison, MI), but are also applicable to other similar systems. By using either rabbit reticulocyte lysate (posttranslational modification proficient) [8] or wheat germ extract (modification deficient) [9], full-length HIF-1α, HIF-2α, and c-Myc can be translated in vitro in the absence or presence of [^{35}S]methionine. Similarly, HIF-1α fragment spanning codons 194–329 (referred to as PAS1B) and an equivalent HIF-2α fragment spanning codons 195–331 (referred to as PAS2B) are prepared. These products are subsequently analyzed for protein–protein interactions with Sp1 co-immunoprecipitation (*see* Subheadings 3.2, 3.3, and 3.4):

1. Thaw the Quick Master Mix quickly (<5 min) by hand-warming and place on ice. Unused master mix can be refrozen once. The other components can be thawed at room temperature and kept on ice (*see* **Note 2**).
2. In a microcentrifuge tube, assemble the standard reaction mixture including a plasmid DNA encoding the protein of interest and plus/minus [^{35}S]methionine. Mix by pipetting up and down gently. Centrifuge briefly to bring the mixture down to the bottom of the tube if necessary (*see* **Notes 3** and **4**).
3. Incubate the reaction at 30 °C for 90 min (*see* **Note 5**).
4. Once the transcription/translation reaction is complete, take an aliquot of 5 μL of the in vitro translated product for co-immunoprecipitation. The remaining reaction mixture may be stored at −20 °C for short-term or −70 °C for long-term periods.

3.2 Protein–Protein Interactions

Co-immunoprecipitation (Co-IP) is an effective assay for studying protein–protein interactions. To show HIF-1α interaction with Sp1, ^{35}S-labeled HIF-1α products from in vitro translation are immunoprecipitated with anti-Sp1 antibody. The captured complexes are then analyzed by SDS-PAGE. The general approach is illustrated in Fig. 1:

1. Prepare Protein A/G agarose for immunoprecipitation by washing the agarose beads twice with phosphate-buffered saline (PBS). Restore to a 50 % slurry with PBS when done (*see* **Note 6**).
2. In a microcentrifuge tube, add 5 μL of ^{35}S-labeled HIF-1α from in vitro translation.
3. Pre-clear the in vitro translated protein by adding 100 μL of Protein A/G agarose bead slurry (50 %) in 1 mL of NETN buffer (100 mM NaCl, 1 mM EDTA, 20 mM Tris (pH 8.0), and 0.5 % Nonidet P-40). Incubate at 4 °C for 1 h on a rotator (*see* **Note 7**).

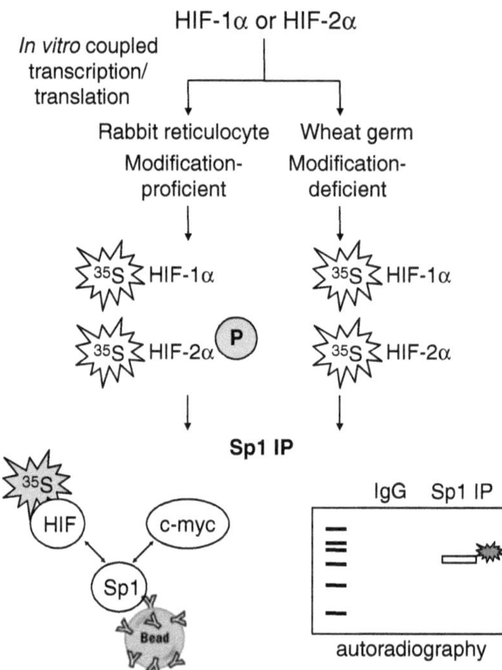

Fig. 1 Study of HIF-1α/Sp1 interaction using in vitro translated proteins. This schematic shows production of HIF-1α or HIF-2α in the presence of radioactive [^{35}S]methionine in an in vitro coupled transcription/translation system. HIF-2α is presumably phosphorylated at a specific residue when prepared in rabbit reticulocyte lysate but not in wheat germ extract. By contrast, HIF-1α is not phosphorylated at the corresponding site. After immunoprecipitation with anti-Sp1 antibody and SDS-PAGE, captured radioactive signal will reveal Sp1 binding activity through autoradiography

4. Spin down the Protein A/G agarose at 10,000 × *g* at 4 °C for 10 min. Transfer the supernatant containing the in vitro translated protein to a fresh microcentrifuge tube.
5. Add the recommended amount of anti-Sp1 antibody to the in vitro translated protein in a final volume of 500 μL in NETN buffer. Incubate on a rotator at 4 °C for 1 h.
6. Capture the immune complex by adding 100 μL of Protein A/G agarose bead slurry (i.e., 50 μL packed beads). Incubate on a rotator at 4 °C overnight.
7. Centrifuge to collect the agarose beads at 10,000 × *g* for 30 s. Discard the supernatant and wash the beads five times with 1 mL ice-cold NETN buffer (*see* **Note 8**).
8. Resuspend the agarose beads in 20 μL 2× SDS sample buffer and mix gently. Heat to dissociate the immune complex from the beads and to denature the bound protein of interest (*see* **Notes 9** and **10**).

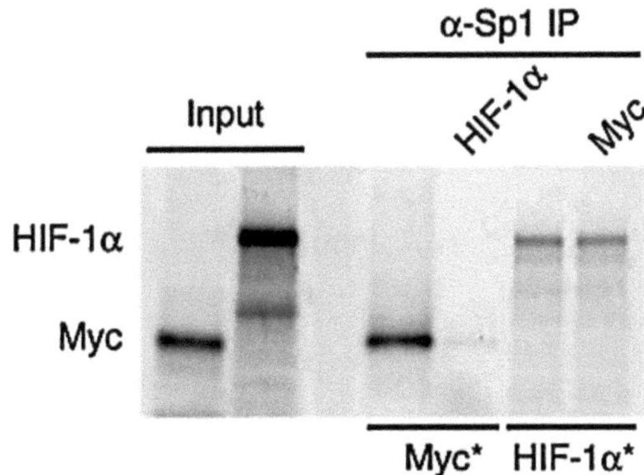

Fig. 2 HIF-1α competes with c-Myc for Sp1 binding. ^{35}S-labeled c-Myc and HIF-1α (with *asterisks*) prepared from rabbit reticulocyte lysate were mixed with equal amount of unlabeled HIF-1α and c-Myc, respectively, before anti-Sp1 co-immunoprecipitation. Input: 10 % of in vitro translated products without immunoprecipitation. Data are adapted with permission from Koshiji et al. (2005)

9. Collect the beads by centrifugation and perform SDS-PAGE with the supernatant (*see* **Note 11**).
10. Dry the gel and perform autoradiograph.

3.3 Competitive Protein–Protein Interactions

Although both HIF-1α and c-Myc interact with Sp1, their binding affinities differ. To test the affinity of HIF-1α and c-Myc binding to Sp1, we ascertain whether HIF-1α binds Sp1 in the presence of c-Myc and vice versa:

1. Prior to co-immunoprecipitation with anti-Sp1 antibody, incubate 5 μL of ^{35}S-labeled c-Myc with 5 μL of unlabeled HIF-1α at 4 °C for 30 min to 1 h. Reciprocally, mix ^{35}S-labeled HIF-1α and unlabeled c-Myc in the same way.
2. Perform co-immunoprecipitation as described above in Subheading 3.2.

As shown in Fig. 2, HIF-1α inhibits c-Myc for Sp1 binding, whereas c-Myc has no effect on HIF-1α, suggesting competitive HIF-1α interaction with Sp1.

3.4 Biochemical Modification of Protein–Protein Interactions

Protein–protein interactions are regulated frequently by biochemical modification. The observation that HIF-2α fails to bind Sp1 has been accounted for by PKD1-mediated phosphorylation [7]. Accordingly, we can alter HIF-2α binding activity by manipulating phosphorylation of in vitro translated products with the treatment of phosphatase or kinase inhibitor and the use of wheat germ extract, which is deficient in posttranslational modification:

1. Use 5 μL of ^{35}S-labeled HIF-2α fragment (PAS2B) prepared from rabbit reticulocyte lysate or wheat germ extract.

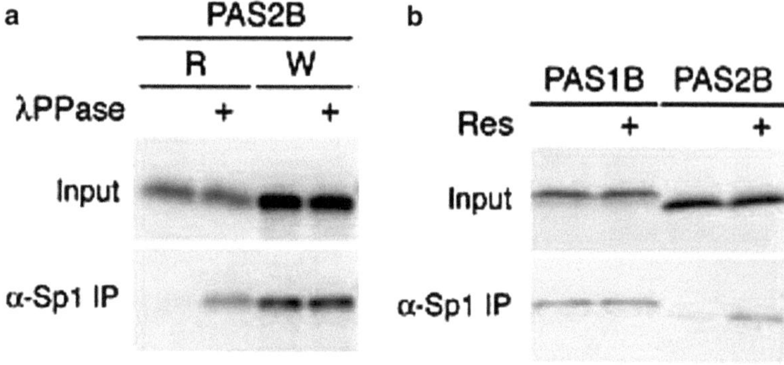

Fig. 3 HIF-2α acquires Sp1 binding activity through biochemical modification. (**a**) HIF-2α fragment PAS2B prepared from rabbit reticulocyte lysate (*R*) or in wheat germ extract (*W*) was treated with λ protein phosphatase (λPPase) before anti-Sp1 co-immunoprecipitation. (**b**) HIF-1α fragment PAS1B and HIF-2α fragment PAS2B were translated in rabbit reticulocyte lysate in the presence of resveratrol (Res), and their Sp1 binding activity was analyzed as above. Input: 10 % of in vitro translated products without immunoprecipitation. Data are adapted with permission from To et al. (2006)

2. For dephosphorylation of HIF-2α, incubate in vitro translated products with 100 U of λ protein phosphatase at 37^{0}C for 30 min.
3. For prevention of HIF-2α phosphorylation, add the PKD1 inhibitor resveratrol at the final concentration of 100 μM during in vitro translation. Alternatively, wheat germ extract can be used for in vitro translation to produce HIF-2α without posttranslational modifications.
4. Perform co-immunoprecipitation as described above in Subheading 3.2.

As shown in Fig. 3, HIF-2α acquires Sp1 binding activity when expressed in wheat germ extract (presumably without any post-translational modification) (*see* **Note 12**). Furthermore, HIF-2α prepared from rabbit reticulocyte lysate also gains Sp1 binding after treatment with the phosphatase or the kinase inhibitor.

4 Notes

1. A "translation" grade of [^{35}S]methionine, such as the one from Amersham Pharmacia (currently under GE Healthcare Life Sciences), is recommended to minimize background labeling of the in vitro translated protein. Background labeling of a 42-kDa protein from the rabbit reticulocyte lysate has been reported when using other grades of radioactive label [10].
2. During assembling of the in vitro transcription/translation reaction, the master mix should be kept on ice whenever

possible. If not consumed completely, the master mix should be refrozen promptly after thawing to reduce loss of translational activity. Do not freeze and thaw the master mix more than twice.

3. Unevenly mixed reaction mixture may result in low yield of translation product. Since the rabbit reticulocyte lysate and wheat germ extract are fairly viscous, it is advised to gently pipette the reaction mixtures during preparation to achieve thorough mixing.
4. Ethanol is known to inhibit translation. Residual ethanol, which may be present in plasmid DNA samples during DNA purification, should be removed before the plasmid is added to the in vitro transcription/translation master mix. Moreover, if the yield of in vitro translation reaction is low, additional RNasin Ribonuclease inhibitor (50 units/reaction) may be added to the reaction to prevent degradation of in vitro transcribed transcripts (http://www.promega.com/resources/articles/pubhub/enotes/optimize-your-tnt-reticulocyte-lysate-systems-reaction/).
5. The optimized temperature for the in vitro coupled transcription/translation reaction is 30 °C. Performing the reaction at 37 °C may decrease the protein yield.
6. It is recommended to cut off the tip of the pipette to avoid disruption of the agarose beads during their pipetting.
7. Pre-clearing can reduce nonspecific binding of in vitro translated proteins to the agarose beads used later for immunoprecipitation.
8. If a significant background is observed, it may be helpful to switch to the RIPA buffer [50 mM Tris, 150 mM NaCl, 0.1 % SDS, 0.5 % sodium deoxycholate, 1 % NP-40] for the washing. However, in our experience, washing with RIPA buffer may strip the HIF-1α/Sp1 immune complex off the agarose beads.
9. In vitro translated protein must be denatured before analysis by SDS-PAGE. However, because of the high protein content in the rabbit reticulocyte lysate and wheat germ extract, proteins are prone to aggregate at the high denaturation temperature. These aggregates may appear as a high-molecular-weight band intensified by radioactive labeling, thereby overshadowing the signal for the protein of interest (i.e., Sp1-bound HIF-1α fragment). Milder denaturation conditions, instead of boiling at 100 °C, may be adopted to reduce the risk of protein aggregation. The immuno-complexes bound to the agarose beads can be eluated with 20 μL 2× sample SDS buffer and denatured at 60 °C for 20 min, 70 °C for 10 min, or 80 °C for 3–4 min (http://www.promega.com/resources/articles/pubhub/enotes/trouble-free-sdspage-analysis-of-proteins-synthesized-in-tnt-cell-free-expression-systems/).

10. Do not load too much protein on a SDS-PAGE gel. It may lead to protein aggregation or band distortion. In general, do not load more than 5 μL of in vitro translated protein per lane. For the "input" lane, usually 1 μL is sufficient.
11. Electrophoresis is usually performed until the bromophenol blue dye has run off the bottom of the gel. However, disposal of unincorporated radioactive label is easier if the gel is stopped while the dye front remains in the gel, as the dye front also contains the unincorporated labeled amino acids.
12. Wheat germ extract has been reported to produce full-length proteins with correct folding and posttranslational modifications—e.g., chicken myoD1 [11]. While this finding may not be universal, an alternative *E. coli* S30 bacterial in vitro translation system, which is known to be devoid of posttranslational modification capability, may be considered.

Acknowledgment

This work was supported in part by the Public Health Service grant CA-131355 from the National Cancer Institute.

References

1. Kaelin WG, Ratcliffe PJ (2008) Oxygen sensing by metazoans: the central role of the HIF hydroxylase pathway. Mol Cell 30:393–402
2. Majmundar AJ, Wong WJ, Simon MC (2010) Hypoxia-inducible factors and the response to hypoxic stress. Mol Cell 40:294–309
3. Gordan JD, Thompson CB, Simon MC (2007) HIF and c-Myc: sibling rivals for control of cancer cell metabolism and proliferation. Cancer Cell 12:108–113
4. Huang LE (2008) Carrot and stick: HIF-alpha engages c-Myc in hypoxic adaptation. Cell Death Differ 15:672–677
5. Koshiji M, Kageyama Y, Pete EA, Horikawa I, Barrett JC, Huang LE (2004) HIF-1alpha induces cell cycle arrest by functionally counteracting Myc. EMBO J 23: 1949–1956
6. Koshiji M, To KK-W, Hammer S, Kumamoto K, Harris AL, Modrich P, Huang LE (2005) HIF-1alpha induces genetic instability by transcriptionally downregulating MutSalpha expression. Mol Cell 17:793–803
7. To KK-W, Sedelnikova OA, Samons M, Bonner WM, Huang LE (2006) The phosphorylation status of PAS-B distinguishes HIF-1alpha from HIF-2alpha in NBS1 repression. EMBO J 25: 4784–4794
8. Beckler GS, Thompson D, Van Oosbree T (1995) In vitro translation using rabbit reticulocyte lysate. Methods Mol Biol 37:215–232
9. Olliver CL, Grobler-Rabie A, Boyd CD (1985) In vitro translation of messenger RNA in a wheat germ extract cell-free system. Methods Mol Biol 2:137–144
10. Jackson RJ, Hunt T (1983) Preparation and use of nuclease-treated rabbit reticulocyte lysates for the translation of eukaryotic messenger RNA. Methods Enzymol 96:50–74
11. Nakamura S (1993) Possible role of phosphorylation in the function of chicken MyoD1. J Biol Chem 268:11670–11677

Chapter 7

Methods for Determining Myc-Induced Apoptosis

Dan Lu and Trevor D. Littlewood

Abstract

Although many oncoproteins promote cell growth and proliferation, some also possess the potential to induce cell death by apoptosis. Deregulated expression of the *myc* oncogene promotes apoptosis in both cultured cells and in some tissues in vivo. Here we describe techniques to detect Myc-induced apoptosis in vitro using flow cytometry and microscopy and in vivo using immunohistochemical staining.

Key words Apoptosis, Flow cytometry, Microscopy, Annexin V, TUNEL, Immunohistochemistry, Cleaved caspase 3, Hematoxylin and eosin

1 Introduction

Oncogenes promote cell growth and proliferation and are often aberrantly expressed in tumor cells. Intriguingly, many oncogenes can also induce cell cycle arrest or apoptosis thus reducing the likelihood of tumor formation and progression, a phenomenon often referred to as intrinsic tumor suppression [1] (Fig. 1). For example, Myc-induced apoptosis suppresses tumorigenesis in various tumor models [2–4]. However, if Myc-induced apoptosis is suppressed by elevated expression of anti-apoptotic proteins such as Bcl_xL, then invasive tumors develop [2]. *c-Myc* was the first oncogene demonstrated to have this innate ability to induce apoptosis [5]. Activation of the ectopically switchable MycER protein in serum-deprived rat fibroblasts induces apoptosis within hours [5, 6].

c-Myc sensitizes cells to various apoptotic stimuli by potentiating both the intrinsic and extrinsic apoptotic pathways rather than inducing apoptosis per se [7]. For example, Myc functionally cooperates with Bax, a central effector of cell death, causing the release of holocytochrome C from the mitochondria and activation of caspase 9 and apoptosis [8]. In addition, deregulated c-Myc expression leads to the stabilization and activation of p53 (which promotes the expression and activity of pro-apoptotic proteins) in both ARF-dependent and independent manners [9–11].

Laura Soucek and Nicole M. Sodir (eds.), *The Myc Gene: Methods and Protocols*, Methods in Molecular Biology, vol. 1012, DOI 10.1007/978-1-62703-429-6_7, © Springer Science+Business Media, LLC 2013

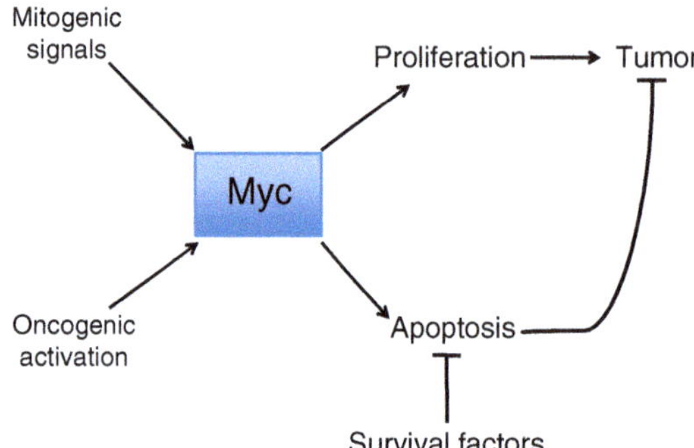

Fig. 1 Intrinsic tumor suppression. Endogenous Myc expression is tightly regulated and coordinates cell cycle progression, whereas deregulated (oncogenic) Myc expression (as observed in many tumors) drives both cell proliferation and, in some circumstances, cell death. In conditions where apoptosis is suppressed by survival factors or cooperating oncogenes (e.g., BclXL), then Myc promotes tumorigenesis

Interestingly, only elevated levels of c-Myc protein induce ARF, implying that low endogenous, but deregulated, levels of Myc may drive tumorigenesis without eliciting an apoptotic response [4]. c-Myc also induces expression of the pro-apoptotic protein Bim independently of p53 and potentiates the extrinsic, death receptor-dependent death [12]. Finally c-Myc also inhibits the synthesis of superoxide dismutases via NFκB leading to an increase in reactive oxygen species that contribute to apoptosis [13].

In addition to death ligands that promote death, cell survival is likewise promoted by survival factors. In culture, the cocktail of factors present in calf serum acts as potent inhibitors of cell death. Factors in serum reduce ROS accumulation in cells [13], abrogate death receptor signaling [12], and stimulate the activity of the serine/threonine kinase Akt that modulates the activity and degradation of several proteins involved in apoptosis. For example, Akt-dependent phosphorylation of the pro-apoptotic Bad protein promotes its degradation [14, 15]. In the absence of Bad, anti-apoptotic proteins such as Bcl-2 and Bcl-$_X$L are free to bind and inhibit other pro-apoptotic mediators including Bax.

Cells undergoing Myc-induced apoptosis in vitro show distinct morphological changes including nuclear condensation, membrane blebbing, and cytoplasmic condensation ending in cell fragmentation [5] (Table 1; Fig. 2a). More quantitative analysis of apoptosis can be achieved using flow cytometry by measuring relative percentage of cells staining positive for both propidium iodide (PI) and Annexin V (Fig. 2b). PI is incorporated into fully membrane porous cells binding to their DNA content therefore staining all dying and dead cells with permeable plasma membranes, whereas Annexin V specifically binds phosphatidylserine, a phospholipid exposed on

Table 1
Features of apoptosis and recommended methods for their detection

Feature	Method of detection
Cell detachment and surface blebbing	Microscopy
Exposure of surface Annexin V	Flow cytometry
Chromatin condensation	Microscopy
DNA fragmentation	TUNEL
Activation of caspase 3 (cleavage of the inactive pro-caspase 3 into the active form)	Western blotting, immunohistochemistry

Please note that this is not an exhaustive list and represents only methods that the authors routinely use

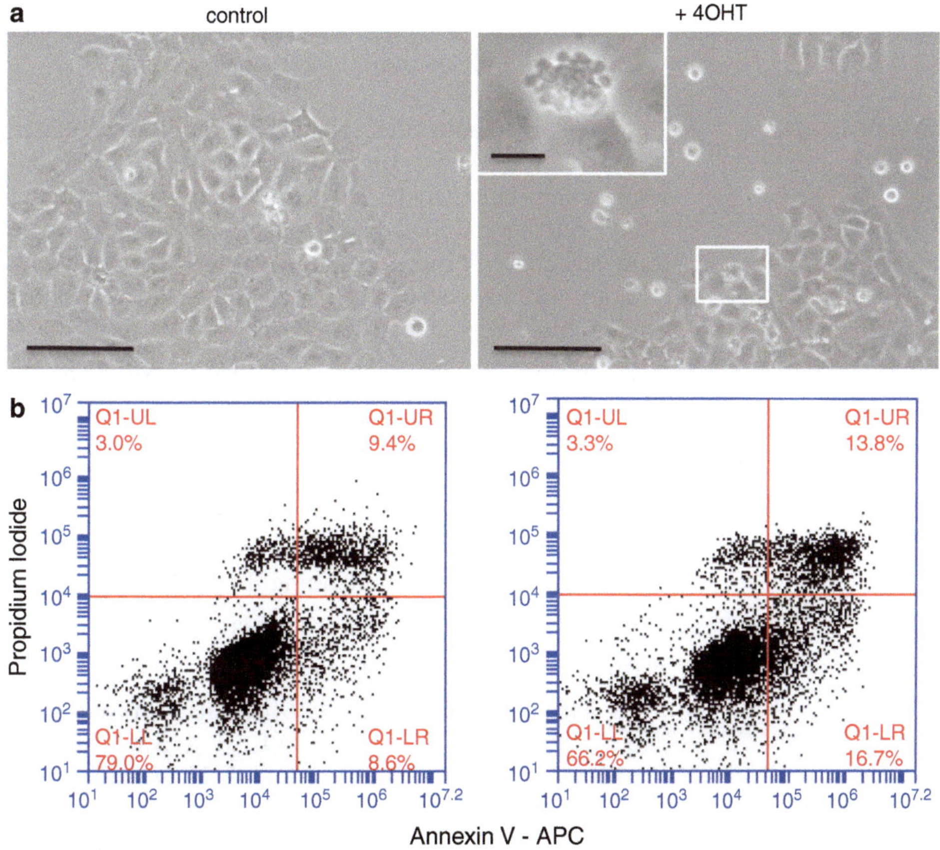

Fig. 2 Myc-induced apoptosis in vitro. (**a**) Representative phase contrast microscopy images of Rat-1/MycER^{Tam} fibroblasts cultured in medium containing 0.1 % BGS in the absence (control) or presence of 100 nM 4OHT (for 48 h) to activate MycER^{Tam}. Apoptosis is characterized by distinct morphological changes including membrane blebbing, cytoplasmic condensation, and cell fragmentation. Scale bars = 100 μm or 50 μm (*inset*). (**b**) Flow cytometric analysis of control and 4OHT-treated cells as above. Apoptotic cells with exposed Annexin V are present in the *lower right* (Q1-LR) quadrant. Late apoptotic and necrotic cells are fully membrane permeable; therefore, both Annexin V- and PI-positive cells are present in the *upper right* (Q1-UR) quadrant

the exterior of apoptotic cells only. Activation of the executioner proteases (caspases) requires their cleavage and assembly into specific protein species. Thus an antibody specific for the cleaved (activated) form of caspase 3 can be used to determine apoptosis by western blotting of protein lysates and in tissue sections by immunohistochemistry. In addition, histochemical detection of pyknotic nuclei (those with condensed chromatin) and terminal deoxynucleotidyl transferase dUTP nick end labeling (TUNEL) that reveals fragmented DNA are characteristic of apoptotic cells (Fig. 3). These techniques are described below.

2 Materials

Prepare all solutions using ultrapure water (DW, prepared by purifying deionized water to attain a sensitivity of 18 MΩ cm at 25 °C) and analytical grade reagents. Prepare and store all reagents at room temperature unless indicated otherwise.

2.1 *In Vitro* Assays

2.1.1 Cell Culture and Microscopy

1. In theory any mammalian cell line exhibiting deregulated Myc expression is likely to undergo apoptosis when deprived of survival signals. Here we describe experiments with rat fibroblasts (Rat-1 cell line) engineered to constitutively express the ectopically switchable fusion protein $MycER^{Tam}$ [6]. In these cells, addition of the specific ER^{Tam} ligand, 4-hydroxytamoxifen, renders $MycER^{Tam}$ active.
2. Cell culture growth media: Dulbecco Modified Eagle's Medium (DMEM) (PAA Laboratories) supplemented with 2 mM L-glutamine and either 10 % heat-inactivated fetal bovine serum (FBS) (Life Technologies) for general cell maintenance or 0.1 % for apoptosis assays (*see* **Note 1**).
3. Cell culture incubator capable of maintaining 37 °C and a humidified atmosphere containing 5 % CO_2.
4. Class II microbiological safety cabinet.
5. Plasticware appropriate for cell culture.
6. Phosphate-buffered saline (PBS) buffer without Ca^{2+} and Mg^{2+}: 0.2 g KCl, 0.2 g KH_2PO_4, 8 g NaCl, and 1.15 g Na_2HPO_4 anhydrous made up in 1 L H_2O made to pH 7.0 (*see* **Note 2**).
7. 1× Trypsin: Add 1 mL 10× 0.5 % Trypsin–EDTA to 9 mL PBS.
8. 4-hydroxytamoxifen (4OHT): Dissolve 4-hydroxytamoxifen (Sigma) to a final concentration of 100 μM in 100 % ethanol. Store at −20 °C (*see* **Note 3**).
9. Inverted microscope for live cell imaging: Any model with suitable bright field (e.g., phase or DIC) and image capture ability.

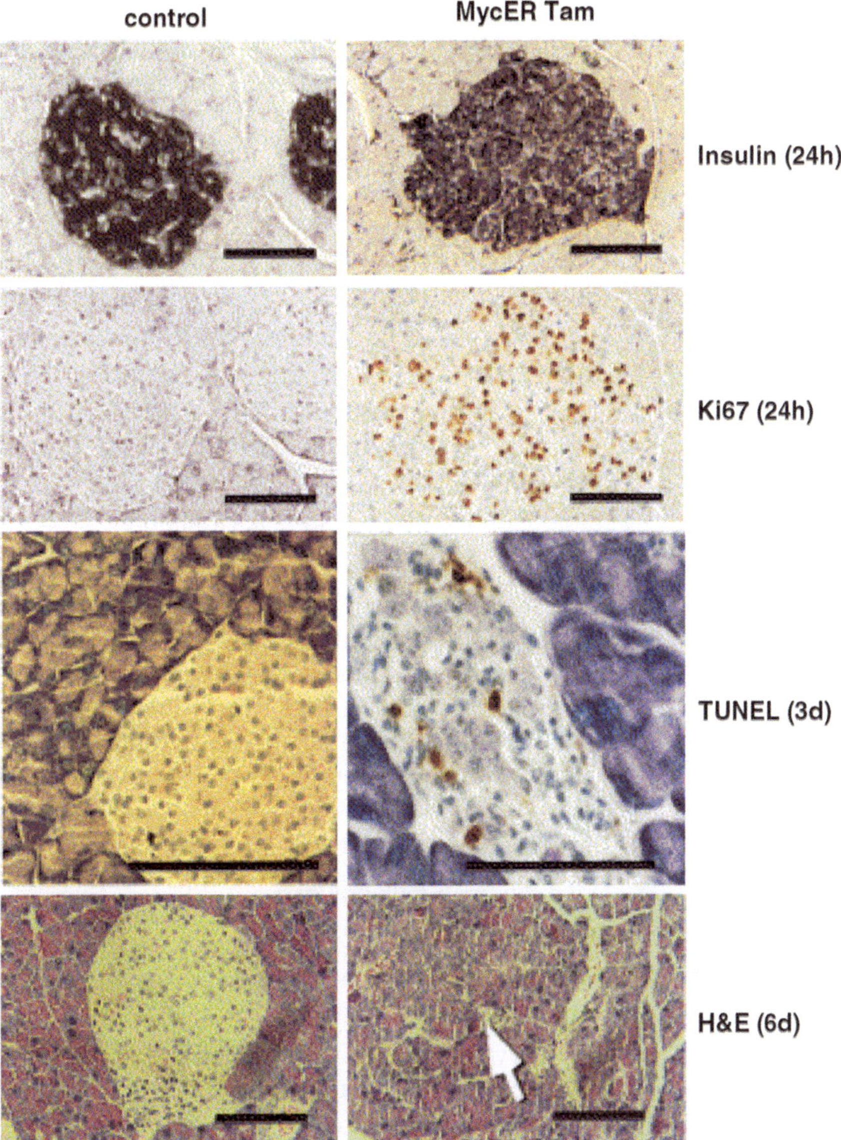

Fig. 3 Myc-induced apoptosis in vivo. Expression of the switchable MycER^{Tam} protein was directed to the insulin-secreting pancreatic β-cells of mice and activated by injection with 4-OHT. Control mice were untreated. Sections demonstrate Myc-induced β-cell proliferation (Ki67) 24 h after 4OHT injection and apoptosis at 3 days (TUNEL). Note that 6 days after 4OHT administration, the pancreatic islet has involuted due to Myc-induced apoptosis (H&E staining). Scale bar = 50 μm. Reproduced from [2] with permission from Cell Press

2.1.2 Flow Cytometry for Annexin V

1. 1× Annexin-V binding buffer: Dilute 10× binding buffer (eBioscience) in PBS to provide a 1× working solution. Store at 4 °C.
2. Propidium iodide (PI) staining solution: Commercially available from eBioscience. No dilution required. Store at 4 °C.

3. APC-conjugated Annexin V: Allophycocyanin (APC) is a fluorescent molecule. Annexin V Apoptosis Detection Kit APC can be purchased from eBioscience (*see* **Note 4**). No dilution required. Store at 4 °C.
4. Flow cytometer: for example, BD Accuri® C6 flow cytometer (*see* **Note 5**).

2.1.3 Western Blotting for Cleaved Caspase 3

1. Vertical gel electrophoresis apparatus and suitable blotting module.
2. Dark room equipped with facilities for developing X-ray film (either manually or via automatic film processor).
3. Protein lysis buffer: 50 mM Tris–Cl pH 7.4, 150 mM NaCl, 1 mM EDTA, 0.5 % Triton X-100, 1 mM NaF, 1 mM Na_3VO_4 protease inhibitor cocktail set III (Calbiochem).
4. Immobilon P PVDF membrane (Millipore).
5. Invitrogen NuPAGE® 4–12 % precast Bis Tris Gels (Life Technologies).
6. Invitrogen NuPAGE 20× running buffer (Life Technologies), 20× blotting buffer, and 4× LDS sample buffer. Dilute running buffer and transfer buffer to 1× using DW before use. In addition, methanol to a final concentration of 20 % should be added to the transfer buffer (e.g., 200 mL methanol, 50 mL 20× blotting buffer, 750 mL DW for 1 L of transfer buffer).
7. 1 M dithiothreitol (DTT).
8. Dried skimmed milk powder.
9. Tween-20.
10. Primary antibodies specific for cleaved caspase 3 (Cell Signaling) (*see* **Note 6**).
11. Secondary antibodies: for example, horseradish peroxidase (HRP)-conjugated goat anti-rabbit IgG.
12. ECL detection kit (Pierce ECL western blotting substrate, Thermo Scientific).
13. Fuji SuperRX X-ray film (or similar).

2.2 In Vivo Assays

1. Formalin-fixed and paraffin-embedded tissue sections ideally 4–6 μm thick and mounted onto glass slides. Formalin-fixed, paraffin-embedded human tonsil sections can be used as a positive control for apoptosis—apoptotic bodies can be seen in tingle-body macrophages within germinal centers.
2. Slide staining apparatus (e.g., Coplin jars, *see* **Note 7**).
3. Hydrophobic barrier pen.
4. Microscope preferably with image analysis software capable of simple measurements (area, pixel density, etc.)

2.2.1 Hematoxylin and Eosin (H&E) Staining

1. Hematoxylin solution, Harris modified (Sigma).
2. Scott's solution: 20 mg magnesium sulfate and 2 mg sodium bicarbonate in 1 L DW.
3. Eosin Y solution (Sigma).
4. Destain solution: 250 mL 50 % methanol, 250 mL DW, and 5 mL HCl.
5. Mounting medium, DPX.

2.2.2 TUNEL Staining

1. Xylene and ethanol at various concentrations.
2. 50 μg/mL proteinase K (Roche) prepared in 10 mM Tris–Cl, pH 7.5 (*see* **Note 8**).
3. Terminal transferase (TdT) enzyme (Roche).
4. Digoxigenin-dUTP (Roche).
5. TdT buffer: 200 mmol/L potassium cacodylate, 25 mmol/L Tris–HCI, pH 6.6 at 25 °C, 0.25 mg/mL bovine serum albumin (BSA), 5 mmol/L cobalt chloride (Roche).
6. TB buffer: 300 mM sodium chloride, 30 mM sodium citrate.
7. 0.1 M Tris–Cl buffer, pH 7.5 (TBS).
8. 10 % bovine serum albumin (BSA) in TBS.
9. Anti-digoxigenin alkaline phosphatase (Fab fragments) (Roche).
10. Substrate solution. 0.19 mg/mL 5-bromo-4-chloro-3-indolyl phosphate (BCIP), 0.4 mg/mL nitroblue tetrazolium (NBT) in 100 mM Tris–Cl, pH 9.5, 50 mM $MgSO_4$ (Roche) (*see* **Note 9**).
11. Mounting medium, e.g., Glycergel (Dako).

2.2.3 Immunohistochemistry for Cleaved Caspase 3

1. Formalin-fixed, paraffin-embedded human tonsil section as a positive control.
2. 0.1 M citrate buffer, pH 6.
3. PBS (see above).
4. Goat serum.
5. Rabbit polyclonal primary antibody specific for cleaved caspase 3 (Asp175) (Cell Signaling) (*see* **Note 10**).
6. Biotinylated goat anti-rabbit secondary antibody.
7. Vectastain ABC kit (Vector Laboratories).
8. 3, 3′-diaminobenzidine (DAB) substrate kit (Vector Laboratories) (*see* **Note 11**).
9. 0.3 % (w/v) $NiCl_2$ in DW.
10. Mounting medium, DPX.

3 Methods

Carry out all procedures at room temperature unless otherwise specified.

3.1 Phase Contrast Microscopy

1. Plate 5×10^4 Rat1 MycER^{Tam} cells [6] into each well of a 6-well plate in 3 mL of DMEM containing 10 % BGS and incubate overnight at 37 °C (*see* **Note 12**).
2. Aspirate the media and wash twice with 3 mL PBS.
3. Replace the culture media with the following:
   a. 3 mL of DMEM containing 0.1 % BGS and 3 μL ethanol.
   b. 3 mL of DMEM containing 0.1 % BGS and 3 μL of 100 μM 4OHT (i.e., final concentration of 100 nM 4OHT). Incubate at 37 °C.
4. Apoptotic cells are observed by microscopy after a few hours in the presence of 4OHT (stochastic 50 % death in 24 h). Apoptotic cells show a distinct morphology characterized by detachment from the substratum, membrane blebbing, and cell fragmentation (Fig. 2a) (*see* **Note 13**).

3.2 Detection of Exposed Annexin V by Flow Cytometry

1. Culture cells as described in **steps 1–3** of Subheading 3.1 above.
2. Depending on the kinetics of apoptosis induction, collect culture media containing detached cells and harvest the adherent cells by trypsinization (*see* **Note 14**) from each sample at various times after the addition of 4OHT and pool.
3. Collect cell pellet by centrifugation at $250 \times g$ for 5 min at 4 °C. Do not exceed $300 \times g$ to avoid cell damage.
4. Discard supernatant, and wash cells once with 5 mL 1× binding buffer and repeat **step 3**.
5. Discard supernatant, and resuspend cells in 1× binding buffer at 1×10^6 cells/mL. Place in icebox at 2–8 °C and store in the dark.
6. Dispense cell suspension into 4×100 μL aliquots labeled as (a) unstained, (b) PI only, (c) Annexin V APC only, and (d) PI and Annexin V APC (*see* **Note 15**).
7. Store (a) untreated at 2–8 °C in the dark.
8. To (c) and (d), add 5 μL of APC-conjugated Annexin V (at supplied concentration), and incubate 10–15 min at room temperature.
9. Add 500 μL of 1× binding buffer to all samples (a–d), and centrifuge at $250 \times g$ for 5 min to pellet cells.
10. Resuspend each cell pellet in 200 μL of 1× binding buffer and store at 2–8 °C in the dark.
11. To (b) and (d) add 5 μL PI staining solution (at supplied concentration). Store at 2–8 °C in the dark for 5 min.

12. Analyze by flow cytometer using a flow rate of 35 μL/min with a core size of 16 μm sampling 10,000 events for each sample.
13. Gating should be performed on unstained samples (a) using forward scatter (cell size) versus side scatter (cell density) to eliminate small cell debris.
14. PI only (b) and Annexin V only (c) single stained should be detected in FL2 and FL4 channels, respectively. Normalization should be performed to avoid signal spillage into adjacent channels (consult flow cytometer manufacturer's manuals).
15. Annexin V- and/or PI-positive cells distinguish between viable, early apoptotic, and late apoptotic/necrotic cells (Fig. 2b) (*see* **Note 4**).

3.3 Western Blotting for Cleaved Caspase 3

1. Harvest cells as described in **steps 1–3** in Subheading 3.2 above.
2. Resuspend cell pellet in suitable volume (e.g., 100 μL per 10^6 cells, *see* **Note 16**) of lysis buffer and incubate on ice for 10 min (*see* **Note 17**).
3. Add appropriate volume of 4× LDS sample buffer and DTT to a final concentration of 100 mM.
4. Heat at 95 °C for 3 min and allow to cool down.
5. Briefly centrifuge (10,000×*g* for 2 min) samples, and load 10–15 μL per well of a precast gel.
6. Apply 200 V for 35 min.
7. Blot onto PVDF membrane (35 V for 40 min) with NuPage transfer buffer.
8. Incubate membrane in 4 % dried skimmed milk in PBS for 1 h.
9. Incubate membrane in an appropriate dilution of primary antibody specific for cleaved caspase 3 in PBS supplemented with 0.1 % Tween-20 (PBST) for 1 hour at room temperature with gentle rocking.
10. Wash three times (3 min per wash) with PBST.
11. Incubate with appropriate dilution of secondary horseradish peroxidase (HRP)-conjugated antibody (*see* **Note 18**) in PBST for 1 hour at room temperature with gentle rocking.
12. Wash three times (3 min per wash) with PBST.
13. Detect binding of secondary antibody with enhanced chemiluminescence (ECL) (see supplier's instructions) and apply to membrane for 1 min.
14. Expose membrane to X-ray film for between 10 s and 10 min depending on signal intensity and develop (*see* **Note 19**).

3.4 Hematoxylin and Eosin (H&E) Stain

1. Embed samples in paraffin wax.
2. 4–6 μm paraffin sections are adhered to subbed slides.
3. Deparaffinize in xylene for 10 min.

4. Rehydrate in descending series of ethanol solutions (100 %, 70 %, 5 min each).
5. Rinse in tap water, then in DW.
6. Place slides in hematoxylin solution for 5 min.
7. Briefly dip slides eight times in destain solution.
8. Place slides in Scott's solution to "blue" for 5 min.
9. Place slides in eosin solution for 6 min.
10. Quickly rinse in DW.
11. Rapidly dehydrate through graded alcohols to two changes of xylene.
12. Cover slip and mount with DPX.

3.5 Immunohistochemistry for Fragmented DNA (TUNEL)

1. Prepare slides as described in **steps 1–5** in Subheading 3.4 above.
2. Circumscribe tissue section with a hydrophobic barrier pen.
3. Incubate in 50 μg/mL proteinase K for 3–5 min at room temperature (*see* **Note 20**). If human tonsil sections are not available, then a positive control can be generated by treating one tissue section with micrococcal nuclease or DNase I (grade 1) to induce DNA strand breaks before proceeding (*see* **Note 21**).
4. Wash in DW four times (2 min each time).
5. Immerse in TdT buffer for 5 min at room temperature.
6. Tip off excess TdT buffer.
7. Incubate with TdT enzyme (0.05–0.2 U/μL) (*see* **Note 22**) and 0.005 nmol/μL digoxigenin-dUTP in TdT buffer (prepared immediately before use) in humid atmosphere at 37 °C for 30 min (*see* **Note 23**). Generally 50 μL of solution is adequate for each slide providing that they are kept in a humidified chamber (*see* **Note** 7).
8. Terminate the reaction by transferring the slides to TB buffer for 15 min at RT.
9. Gently rinse in DW.
10. Place slides in 0.1 M Tris buffer, pH 7.6 (TBS) for 5 min.
11. Cover sections with 10 % BSA in TBS (blocking buffer) for 10 min at room temperature (*see* **Note 24**).
12. Incubate with 1 U/mL anti-digoxigenin alkaline phosphatase (Fab fragments) in blocking buffer for 1 h at room temperature.
13. Rinse with TBS for 5 min.
14. Incubate with the BCIP/NBT chromogenic substrate solution. The intensity of the color reaction should be monitored under the microscope and terminated when desired (*see* **Note 25**).
15. Terminate the reaction by rinsing in DW.

16. Counterstain if desirable (with 1 % eosin for BCIP/NBT or 1 % light green or Fast Red).
17. Mount sections using aqueous mounting medium, e.g., Glycergel.
18. Examine under microscope (*see* **Note 26**).

3.6 Immuno-histochemistry for Activated Caspase 3

1. Prepare formalin-fixed, paraffin-embedded sections as described in **steps 1–5** in Subheading 3.4 above.
2. Circumscribe tissue section with a hydrophobic barrier pen.
3. Microwave sections in preheated citrate buffer pH 6 for 1 min on full power (850 W) followed by 9 min at medium power.
4. Cool to room temperature for 20 min.
5. Wash sections in milli-Q DW twice for 5 min each.
6. Incubate sections in 3 % hydrogen peroxide in PBS for 10 min.
7. Wash sections twice for 5 min each in PBS.
8. Incubate sections with 5 % goat serum in PBS for 1 h at room temperature.
9. Tip off excess serum, and incubate sections with an appropriate dilution of cleaved caspase 3 antibody (*see* **Notes 10** and **18**) in PBS containing 5 % goat serum overnight at 4 °C.
10. Wash twice for 5 min each in PBS.
11. Incubate sections with 1:400 dilution of secondary antibody (biotinylated goat anti-rabbit immunoglobulins) in PBS containing 5 % goat serum for 30 min at room temperature.
12. Wash twice for 5 min each time in PBS.
13. Incubate sections with ABC complex for 30 min at room temperature.
14. Wash twice for 5 min each with PBS.
15. Apply DAB solution and monitor reaction closely using a microscope (*see* **Notes 11** and **27**).
16. Wash thoroughly with DW for 5 min.
17. Counterstain as required.
18. Dehydrate, clear in two changes of xylene, cover slip, and mount as **step 12** in Subheading 3.4 and examine under the microscope.

4 Notes

1. Other commercially available serum could also be used such as bovine growth serum (BGS) (Life Technologies).
2. PBS must be sterilized by filtering through 0.45 μm filter. Alternatively sterile PBS is commercially available from PAA

Laboratories. The use of calcium and magnesium ion free PBS is recommended for harvesting of cells with trypsin.

3. The z isomer of 4OHT (Sigma) should be used.
4. Alternate fluorophore such as Annexin V-FITC conjugate can also be used however this may lead to signal overflow into FL2 channel in the flow cytometer contaminating PI absorbance. Normalization of samples will then be required (see flow cytometer manufacturer's protocols). Cells exhibiting Annexin V only are considered to be "early apoptotic" cells, whereas those exhibiting both PI and Annexin V are considered to be "late apoptotic" undergoing secondary necrosis. Therefore, the kinetics of apoptosis induction should be taken into account when analyzing such data.
5. Other flow cytometers can be used, but protocol may need to be adapted to suit manufacturer's guidelines.
6. Caspase 3 (like all caspases) is present as a larger inactive pro-caspase of ~32 kD and on activation is cleaved to smaller subunits—antibodies specific for the larger of the cleaved subunits (17–21 kDa) can therefore be used to detect its activation.
7. A simple humidified chamber can be constructed by fixing plastic pipettes to the bottom of a 245 × 245 mm assay dish (Corning) and placing moistened tissue paper inside the chamber.
8. Other sources of proteinase K can be used, but they should be certified nuclease-free so as not to generate false-positive results.
9. Alternatively the BCIP/NBT substrate kit from Vector Laboratories is also suitable.
10. Alternatively specific cleaved caspase 3 (Asp175) antibodies directly coupled to various conjugates are available. For example, Alexa Fluor 488 (Cell Signaling) or biotinylated (Cell Signaling) conjugates can be used.
11. Alternatively dissolve 6 mg of DAB in 9 mL of 50 mM Tris–Cl, pH 7.5, and add 1 mL 0.3 % $NiCl_2$ if required.
12. If using a greater/reduced culture area size, adjust cell numbers linearly proportionally (e.g., for 25 cm^2 flask, plate 125,000 cells).
13. One of the first manifestations of apoptosis in vitro is the rounding up and detachment of the cell from the substratum. In many cases this cannot be distinguished from the early stages of cell division. Time-lapse microscopy is extremely useful in distinguishing dividing cells (that readhere to the substratum) and those that are apoptotic (and do not reattach) and exhibit membrane blebbing.
14. Trypsinization must be performed with care as it could disrupt the plasma membrane causing phosphatidylserine to flip, allowing

Annexin V stains to bind giving false readings. Cell scraping could also be used instead of trypsinization using 2 mL of PBS as buffer. Care must be taken to avoid damage to cells, which could lead to increased plasma membrane permeability.

15. Annexin V binding is reversible, therefore perform staining and flow cytometry procedures as quickly as possible to reduce errors. Analyze by flow cytometry within 4 h of initial collection of cells.
16. Count cells with a simple hemocytometer counting chamber at this stage in order to calculate volume of binding buffer to add. Alternatively, the protein concentration of the lysate can be determined using one of the commercially available protein assay kits (e.g., Thermo Scientific Pierce BCA protein assay, Rockford).
17. Alternatively cells can be lysed with other methods prior to protein analysis. For example, cells can by lysed directly in 125 mM Tris–Cl, pH 6.8, 2 % SDS, 10 % β-mercaptoethanol, although this lysis buffer may not be compatible with the NuPage gels described nor protein assay reagents.
18. Appropriate antibody concentrations are normally indicated on the supplied data sheet.
19. Equal loading of the protein gel and transfer to the membrane can be assessed by Ponceau S staining of the membrane or reprobing with an antibody to a "housekeeping gene" where the protein is assumed to be expressed equally in all cells under all conditions (e.g., β-actin or γ-tubulin).
20. The optimum incubation time and concentration of proteinase K may need to be determined empirically. Other methods employ pepsin or trypsin or, for tissues with extensive extracellular matrix, microwave treatment in 0.1 M citrate buffer, pH 6.0 instead of proteinase K treatment.
21. Incubate the tissue in 3 U/mL DNase I in 50 mM Tris–Cl, pH 7.5, 10 mM $MgCl_2$, 1 mg/mL BSA for 10 min at room temperature.
22. Pilot experiments in which the concentration of TdT is titrated between 0.05 and 0.2 U/μL should be conducted.
23. A negative control should be included in which TdT is omitted.
24. BSA can be substituted with 10 % FCS.
25. BCIP/NBT gives a purple/bluish color. Other substrates such as Fast Red may be used. Intensity of color can be optimized by varying incubation time and can be modified by user accordingly.
26. Troubleshooting: (a) false-negative results may be due to restricted access of TdT enzyme due to extensive extracellular matrix, and (b) false positives may be due to extensive DNA

damage occurring in late stages of necrosis or DNA strand breaks in cells exhibiting high proliferative rates and/or high metabolic activity. Accompanying morphological (e.g., pyknotic nuclei) changes should be used to confirm apoptosis.

27. DAB produces a brown stain; if a more intense color is required, then nickel chloride may be added to the substrate to produce a grey/black color—see kit instructions. *See* also **Note 11**.

Acknowledgments

This work was supported by Cancer Research UK. We are indebted to our colleagues for advice on the methods presented in this chapter.

References

1. Lowe SW, Cepero E, Evan G (2004) Intrinsic tumour suppression. Nature 432:307–315
2. Pelengaris S, Khan M, Evan GI (2002) Suppression of Myc-induced apoptosis in beta cells exposes multiple oncogenic properties of Myc and triggers carcinogenic progression. Cell 109:321–334
3. Dansen TB, Whitfield J, Rostker F et al (2006) Specific requirement for Bax, not Bak, in Myc-induced apoptosis and tumor suppression in vivo. J Biol Chem 281:10890–10895
4. Murphy DJ, Junttila MR, Pouyet L et al (2008) Distinct thresholds govern Myc's biological output in vivo. Cancer Cell 14:447–457
5. Evan GI, Wyllie AH, Gilbert CS et al (1992) Induction of apoptosis in fibroblasts by c-myc protein. Cell 69:119–128
6. Littlewood TD, Hancock DC, Danielian PS et al (1995) A modified oestrogen receptor ligand-binding domain as an improved switch for the regulation of heterologous proteins. Nucleic Acids Res 23:1686–1690
7. Iaccarino I, Hancock D, Evan G et al (2003) c-Myc induces cytochrome c release in Rat1 fibroblasts by increasing outer mitochondrial membrane permeability in a Bid-dependent manner. Cell Death Differ 10:599–608
8. Juin P, Hunt A, Littlewood T et al (2002) c-Myc functionally cooperates with Bax to induce apoptosis. Mol Cell Biol 22:6158–6169
9. Wagner AJ, Kokontis JM, Hay N (1994) Myc-mediated apoptosis requires wild-type p53 in a manner independent of cell cycle arrest and the ability of p53 to induce p21waf1/cip1. Genes Dev 8:2817–2830
10. Zindy F, Eischen CM, Randle DH et al (1998) Myc signaling via the ARF tumor suppressor regulates p53-dependent apoptosis and immortalization. Genes Dev 12: 2424–2433
11. Hermeking H, Eick D (1994) Mediation of c-Myc-induced apoptosis by p53. Science 265:2091–2093
12. Hueber A, Zörnig M, Lyon D et al (1997) Requirement for the CD95 receptor-ligand pathway in c-Myc-induced apoptosis. Science 278:1305–1309
13. Tanaka H, Matsumura I, Ezoe S et al (2002) E2F1 and c-Myc potentiate apoptosis through inhibition of NF-kappaB activity that facilitates MnSOD-mediated ROS elimination. Mol Cell 9:1017–1029
14. Kauffmann-Zeh A, Rodriguez-Viciana P, Ulrich E et al (1997) Suppression of c-Myc-induced apoptosis by Ras signalling through PI(3)K and PKB. Nature 385:544–548
15. Juin P, Hueber AO, Littlewood T et al (1999) c-Myc-induced sensitization to apoptosis is mediated through cytochrome c release. Genes Dev 13:1367–1381

Chapter 8

Methods to Study MYC-Regulated Cellular Senescence

Vedrana Tabor, Matteo Bocci, and Lars-Gunnar Larsson

Abstract

Studies in primary and tumor cells suggest that MYC plays an important role in regulating cellular senescence, thereby impacting on tumor development. Here we describe different common methods to measure senescence in cell cultures and in tissues. These include measurement of senescence-associated β-galactosidase activity (SA-β-gal), senescence-associated heterochromatin foci (SAHFs), proliferative arrest, morphological changes, and expression and activity of proteins involved in the senescence process, such as p53 and Rb pathway proteins and secretory proteins. It is important to note that there is no unique marker that unambiguously defines a senescent state, and it is therefore necessary to combine measurements of several different markers that together determine whether cells are senescent or not. Measurement of senescence is an important aspect of studies of MYC biology and will improve our understanding of MYC function and regulation both in preclinical and clinical settings. This may form the basis for new concepts of pro-senescence therapy to combat MYC in cancer.

Key words MYC, Senescence, SA-β-gal, SAHF, Proliferation, RAS, BRAF, p53, Rb

1 Introduction

Cellular senescence is defined as a state of irreversible cell cycle arrest and was first described by Hayflick and Moorhead 50 years ago [1]. They observed that normal human fibroblasts in culture had a limited proliferative capacity, a phenomenon that was thought to reflect the normal cellular aging process. This view was later supported by in vivo studies of senescence in aging primates [2]. This age-related replicative senescence is linked to telomere erosion, but it is now clear that cellular senescence can be induced prematurely by various types of stresses, including DNA damage, oxidative stress, therapeutic drugs, cytokine signaling, and oncogenic stress [3–6]. The latter was first described by Serrano and colleagues [7], showing that mutated, oncogenic RAS triggers senescence when introduced in primary fibroblasts, a phenomenon referred to as "oncogene-induced senescence" (OIS). During the last decade, it has become evident that senescence together with apoptosis represents two main barriers for oncogenic transformation

Laura Soucek and Nicole M. Sodir (eds.), *The Myc Gene: Methods and Protocols*, Methods in Molecular Biology, vol. 1012, DOI 10.1007/978-1-62703-429-6_8, © Springer Science+Business Media, LLC 2013

of normal cells and that these barriers need to be overcome or bypassed for tumor development to occur [3–6]. The p53 and RB pathways, two major tumor suppressor pathways in cells, are both crucial for the induction and maintenance of senescence in cells in culture and in vivo [3–6].

It has become increasingly clear that MYC plays an important role in the regulation of cellular senescence. MYC overexpression contributes to immortalization of many types of primary cells in culture, thereby bypassing replicative senescence [8, 9]. Data accumulated during recent years suggest that MYC contributes to tumorigenesis by suppressing also oncogene-induced senescence triggered, for instance, by RAS, BRAF, STAT5, and mTOR [6, 10–13], as well as senescence induced by the growth inhibitory cytokine TGF-β [14]. At least for suppression of RAS-induced senescence, MYC is dependent on CDK2-mediated phosphorylation at the serine 62 residue in MYC-box 1 [10]. Studies using transgenic MYC mouse tumor models with regulatable MYC or a dominant-negative MYC construct have shown that inactivation of MYC in different tissues results in tumor regression through either apoptosis, differentiation, or senescence, depending on the context [15, 16]. However, under certain circumstances, for instance in cells deficient in CDK2 or WRN, MYC triggers, rather than inhibits, senescence [6, 17–21]. This effect involves MYC-induced DNA damage and is dependent on intact p53 and Rb pathways [17–20]. At least in MYC-driven B lymphoma, senescence induction requires non-cell-autonomous TGF-β signaling and the histone methyl transferase SUV39H1 [21]. Based on the insights into the role of MYC and other oncogenes in senescence regulation, the concept of "pro-senescence therapy" for combating tumors with deregulated MYC has emerged as a potential new strategy for treatment of cancer [6, 22].

This article addresses the issue of how to detect and measure cellular senescence. A number of different markers of senescence have been described in the literature [23], but unfortunately there is no single marker that unambiguously defines a senescent state. It is therefore necessary to combine measurements of several different markers that taken together increase the fidelity of the senescence analysis. In the Methods section, we will go through some of the more common methods and markers to measure senescence in cells in culture and in tissues.

2 Materials

2.1 Solutions for SA-β-gal Detection

1. Phosphate-buffered saline/magnesium chloride (PBS/$MgCl_2$): 1 mM $MgCl_2$ in PBS, pH 5.5 for rodent, and pH 6.0 for human cells.

2. Fixation solution: 4 % paraformaldehyde, 0.25 % glutaraldehyde in PBS/$MgCl_2$ (*see* **Notes 1** and **2**).
3. 20× potassium cyanide (KCN) solution: 820 mg potassium ferricyanide $K_3Fe(CN)_6$, 1 mg potassium ferrocyanide $K_4Fe(CN)_6 \times 3H_2O$, 25 ml PBS, store at 4 °C (*see* **Note 3**).
4. 40× X-Gal solution: 40 mg 5-bromo-4-chloro-3-indolyl-beta-D-galactopyranoside (X-Gal) (Sigma) and 1 ml dimethylformamide (DMF) (*see* **Notes 4** and **5**).
5. Staining solution: 9.25 ml PBS/$MgCl_2$, 0.5 ml KCN solution, 0.25 ml 40× X-Gal solution (*see* **Note 5**).
6. Mounting medium: 70 % glycerol.

2.2 Reagents and Solutions for SAHF Detection

1. DAPI, diluted in PBS, 1 μg/ml.
2. Fixation solution: PBS, 4 % paraformaldehyde (*see* **Note 6**).
3. Permeabilization solution (PBS-T): PBS, 0.1 % TritonX-100.
4. Blocking solution: PBS-T, 3 % BSA.
5. H3K9me3 antibody (Millipore) mix: blocking solution, 1:500 H3K9me3 antibody.
6. Mounting medium: ProLong gold antifade with or without DAPI (Invitrogen).

2.3 Reagents and Solutions for Measurement of Proliferative Arrest

1. Trypan blue 0.4 % solution.
2. Trypsin.
3. Ca^{2+} and Mg^{2+} free PBS (PBS-CM free).
4. FITC BrdU flow kit (cell proliferation kit), BD Pharmingen.
5. Ki67 antibody (DAKO) mix: blocking solution, 1:200 Ki67 antibody.

2.4 Reagent for Detection of Morphological Changes in Senescent Cells

1. Phalloidin (Sigma), 10 μM diluted in PBS.

2.5 Reagents and Solutions for Detection of Protein Expression in Senescent Cells

1. Protein lysis buffer: 50 ml NP-40, 50 ml Tris 1 M pH 8, 15 ml NaCl 5 M, 5 ml EDTA 0.5 M, 2.5 ml aprotinin, ultrapure water to 500 ml.
2. PhosSTOP (phosphatase inhibitors, Roche): 1 tablet in 10 ml of protein lysis buffer.
3. Complete EDTA-free (protease inhibitors, Roche): 1 tablet in 10 ml of protein lysis buffer.
4. PBS-T: PBS, 0.1 % Tween-20.
5. Blocking buffer: PBS-T, 5 % BSA.

6. NuPAGE MOPS/MES.
7. NuPAGE antioxidant.
8. NuPAGE sample buffer 4×.
9. NuPAGE reducing agent 10×.
10. NuPAGE 4–12 % Bis–Tris gel.
11. Ponceau red solution.

3 Methods

3.1 Cytochemical or Histochemical Detection of SA-β-gal Activity

Senescence-associated β-galactosidase (SA-β-gal) activity (Fig. 1) is considered a "golden standard" of senescence detection. It was first described by Dimri et al. [24], showing the presence of β-gal enzymatic activity measured at pH 6.0 in human cells (pH 5.5 in mice) as a marker of replicative senescence both in vitro and *in vivo*. The optimal pH for β-gal enzymatic activity is 4.0, when all of the cells, irrespective of their proliferating status, will score positive. However, one has to be aware that β-gal activity can also increase in quiescence, as a consequence of cell confluency as shown by Severino and co-workers [25]. This notion, taken together with the fact that β-gal enzymatic activity can be induced by many factors, indicates that SA-β-gal cannot be used as a single marker for determining the senescent state of the cell. However, it can be a very strong indicator of a senescent state when used in combination with other markers of senescence. SA-β-gal activity can be detected using several different techniques: here we will describe the chromogenic detection method (for description of additional SA-β-gal-techniques, see ref. 26):

1. Wash adherent cells/frozen tissue sections (*see* **Note 7**) in PBS to remove leftover media, 2 min per wash.
2. Fix cells/frozen tissue sections immediately with freshly prepared fixative solution for 15 min at room temperature (RT), avoid drying out of the slides (*see* **Note 8**).
3. Remove fixative, wash cells/tissue sections twice in PBS/$MgCl_2$.
4. Add the staining solution (50–100 μl per tissue section/10 mm glass slide) and incubate in a humidified atmosphere for 12–16 h in the dark at 37 °C (*see* **Note 9**).
5. Remove the staining solution and wash slides three times in PBS, each time 3–5 min.
6. Mount slides with glycerol and store at 4 °C protected from light (*see* **Note 10**).
7. Assess the blue-positive staining of cells under the microscope (*see* **Note 11**).

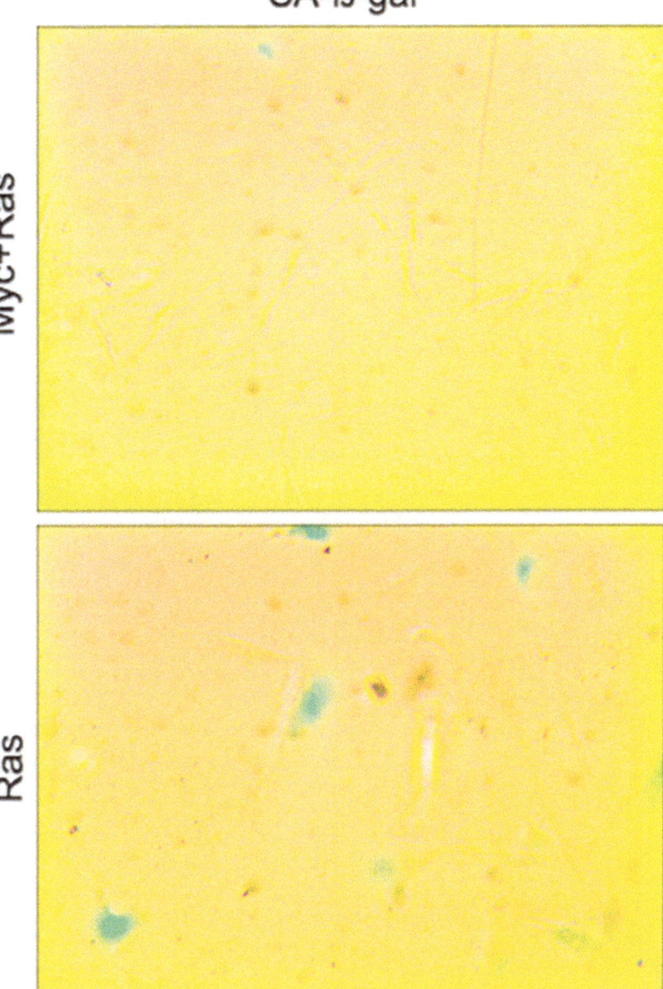

Fig. 1 Murine embryonic fibroblasts (MEFs) stably transduced with RAS^{G12D} or MYC and RAS^{G12D} together were grown for 5 days, after which senescence was assessed by β-gal activity

3.2 Senescence-Associated Heterochromatic Foci (SAHF) Detection

Narita and co-workers described the phenomenon of a distinct nuclear DNA pattern observed in cells undergoing senescence [27]. This is visualized by the DNA dye DAPI and can be seen as distinct foci present in the nucleus of the cell. Furthermore, this distinct DNA focal pattern colocalizes with heterochromatic markers, such as tri-methylated lysine 9 of histone H3 (H3K9me3) (Fig. 2) and HP1 proteins (α, β, and γ). SAHFs form during senescence (both replicative and premature), but not quiescence, hence although they are not markers of cell cycle arrest in general, they are markers of certain types of senescent cells [27]. One has to be aware that SAHFs, which usually form during oncogene-induced senescence, do not necessarily appear upon other types of senescence-inducing

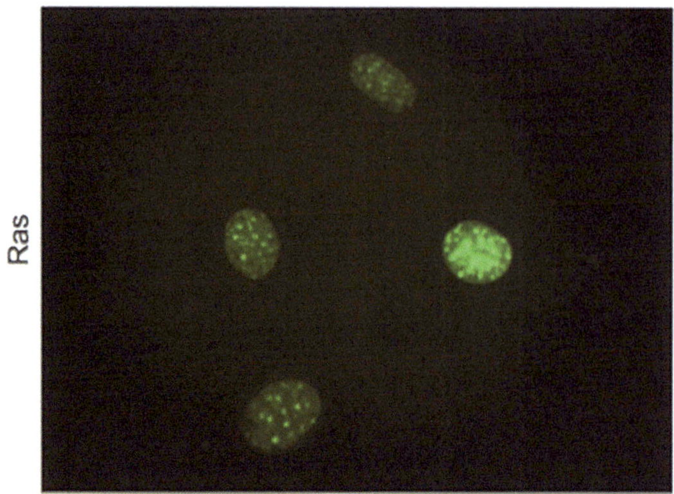

Fig. 2 MEFs stably transduced with RAS^{G12D} were grown for 5 days, after which H3K9me3 foci were evaluated by immunofluorescence

stress depending on the cell type [28, 29]. Therefore, cells may show senescent-like arrest even in the absence of SAHF formation, and consequently lack of SAHFs is not always representing the lack of senescence response. On the other hand, tumor cells bypassing senescence may also display SAHFs, and therefore, the presence of SAHFs does not necessarily represent a senescent state [28, 29]. In conclusion, SAHFs can be used as markers of senescent arrest in combination with the other before-mentioned markers of this state.

3.2.1 SAHF Detection by DAPI

This staining can be used as a last step in any of the immunofluorescence stainings, such as H3K9me3, HP1-γ, or in a simpler combination with phalloidin actin staining. If used alone:

1. Wash cells/tissue sections in PBS to remove leftover media, 2 min per wash.
2. Fix cells/tissue sections in fixation solution, 15 min at room temperature.
3. Wash three times in PBS.
4. Permeabilize in permeabilization solution for 5 min at room temperature, wash one time in PBS for 5 min.
5. Incubate cells/tissue sections with DAPI for 5 min in the dark at room temperature, wash three times with PBS, 10 min each wash.
6. Mount for microscopy using ProLong gold antifade, store in the dark at 4 °C.

3.2.2 SAHF Detection by H3K9me3

1. Wash cells/tissue sections (*see* **Note 12**) in PBS to remove leftover media, 2 min per wash.

2. Fix cells/tissue sections in fixation solution, 15 min at room temperature; wash three times in PBS.
3. Permeabilize cells/tissue sections (*see* **Note 13**) in permeabilization solution for 5 min at room temperature, wash one time in PBS.
4. Block cells/tissue sections in blocking solution for 30 min at 37 °C.
5. Incubate cells/tissue sections in primary H3K9me3 antibody for 12–14 h at 4 °C.
6. Wash three times with PBS, 10 min each wash.
7. Incubate cells/tissue sections with secondary fluorophore-conjugated antibody (e.g., Alexa-488) for 1 h at RT in the dark.
8. Wash three times in PBS, 10 min each wash.
9. Mount for microscopy using ProLong gold antifade, store in the dark at 4 °C (*see* **Note 14**).

3.3 Assessment of Proliferative Arrest

Proliferative arrest is the most typical marker of senescence, as all of the senescent cells cease to proliferate. However, due to the lack of growth factor stimulation, cells can enter a state of quiescence, which is also characterized by the cell growth arrest. Proliferative arrest can be assessed in several different ways. In the case of growth curves spanning over several days (usually 5–7 days), trypan blue dye exclusion is the most common way. One can opt to count in a counting chamber or use any of the automated cell counters presently available on the market. Proliferative arrest can also be measured by lack of BrdU incorporation (Fig. 3). The absence of DNA replication can be shown by a failure to incorporate BrdU, a nucleotide analogue. Absence of BrdU incorporation might not be a sign of cells entering cellular senescence *per se*, as cells might temporarily exit the cell cycle, as it is the case with quiescence. Additional marker of proliferative arrest is diminished Ki67 positivity. Ki67 is a broad cell cycle marker and might be used as a stand-alone staining or in combination with a more specific mitotic marker such as histone H3 phosphorylated at Ser 10 (H3S10P) in order to determine number of cells engaged in the cell cycle as well as in mitosis.

3.3.1 Trypan Blue Exclusion Test

1. Wash adherent cells in PBS and trypsinize them (omit this step for nonadherent cells).
2. Centrifuge cells for 5 min at 100 g, remove supernatant, and resuspend the cell pellet in 500 μl PBS (*see* **Notes 15** and **16**).
3. Mix 1 part of trypan blue and 1 part of the cell suspension (mixing can be performed in a 96-well microtiter plate), and incubate the mixture for 1 min at room temperature (*see* **Note 17**).
4. Apply a drop of the mixture to a hemocytometer.

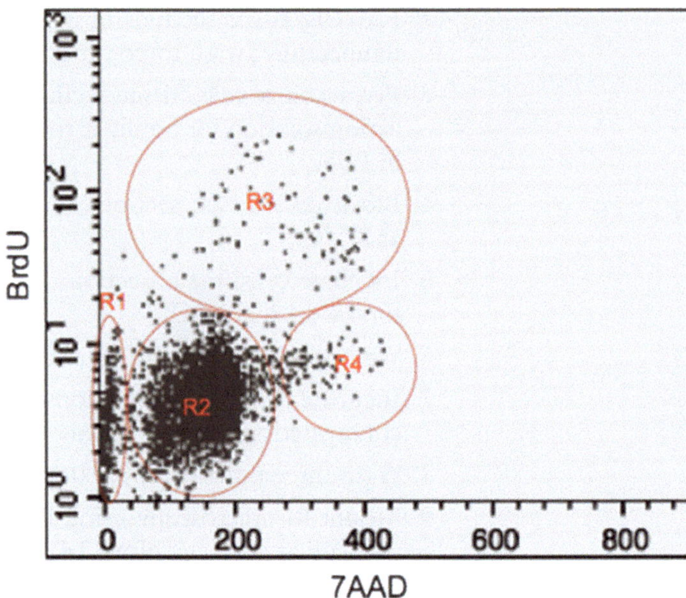

Fig. 3 BrdU/7AAD cell cycle profile of lungs from transgenic mice expressing activated $BRAF^{E600}$ with manifest adenomas. Regions encircled in red: R1 = sub G0/apoptosis; R2 = G1 arrest/senescence; R3 = S-phase; R4 = G2/M phase

5. Place the hemocytometer on the stage of a microscope, focus on the cells.
6. Count the nonstained (viable) and stained (nonviable) cells.
7. To obtain the total number of cells, please refer to the following link: http://www.animal.ufl.edu/hansen/protocols/hemacytometer.htm
8. To calculate the percentage of viable cells (*see* **Note 18**):

   Viable cells (%) = A/B, where

   A—total number of viable cells per ml of aliquot.

   B—total number of cells per ml of aliquot.

3.3.2 BrdU Incorporation (BrdU/7AAD Profile)

1. Treat exponentially growing cells for a period of 48 h with 20 μM BrdU (*see* **Notes 19** and **20**). Alternatively, if working with animals, inject animal with 1 mg of BrdU for a period of 2–24 h (*see* **Note 21**).
2. Harvest cells from the plate or the tissue and centrifuge at 500 × *g* for 5 min.
3. Wash cells in Ca^{2+} and Mg^{2+} free PBS (PBS-CM free).
4. Cell proliferation kit from BD Pharmingen can be used, and a standard protocol described in that kit works for almost all cell types: http://www.bdbiosciences.com/ptProduct.jsp?prodId=8332, briefly:

- Fix cells in fixation buffer for 15 min at RT.
- Spin down cells and wash with washing buffer.
- Permeabilize cells with permeabilization buffer for 10 min at 4 °C.
- Spin down cells and wash with washing buffer.
- Refix cells with fixation buffer for 5 min at RT.
- Spin down cells and wash with washing buffer.
- Treat with DNase for 1 h at 37 °C.
- Spin down cells and wash with washing buffer.
- Incubate with anti-BrdU antibody for 20 min at RT.
- Spin down cells and wash with washing buffer.
- Resuspend in 7AAD (PI), measure immediately.

5. Analyze cells with excitation at 488 nm (*see* Subheading 3.4.2). The cell population is defined in the FSC vs. SSC dot plot. BrdU incorporation is measured in FL1, while 7AAD (PI) is visualized in FL3 channel.

3.3.3 Ki67 and H3S10P Stainings

Single Ki67 staining can be performed by following the protocol described in Subheading 3.2.2 For the double staining with H3S10P, please follow this procedure:

1. Wash cells/tissue sections (*see* **Note 22**) in PBS to remove leftover media, 2 min per wash.
2. Fix cells in fixation solution, 15 min at room temperature; wash three times in PBS.
3. Permeabilize cells/tissue sections in permeabilization solution for 5 min at room temperature, wash one time in PBS.
4. Block cells/tissue sections in blocking solution for 30 min at 37 °C.
5. Incubate cells/tissue sections in primary H3S10P AlexaFluor488-conjugated antibody for 12–14 h at 4 °C in the dark.
6. Wash three times with PBS, 10 min each wash.
7. Incubate cells/tissue sections in primary Ki67 antibody for 12–14 h at 4 °C in the dark.
8. Wash three times with PBS, 10 min each wash.
9. Incubate cells/tissue sections with secondary fluorophore-conjugated antibody (e.g., Alexa-596) for 1 h at room temperature in the dark.
10. Wash three times in PBS, 10 min each wash.
11. Mount for microscopy using ProLong gold antifade; store in the dark at 4 °C.

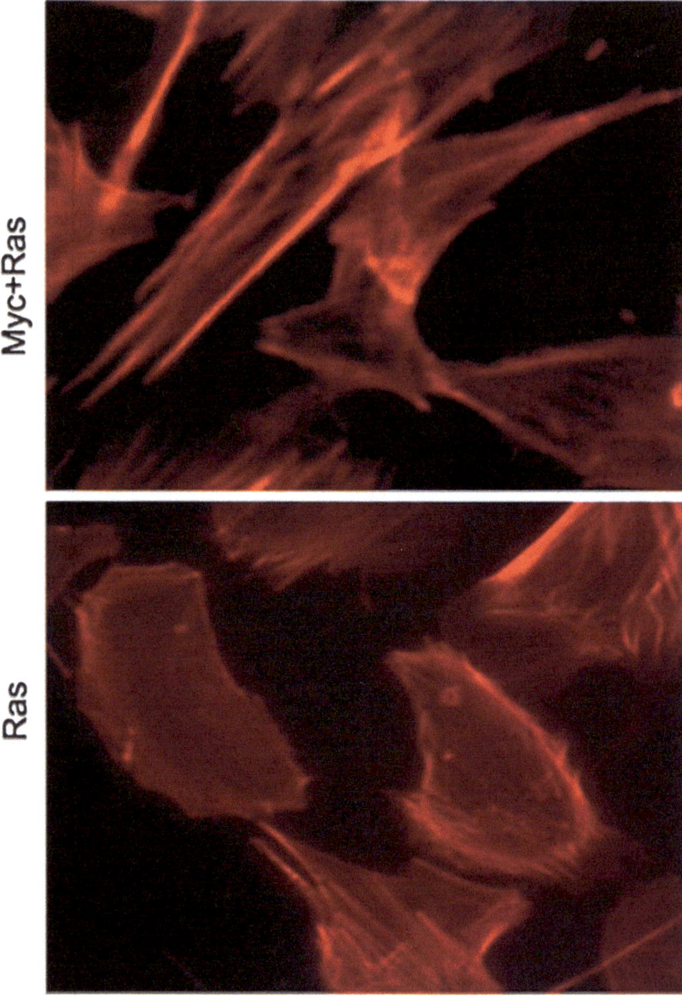

Fig. 4 MEFs stably transduced with RAS^{G12D} or MYC and RAS^{G12D} together were grown for 5 days, after which morphological changes were assessed by F-actin staining

3.4 Detection of Morphological Changes by Visualization of Actin Filaments

Another feature of adherent senescent cells is that they enlarge in size and flatten out and also show extensive vacuolization. Differences in the morphology can be observed by the normal bright light transmitted microscopy, where senescent MEF, BJ, IMR90 cells, BJs, IMR90s, or other fibroblasts appear more enlarged and flattened. Phalloidin staining, where actin filaments are stained, can be also used (Fig. 4). For nonadherent cells, a simple FACS analysis with FSC and SSC can be used to discriminate a population of senescent from normally proliferating cells. Increase in size is one of the characteristics of senescent cells, as well as elevated granularity, due to the increased activity of the lysosomal compartment [30]. This increase can be studied by measuring forward and side scatter using flow cytometry.

3.4.1 Phalloidin Staining

This staining can be used as a last step in any immunofluorescence stainings, such as H3K9me3 and HP1γ, or in a simpler combination with DAPI nuclear staining:

1. Wash cells/tissue sections in PBS to remove leftover media.
2. Fix cells in fixation solution (*see* **Notes 23** and **24**), 15 min at room temperature (RT), wash three times in PBS.
3. Permeabilize cells in permeabilization solution for 5 min at RT, wash one time in PBS.
4. Incubate cells in rhodamine phalloidin for 15 min at RT.
5. Wash three times in PBS.
6. Mount for microscopy using ProLong gold antifade with DAPI, store at 4 °C in the dark.

3.4.2 Forward/Side Scatter (FSC/SSC) Measurement by Flow Cytometry

1. Trypsinize cells (*see* **Note 25**), resuspend them in ice cold PBS, and use them immediately for flow cytometry analysis (*see* **Note 26**).
2. The population of live cells is measured in FSC/SSC dot plot, with apoptotic cells as well as cell debris gated out. Senescent cells are visualized as enlarged in size and increased granularity.

3.5 Detection of Protein Expression in Senescent Cells

Senescent cells are also characterized by changes in mRNA and protein expression patterns. As mentioned in the introduction, the p53 and RB pathways are both crucial for the induction of senescence in vitro and in vivo. In human fibroblasts undergoing premature senescence, p53 displays both elevated levels and increased activity (as determined by its phosphorylation state or the expression of its downstream target genes), and RB is present in its active, hypophosphorylated form. Expression of the cyclin-dependent kinase inhibitors $p16^{INK4A}$ and $p21^{CIP}$ (which are engaged in the p53 and RB pathways) is another hallmark of senescence [3–6] (Fig. 5). The senescence-associated secretory phenotype (SASP) or senescence-messaging secretome (SMS), which is characteristic of cells undergoing senescence [3–6], can also be used as marker of senescence. Cytokines and their receptors such as IL-6, IL-8, CXCR2, TGF-β, and IGFBP7 can be quantified at the protein or mRNA levels. Other senescence markers such as expression of the decoy receptor DcR2 and the transcription factor DEC1 are also frequently reported in the literature [23]. Comprehensive list of antibodies commonly used can be found in Table 1.

The above-mentioned markers can be used for both total protein abundance analysis as described below (western blotting) or for the protein localization by immunohistochemistry or immunofluorescence (please refer Subheading 3.2.2 for the general protocol for immunofluorescence):

1. Grind 200 mg of tissue in 400 μl of protein lysis buffer with protease and phosphatase inhibitors (*see* **Notes 27** and **28**).

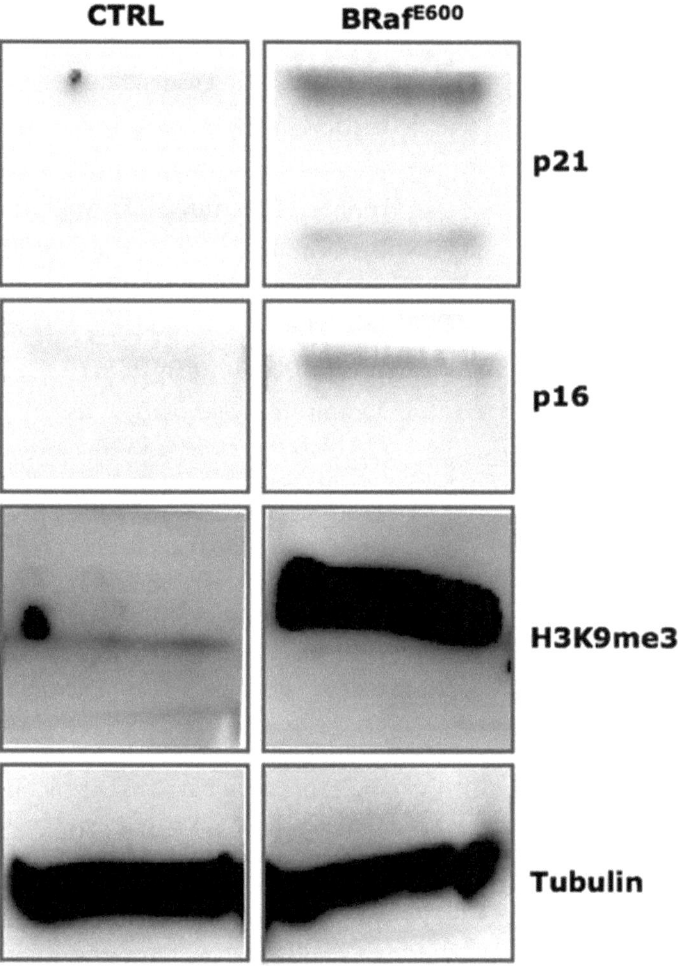

Fig. 5 Whole protein lysates from mouse lungs were blotted for the detection of senescence markers p21^{Cip1}, p16^{INK4A}, and H3K9me3

2. Sonicate protein lysate (*see* **Note 29**) and spin down, 12,000 × *g* for 10 min at 4 °C.
3. Determine protein concentration with BSA Bradford protein assay.
4. Load 50 μg of total protein (*see* **Note 30**) containing sample and reducing buffer and denature for 10 min at 95 °C; place on ice to cool down.
5. Load samples onto a precast gel and run at 115 mA for 1 h (*see* **Note 31**).
6. Transfer the gel on a PVDF membrane within the iBlot dry system blotting (Invitrogen).
7. Check transfer with reversible Ponceau red staining; wash excess Ponceau in dH_2O.
8. Block membrane in blocking buffer for 30 min at RT, shaking.

Table 1
Typical antibodies used in studies involving senescence

Antibody	Tested on
Anti-p15^{INK4B} (sc-612, Santa Cruz)	Human diploid fibroblasts [31]
Anti-p16^{INK4A} (sc-1207, Santa Cruz)	Mouse lung adenocarcinomas [35]
Anti-p16^{INK4A} (M-156, Santa Cruz)	T-cell lymphomas [35] B-cell lymphomas [21]
Anti-p16^{INK4A} (DCS-50, Novocastra)	Human diploid fibroblasts [31]
Anti-p16^{INK4A} JC8 (MS-889, NeoMarkers)	Human melanocytes/nevi [36]
Anti-p19^{ARF} (Ab80, Abcam)	Mouse lung adenocarcinomas [31] T-cell lymphomas [35] B-cell lymphomas [37]
Anti-p21^{Cip1} (c-19, Santa Cruz)	B-cell lymphomas [37] Prostate cancer cells [38]
Anti-p53 (DO1, Santa Cruz)	Prostate cancer cells [38]
Anti-p53 (CM5, Novocastra)	B-cell lymphomas [21, 37]
Anti-p53 Ser 15P (Cell Signaling)	B-cell lymphomas [21, 37]
Anti-DcR2 (AAP-371, Stressgen)	Mouse lung adenocarcinomas [31]
Anti-H3K9me3 (Millipore)	T-cell lymphomas [35] B-cell lymphomas [21]
Anti-MYC (N262, Santa Cruz)	Human leukemic cells [10] B-cell lymphomas [21, 37]
Anti-MYC Thr 58P/Ser 62P (Cell Signaling) (A300-206A, Bethyl Laboratories, Inc.)	Human leukemic cells [10]
Anti-Rb (PMG3-245, Pharmingen)	Human leukemic cells [10] B-cell lymphomas [21] T-cell lymphomas [35]
Anti-Rb Thr 356P (Santa Cruz)	Human leukemic cells [10]

9. Incubate membrane with primary antibody (*see* **Note 32**), overnight, with constant rotation/tilting at 4 °C.
10. Wash membrane in PBS-T for 1 h at RT, shaking.
11. Incubate membrane with secondary HRP-conjugated whole antibody in blocking buffer for 1 h at RT, shaking.
12. Wash membrane in PBS-T, 3 times, 10 min each wash at RT, shaking.
13. Develop membrane with chemiluminescent HRP substrate.

Levels of expression of the above-mentioned players can also be determined by mRNA-based methods [10, 31].

4 Notes

1. Fixation solution needs to be prepared fresh.
2. Caution: Hazardous. Paraformaldehyde and glutaraldehyde are toxic and corrosive solutions. Wear personal protective clothing when handling solution and use in a fume hood.
3. Caution: Potassium cyanide solution is dangerous for the environment. Wear personal protective clothing when handling. Discard in an appropriate manner.
4. Caution: Dimethylformamide is toxic and harmful. Wear personal protective clothing when handling solution. Use a fume hood.
5. X-Gal and staining solution are light sensitive. Protect from light.
6. Prepare fresh and store at −20 °C.
7. If working with nonadherent cells, one should cytospin them onto slides, 74 g for 5 min, and then proceed with the protocol. *Important*: proceed immediately with the fixation and subsequent staining, since storing sections at temperatures above −80 °C will result in a diminished enzymatic activity.
8. Add enough fixation solution to submerge the cells or tissue, avoid drying out of the slide. Avoid prolonged fixation, as it will diminish enzymatic activity.
9. Avoid incubating the samples in a CO_2 incubator. The 5–10 % CO_2 will lower the pH of the buffer.
10. Alternatively, before mounting, proceed with an additional immunohistochemical staining (e.g., Ki67) which is described in a separate section (please follow all the steps starting with fixation).
11. It is important to use appropriate positive and negative controls. For example, as a positive control in vivo, one can use *Eμ-myc/Bcl-2* lymphomas treated with Adriamycin [32] or for in vitro any cell expressing exogenous RAS or 1-2 Gy irradiated NIH-3T3 cells. As a negative control, one can use *Eμ-myc/p53*$^{-/-}$lymphomas or any other highly proliferating tissue in vivo or any exponentially growing cell line in vitro. Controls are needed in every experiment in order to determine threshold levels of positive blue staining.
12. For tissue sections, epitope recovery is necessary before proceeding with the blocking and incubation. Although citrate buffers of pH 6.0 are widely used antigen retrieval solutions, high pH buffers have also been shown to be applicable for many antibodies. It should be determined by the individual laboratory which of the buffers perform optimally for each antigen [33, 34]. To our experience, epitope recovery using citrate buffer in a pressure cooker gave good results for most of the stainings.

For complete information, please refer to Molecular Cloning: A Laboratory Manual (http://www.cshlpress.com/default.tpl?cart=134166359811501284&action=full&--eqskudatarq=934).

13. The same positive controls as mentioned in Subheading 3.1 for SA-β-gal staining can be used for this staining.
14. In recent years there was growing evidence that DNA damage and DNA damage response (DDR) play important roles during senescence induction [3–6]. One can opt for staining for DDR as visualized by γ-H2AX foci [28, 29] and include them in the senescence marker panel. Staining protocol for γ-H2AX is the same as for H3K9me3. As a positive control for γ-H2AX staining, 1-2 Gy irradiated NIH-3T3 cells can be used.
15. The size of aliquot will depend on the approximate number of cells; optimal number is 5×10^5 cells/ml for counting in hemocytometer.
16. Proteins present in serum can be stained by trypan blue and therefore produce misleading results; hence, all the determinations should be done in a serum-free medium.
17. Cells should be counted within 3–5 min of mixing with trypan blue, as longer incubation periods will lead to cell death and reduce viability counts.
18. A more sophisticated method of measuring cell viability is to determine the cell's light scatter characteristics or propidium iodine uptake.
19. In order to detect senescent cells more efficiently, one can extend the BrdU incorporation time to max 72 h. Longer exposures can be cytotoxic.
20. Cells should be shielded from light while incorporating BrdU, as light can increase the generation of DNA strand breaks.
21. Depending on the tissue of interest, one might opt for shorter BrdU exposure times of 2–4 h to get a snapshot of ongoing replication, or for tissues comprised of slowly dividing cells, 24 h exposure would be more desirable.
22. Please observe the notes in Subheading 3.3 as they are applicable to this procedure as well.
23. The time and concentration of fixative can vary, depending on the thickness of the specimen (time for fixative to diffuse into the sample) and on how sensitive the antigen may be to the fixative. For an unknown sample, it is worthwhile to vary the conditions widely in a pilot experiment.
24. Phalloidin binding requires the F-actin to have a protein structure near to native. Methanol or acetone used to fix and/or permeabilize essentially abolishes phalloidin binding.

25. Use cells in exponential phase of growth.
26. Each cell type has specific size and morphology; therefore, before performing this measurement, please refer to the manual of flow cytometer that you are using in order to determine voltage and amplification gain for your desired cell type. In general, as a control, we recommend Ras-expressing growth arrested cells (positive control) and exponentially growing cells (negative control).
27. Different protein extraction protocols may apply depending on the tissue/cell type.
28. All the steps of protein isolation should be carried out on ice.
29. One time 10 pulses, 70 % power.
30. Certain less-abundant proteins might require loading of more than 50 μg of total protein.
31. Depending on the proteins one wishes to detect, it might let be run for longer times.
32. Dilution according to manufacturers' datasheet.

Acknowledgments

This work was supported by the Swedish Cancer Society (V.T., L.G.L.), the Swedish Childhood Cancer Society (L.G.L.), the Swedish Research Council (L.G.L.), Olle Engkvist Foundation (L.G.L.), the Karolinska Institutet Foundations (V.T., L.G.L.), and the Advanced Cancer Therapies (ACT!) consortium (L.G.L.).

References

1. Hayflick L (1965) The limited in vitro lifetime of human diploid cell strains. Exp Cell Res 37:614–636
2. Herbig U, Ferreira M, Condel L, Carey D, Sedivy JM (2006) Cellular senescence in aging primates. Science 311(5765):1257
3. Campisi J, d'Adda di Fagagna F (2007) Cellular senescence: when bad things happen to good cells. Nat Rev Mol Cell Biol 8(9):729–740
4. Collado M, Serrano M (2010) Senescence in tumours: evidence from mice and humans. Nat Rev Cancer 10(1):51–57
5. Kuilman T, Michaloglou C, Mooi WJ, Peeper DS (2010) The essence of senescence. Genes Dev 24(22):2463–2479
6. Larsson LG (2011) Oncogene- and tumor suppressor gene-mediated suppression of cellular senescence. Semin Cancer Biol 21(6):367–376
7. Serrano M, Lin AW, McCurrach ME, Beach D, Lowe SW (1997) Oncogenic ras provokes premature cell senescence associated with accumulation of p53 and p16INK4a. Cell 88(5): 593–602
8. Gil J, Kerai P, Lleonart M, Bernard D, Cigudosa JC, Peters G, Carnero A, Beach D (2005) Immortalization of primary human prostate epithelial cells by c-Myc. Cancer Res 65(6): 2179–2185
9. Guney I, Wu S, Sedivy JM (2006) Reduced c-Myc signaling triggers telomere-independent senescence by regulating Bmi-1 and p16(INK4a). Proc Natl Acad Sci U S A 103(10):3645–3650
10. Hydbring P, Bahram F, Su Y, Tronnersjo S, Hogstrand K, von der Lehr N, Sharifi HR, Lilischkis R, Hein N, Wu S, Vervoorts J, Henriksson M, Grandien A, Luscher B, Larsson LG (2010) Phosphorylation by Cdk2 is

required for Myc to repress Ras-induced senescence in cotransformation. Proc Natl Acad Sci U S A 107(1):58–63

11. Mallette FA, Gaumont-Leclerc MF, Huot G, Ferbeyre G (2007) Myc down-regulation as a mechanism to activate the Rb pathway in STAT5A-induced senescence. J Biol Chem 282(48):34938–34944
12. Ruggero D, Montanaro L, Ma L, Xu W, Londei P, Cordon-Cardo C, Pandolfi PP (2004) The translation factor eIF-4E promotes tumor formation and cooperates with c-Myc in lymphomagenesis. Nat Med 10(5):484–486
13. Zhuang D, Mannava S, Grachtchouk V, Tang WH, Patil S, Wawrzyniak JA, Berman AE, Giordano TJ, Prochownik EV, Soengas MS, Nikiforov MA (2008) C-MYC overexpression is required for continuous suppression of oncogene-induced senescence in melanoma cells. Oncogene 27(52):6623–6634
14. van Riggelen J, Muller J, Otto T, Beuger V, Yetil A, Choi PS, Kosan C, Moroy T, Felsher DW, Eilers M (2010) The interaction between Myc and Miz1 is required to antagonize TGFbeta-dependent autocrine signaling during lymphoma formation and maintenance. Genes Dev 24(12):1281–1294
15. Soucek L, Whitfield J, Martins CP, Finch AJ, Murphy DJ, Sodir NM, Karnezis AN, Swigart LB, Nasi S, Evan GI (2008) Modelling Myc inhibition as a cancer therapy. Nature 455(7213):679–683
16. Wu CH, van Riggelen J, Yetil A, Fan AC, Bachireddy P, Felsher DW (2007) Cellular senescence is an important mechanism of tumor regression upon c-Myc inactivation. Proc Natl Acad Sci U S A 104(32):13028–13033
17. Campaner S, Doni M, Hydbring P, Verrecchia A, Bianchi L, Sardella D, Schleker T, Perna D, Tronnersjo S, Murga M, Fernandez-Capetillo O, Barbacid M, Larsson LG, Amati B (2010) Cdk2 suppresses cellular senescence induced by the c-myc oncogene. Nat Cell Biol 12(1):54–59, sup pp 51–14
18. Grandori C, Wu KJ, Fernandez P, Ngouenet C, Grim J, Clurman BE, Moser MJ, Oshima J, Russell DW, Swisshelm K, Frank S, Amati B, Dalla-Favera R, Monnat RJ Jr (2003) Werner syndrome protein limits MYC-induced cellular senescence. Genes Dev 17(13):1569–1574
19. Moser R, Toyoshima M, Robinson K, Gurley KE, Howie HL, Davison J, Morgan M, Kemp CJ, Grandori C (2012) MYC-driven tumorigenesis is inhibited by WRN syndrome gene deficiency. Mol Cancer Res 10(4):535–545
20. Post SM, Quintas-Cardama A, Terzian T, Smith C, Eischen CM, Lozano G (2010) p53-Dependent senescence delays Eμ-myc-induced B-cell lymphomagenesis. Oncogene 29(9):1260–1269
21. Reimann M, Lee S, Loddenkemper C, Dorr JR, Tabor V, Aichele P, Stein H, Dorken B, Jenuwein T, Schmitt CA (2010) Tumor stroma-derived TGF-beta limits myc-driven lymphomagenesis via Suv39h1-dependent senescence. Cancer Cell 17(3):262–272
22. Nardella C, Clohessy JG, Alimonti A, Pandolfi PP (2011) Pro-senescence therapy for cancer treatment. Nat Rev Cancer 11(7):503–511
23. Collado M, Serrano M (2006) The power and the promise of oncogene-induced senescence markers. Nat Rev Cancer 6(6):472–476
24. Dimri GP, Lee X, Basile G, Acosta M, Scott G, Roskelley C, Medrano EE, Linskens M, Rubelj I, Pereira-Smith O, Peacocke M, Campisi J (1995) A biomarker that identifies senescent human cells in culture and in aging skin in vivo. Proc Natl Acad Sci U S A 92(20):9363–9367
25. Severino J, Allen RG, Balin S, Balin A, Cristofalo VJ (2000) Is beta-galactosidase staining a marker of senescence in vitro and in vivo? Exp Cell Res 257(1):162–171
26. Debacq-Chainiaux F, Erusalimsky JD, Campisi J, Toussaint O (2009) Protocols to detect senescence-associated beta-galactosidase (SA-βgal) activity, a biomarker of senescent cells in culture and in vivo. Nat Protoc 4(12):1798–1806
27. Narita M, Nunez S, Heard E, Narita M, Lin AW, Hearn SA, Spector DL, Hannon GJ, Lowe SW (2003) Rb-mediated heterochromatin formation and silencing of E2F target genes during cellular senescence. Cell 113(6):703–716
28. Di Micco R, Sulli G, Dobreva M, Liontos M, Botrugno OA, Gargiulo G, dal Zuffo R, Matti V, d'Ario G, Montani E, Mercurio C, Hahn WC, Gorgoulis V, Minucci S, d'Adda di Fagagna F (2011) Interplay between oncogene-induced DNA damage response and heterochromatin in senescence and cancer. Nat Cell Biol 13(3):292–302
29. Kosar M, Bartkova J, Hubackova S, Hodny Z, Lukas J, Bartek J (2011) Senescence-associated heterochromatin foci are dispensable for cellular senescence, occur in a cell type- and insult-dependent manner and follow expression of p16(ink4a). Cell Cycle 10(3):457–468
30. Bayreuther K, Rodemann HP, Francz PI, Maier K (1988) Differentiation of fibroblast stem cells. J Cell Sci Suppl 10:115–130
31. Collado M, Gil J, Efeyan A, Guerra C, Schuhmacher AJ, Barradas M, Benguria A, Zaballos A, Flores JM, Barbacid M, Beach D, Serrano M (2005) Tumour biology: senescence in premalignant tumours. Nature 436(7051):642

32. Schmitt CA, Fridman JS, Yang M, Lee S, Baranov E, Hoffman RM, Lowe SW (2002) A senescence program controlled by p53 and p16INK4a contributes to the outcome of cancer therapy. Cell 109(3):335–346
33. Shi SR, Imam SA, Young L, Cote RJ, Taylor CR (1995) Antigen retrieval immunohistochemistry under the influence of pH using monoclonal antibodies. J Histochem Cytochem 43(2):193–201
34. Bankfalvi A, Navabi H, Bier B, Bocker W, Jasani B, Schmid KW (1994) Wet autoclave pretreatment for antigen retrieval in diagnostic immunohistochemistry. J Pathol 174(3):223–228
35. Braig M, Lee S, Loddenkemper C, Rudolph C, Peters AH, Schlegelberger B, Stein H, Dorken B, Jenuwein T, Schmitt CA (2005) Oncogene-induced senescence as an initial barrier in lymphoma development. Nature 436(7051):660–665
36. Michaloglou C, Vredeveld LC, Soengas MS, Denoyelle C, Kuilman T, van der Horst CM, Majoor DM, Shay JW, Mooi WJ, Peeper DS (2005) BRAFE600-associated senescence-like cell cycle arrest of human naevi. Nature 436(7051):720–724
37. Reimann M, Loddenkemper C, Rudolph C, Schildhauer I, Teichmann B, Stein H, Schlegelberger B, Dorken B, Schmitt CA (2007) The Myc-evoked DNA damage response accounts for treatment resistance in primary lymphomas in vivo. Blood 110(8):2996–3004
38. Chen Z, Trotman LC, Shaffer D, Lin HK, Dotan ZA, Niki M, Koutcher JA, Scher HI, Ludwig T, Gerald W, Cordon-Cardo C, Pandolfi PP (2005) Crucial role of p53-dependent cellular senescence in suppression of Pten-deficient tumorigenesis. Nature 436(7051):725–730

Chapter 9

Chromatin Immunoprecipitation Assays for Myc and N-Myc

Bonnie L. Barrilleaux, Rebecca Cotterman, and Paul S. Knoepfler

Abstract

Myc and N-Myc have widespread impacts on the chromatin state within cells, both in a gene-specific and genome-wide manner. Our laboratory uses functional genomic methods including chromatin immunoprecipitation (ChIP), ChIP-chip, and, more recently, ChIP-seq to analyze the binding and genomic location of Myc. In this chapter, we describe an effective ChIP protocol using specific validated antibodies to Myc and N-Myc. We discuss the application of this protocol to several types of stem and cancer cells, with a focus on aspects of sample preparation prior to library preparation that are critical for successful Myc ChIP assays. Key variables are discussed and include the starting quantity of cells or tissue, lysis and sonication conditions, the quantity and quality of antibody used, and the identification of reliable target genes for ChIP validation.

Key words ChIP-seq, Chromatin immunoprecipitation, Myc, N-Myc, Epigenetics, Histone modifications

1 Introduction

Myc and MycN (referred hitherto as "c-Myc" and "N-Myc") are basic helix-loop-helix transcription factors that, when dimerized with the binding partner Max, bind E-box sequences in a specific chromatin context and thereby activate expression of target genes [1]. However, Myc can also repress transcription with its partner Miz-1 (ZBTB17) in both cancer cells [2] and human embryonic stem cells (hESCs) [3]. In addition, Myc's impact on chromatin can also manifest in a much more widespread manner, with Myc binding and affecting histone acetylation and methylation across the genome [4, 5]. Both the gene-specific and genome-wide modes of action by Myc are critical to its biological functions, including its important roles in normal development, tumorigenesis, and cellular reprogramming. However, the relationship between Myc and chromatin remodeling is complex and not fully understood. Myc is known to recruit several histone-modifying enzymes including the serine/threonine kinase PIM1 which phosphorylates serine 10 of

Laura Soucek and Nicole M. Sodir (eds.), *The Myc Gene: Methods and Protocols*, Methods in Molecular Biology, vol. 1012, DOI 10.1007/978-1-62703-429-6_9,

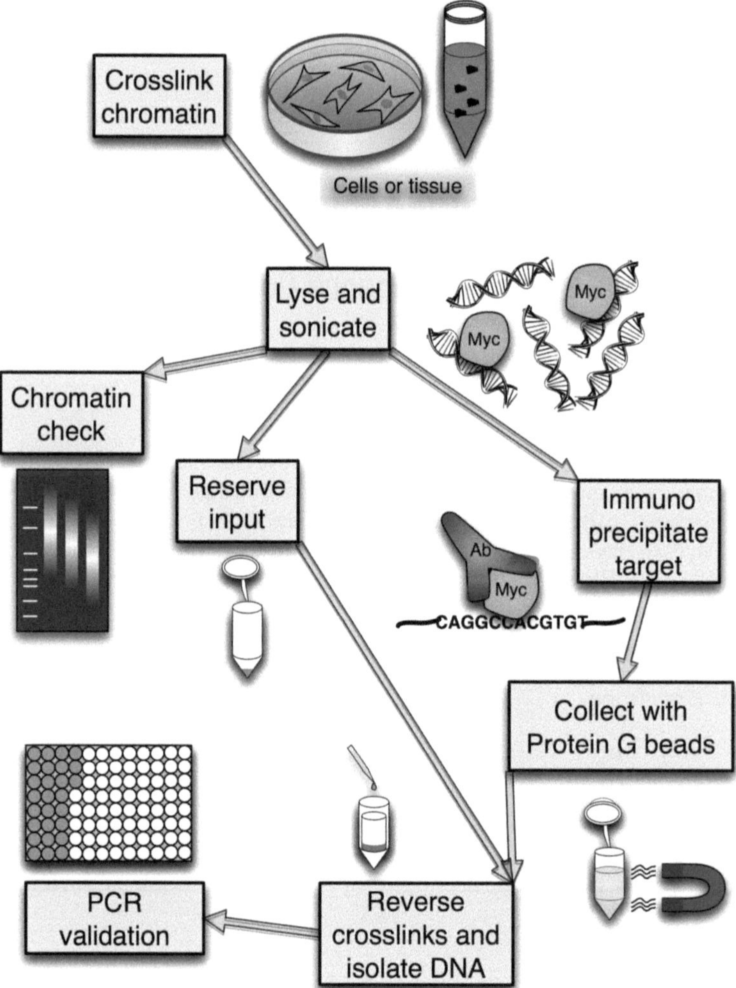

Fig. 1 Schematic diagram illustrating major steps in the Myc ChIP process

histone H3 [6] and the histone acetyltransferases GCN5 [7] and TIP60 [8]. This is a dynamic process possibly involving cross talk between Myc and histone modifications, whereby Myc recruits chromatin-modifying enzymes which lay down marks that subsequently influence transcription, but also direct the localization of Myc. For example, the trimethylation of histone H3 at lysine 4 (H3K4me3) is involved in recruiting Myc to specific locations in the genome [1], and the persistence of these transcriptionally activating H3K4me3 marks is dependent on the presence of Myc [9].

To understand the global function of Myc on chromatin, our laboratory has been using functional genomic methods for years including chromatin immunoprecipitation (ChIP) with DNA hybridization arrays (ChIP-chip). More recently, we have changed to predominantly employ ChIP coupled to high-throughput sequencing (ChIP-seq) to analyze the genomic location of c-Myc and N-Myc (Fig. 1). ChIP-seq has key advantages over ChIP-chip, including its greater dynamic range, higher resolution, and more complete genome-wide coverage. The ChIP-seq technique must generally be optimized for each cell type and each antibody used; therefore in this chapter, we will describe an effective protocol using specific validated antibodies to c-Myc and N-Myc and discuss its application to several types of stem and cancer cells. Key variables include the starting quantity of cells or tissue, sonication conditions, as well as quantity and quality of antibody used. Performing ChIP assays for transcription factors such as Myc has specific additional considerations compared to ChIP assays for histone modifications. Myc is less abundant, less tightly bound to chromatin, and has a shorter half-life in the cell compared to histones. This means that a much higher starting cell number is required, and protein breakdown during sample preparation is a major concern. Additionally, since c-Myc and N-Myc share significant sequence similarity, it is particularly important to test for antibody cross-reactivity.

Finding reliable Myc target genes for ChIP validation is another significant factor. The binding of Myc to E-box sequences is dependent on the chromatin context, which is cell type specific. It can be difficult to determine what genes Myc may be binding in the genome before doing the sequencing, yet the ChIP must be validated before proceeding with library preparation and sequencing, necessitating the use of PCR for specific candidate targets. In this protocol, we suggest Myc target genes that, from our extensive experience, can be effective in several cell types that we have studied. Other recent protocols have discussed ChIP-seq library preparation in some detail [10]. In this protocol we will focus on the aspects of sample preparation prior to library preparation, which are the most critical steps for successful immunoprecipitation of transcription factors such as Myc.

2 Materials

Prepare all solutions with ultrapure water and molecular biology grade reagents (*see* **Note 1**).

2.1 Cross-linking

1. Formaldehyde solution: 37 % w/w (*see* **Note 2**).
2. Quenching solution: 1.25 M glycine.
3. Wash solution: Phosphate Buffered Saline (PBS), pH 7.4. Protease inhibitors are optional (*see* **Note 3**).

2.2 Chromatin Preparation

1. Protease inhibitor tablets: Roche Complete Mini, EDTA free. Use 1 tablet per 10 ml of lysis buffer, wash buffer, or IP dilution buffer (cut to needed size).
2. Cell lysis buffer: 5 mM PIPES pH 8.0 and 85 mM KCl. Before each use, add Igepal CA-630 (Sigma) to a final concentration

of 1 % (10 μl/ml) and protease inhibitors. Store at room temperature.

3. Nuclear lysis buffer: 50 mM Tris-Cl, pH 8.0, 10 mM EDTA pH 8.0, 1 % (w/v) SDS. Before use, add protease inhibitors. Store at room temperature.

2.3 Immunoprecipitation

(*See* **Note 4**).

1. IP dilution buffer: 50 mM Tris-Cl pH 7.4, 1 mM EDTA pH 8.0, 150 mM NaCl, 0.25 % (w/v) deoxycholic acid (Fisher Scientific). Before each use, add Igepal CA-630 to a final concentration of 1 % and protease inhibitors. Store at 4 °C.
2. IP wash buffer 1: same as IP dilution buffer. Protease inhibitors are optional (*see* **Note 3**). Store at 4 °C.
3. IP wash buffer 2: 100 mM Tris-Cl, pH 9.0, 500 mM lithium chloride, 1 % (w/v) deoxycholic acid. Before use, add Igepal CA-630 to a final concentration of 1 %. Protease inhibitors are optional. Store at room temperature.
4. IP wash buffer 3: 100 mM Tris-Cl, pH 9.0, 500 mM lithium chloride, 150 mM NaCl, 1 % (w/v) deoxycholic acid. Before use, add Igepal CA-630 to a final concentration of 1 %. Protease inhibitors are optional. Store at room temperature.
5. Elution buffer: 50 mM $NaHCO_3$, 1 % (w/v) SDS. Store at room temperature.
6. 5 M NaCl.
7. Protein G magnetic beads, ChIP grade (Cell Signaling Technology) (*see* **Note 5**).

2.4 Small Equipment and Consumables

1. Magnetic bead rack (*see* **Note 6**).
2. Type B (loose) Dounce homogenizer for cell lysis.
3. Diagenode Bioruptor UCD 200 (allows sonication of multiple samples at once) or equivalent.
4. Sterile 2.0 ml screw cap, o-ring tubes for immunoprecipitations and for heating samples in water bath to reverse cross-links.
5. Low adhesion or siliconized tubes for final storage of samples.
6. Sterile, DNAse-free pipette tips.
7. Water bath or heating block capable of reaching 67 °C.
8. NanoDrop, or equivalent, for determining chromatin and DNA concentrations.
9. Rocker or rotator: Nutator (Clay Adams) or Labquake (Barnstead Thermolyne).

2.5 DNA Purification

1. Qiagen MinElute PCR kit or equivalent (*see* **Note 7**).
2. RNase A (Roche) reconstituted to 10 mg/ml in reagent grade water.

2.6 Antibodies

1. Suggested antibodies against c-Myc and N-Myc, along with normal IgG control antibodies, are given in Table 1.

2.7 Controls

1. If enough chromatin is available, include a nonspecific IgG as one of the antibodies. It helps one judge the background and quality of the ChIP, works as a contamination check (if IgG is too high, solutions or IP may be contaminated), and serves as a reasonable negative control for PCR validation in the absence of negative control primers.
2. A stringent check for DNA contamination of the PCR reaction is to run a 100 μl sample of elution buffer through the elution, cross-link reversal, PCR purification kit, and final PCR evaluation.
3. Input or total chromatin control: there are several options for taking the input control. For regular ChIP, we routinely take input from the leftover IgG supernatant after incubation overnight. Use 10 % of the volume of one IP, including dilution buffer.
4. Alternatively, take 10 % of the volume of one IP before dilution with IP dilution buffer to use as input.
5. For ChIP-seq, save a portion ≥ 500 ng of chromatin before addition of IP dilution buffer to use as input. This quantity can be determined from the NanoDrop reading taken after **step 3.2, item 8**.

3 Methods

3.1 Preparation of Cross-linked Cells or Tissues

1. Tissues for ChIP may be flash frozen in liquid nitrogen and stored at −80 °C until ready to cross-link. On the day of cross-linking, keep tissues on dry ice until ready. Weigh and cut portion (about 30–80 mg per IP for Myc). Quickly chop with a clean scalpel, then immediately transfer to 15 ml polypropylene tube containing 10 ml of PBS and begin cross-linking (*see* **Note 8**).
2. Cultured cells for ChIP should be in log-phase growth and not confluent. Cells may be flash frozen before or after cross-linking. Immunoprecipitation of Myc will require more starting material than for active histone marks. Plan on 3 or more 6-well plates for pluripotent stem cells, and 4 or more 15 cm plates for adherent cell lines, or roughly 5×10^7 cells per IP. Some researchers use up to 1×10^8 cells for transcription factor IP.

Table 1
Antibodies tested for ChIP assays

Antibody	Species reactivity	Vendor	Catalog number	Web link	Laboratory notes
Normal mouse IgG	None	Santa Cruz	sc-2025	www.scbt.com/datasheet-2025-normal-mouse-igg.html	Negative control antibody
Normal rabbit IgG	None	Santa Cruz	sc-2027	www.scbt.com/datasheet-2027-normal-rabbit-igg.html	Negative control antibody
N-Myc (NCM II), ChIP grade	Human, mouse	Abcam	ab16898	www.abcam.com/n-Myc-antibody-NCM-II-100-ChIP-Grade-ab16898.html	Not known to cross-react with c-Myc. Also suitable for IP. Works well for mouse ChIP in our lab
N-Myc (C-19)	Human, mouse	Santa Cruz	sc-791	www.scbt.com/datasheet-791-n-myc-c-19-antibody.html	Works well for mouse ChIP in our lab
N-Myc (B8.4.B)	Human, mouse	Santa Cruz	sc-53993	www.scbt.com/datasheet-53993-n-myc-b8-4-b-antibody.html	A lab favorite for human cells. Verified in our lab to be specific for N-Myc
N-Myc (M-50)	Mouse	Santa Cruz	sc-22836	www.scbt.com/datasheet-22836-n-myc-m-50-antibody.html	Didn't work well in our lab. Not recommended
c-Myc (3C7)	Human, mouse	Chemicon/ Millipore	CBL434	www.millipore.com/catalogue/item/cbl434	Didn't work well in our lab. Not recommended
c-Myc (N262)	Human, mouse	Santa Cruz	sc-764	www.scbt.com/datasheet-764-c-myc-n-262-antibody.html	Antibody strongly cross reacts with N-Myc.
c-Myc [8]	Human, mouse	Abcam	ab17355	www.abcam.com/c-Myc-antibody-8-ChIP-Grade-ab17355.html	Worked fine in our lab, but ab56 works better for hESC
c-Myc (9E11), ChIP grade	Human	Abcam	ab56	www.abcam.com/c-Myc-antibody-9E11-ChIP-Grade-ab56.html	A lab favorite. Verified in our lab to be specific for c-Myc

For all antibodies, we use 3 μg per IP. Santa Cruz refers to Santa Cruz Biotechnology

3. Cross-link cells or tissues on a rocker in 1 % final volume of freshly diluted formaldehyde at room temperature in culture medium or in PBS. Cross-link 10 min for histones and 12–13 min for transcription factors such as Myc (*see* **Note 9**).
4. Quench the reaction by adding 1.25 M glycine (10×) to a final dilution of 0.125 M. Mix on a rocker for 5 min.
5. Wash adherent cells on the plate by decanting the cross-linking solution and rinsing plates twice with cold PBS. (Keep cells at room temperature while exposed to formaldehyde, but after cross-linking, keep PBS wash solution on ice, and spin in a 4 °C centrifuge at 1,000 × *g*. Optionally, add protease inhibitors to the PBS used for washing.) Transfer adherent cells from the plate to a 15 ml polypropylene tube by decanting the second wash, then scraping cells into the residual PBS with a cell lifter. Add more PBS and spin down for the final wash. Do not be surprised when cross-linked cells become hard to pool. They tend to float after cross-linking and may require additional centrifugation to pellet.
6. Tissues and suspension cells can be washed directly in the tube they were cross-linked in, three times in cold PBS in a 4 °C centrifuge at 1,000 × *g*.
7. After the final wash, pellet cells and remove supernatant. Proceed immediately to ChIP or flash freeze cross-linked cells in liquid nitrogen and freeze at −80 °C until needed.

3.2 Cell and Nuclear Lysis and Sonication

1. Lyse cells in cell lysis buffer with protease inhibitors and Igepal on ice for 20 min, using 1 ml lysis buffer per 5×10^7 cells. Reserve 500 μl of cell lysis buffer to rinse tips and tubes if low cell number is a problem. Keep chromatin on ice at all times to avoid degradation.
2. Transfer cells to a type B (loose) Dounce homogenizer, 1 ml at a time. Dounce about 20 strokes on ice for a dilute cell preparation or 20–25 strokes for tissues or cells that are in clumps. Transfer cells to a 1.5 ml tube. Reserve 1 ml of cell lysis buffer to rinse the homogenizer after homogenizing, then add the rinse to the rest of the cells. This will increase yield, especially when there is little material to spare.
3. Spin the lysed cells at 2,500 × *g* for 5 min at 4 °C in a precooled centrifuge to pellet the nuclei.
4. Resuspend the pellet in nuclear lysis buffer with protease inhibitors, using no more than 100 μl per IP sample. Keep in mind that the Bioruptor will not efficiently sonicate more than 300 μl per 1.5 ml tube (*see* **Note 10**).
5. Incubate on ice for 15 min. Flick regularly to mix the nuclei. Proceed to sonication.

Table 2
Representative sonication conditions for Subheading 3.2

Cell type	Example	Starting quantity	Suggested sonication time (min)
Tissue	Mouse testis	90 mg	30–35
Suspension aggregates	Neurospheres, embryoid bodies	5 × 10 cm plates	30–35
Adherent colonies	hESC	5 × 6-well plates	25–30
Adherent monolayers	Tet21N, mouse embryonic fibroblasts	5 × 15 cm plate	20–25

Starting quantities of cells can be divided into multiple IPs with different antibodies after sonication (generally 30–50 mg tissue per IP or 2–3 × 10 cm plates of cells per IP). Times given are optimized for ChIP-seq, which requires sonication to a size of 200–500 bp. ChIP without sequencing (using end-point PCR or qPCR to examine specific target genes) can tolerate somewhat larger fragments, so use about one-third less time. All sonications are performed in 100–300 μl total chromatin volume in a 1.5 ml tube using a Diagenode Bioruptor set to sonicate in cycles of 30 s high, 1 min off. Sonication is performed in 15 min increments, incubating the chromatin on ice for 5 min between increments

6. Sonicate 100–300 μl samples for the appropriate predetermined amount of time (*see* **Note 11** and Table 2) in order to reduce fragments to the 200–500 bp size range, which is critical for efficient sequencing (*see* Fig. 2). Leave cells on ice between rounds of sonication, and replace warm water in the sonicator with cold water between each round of sonication (or use a cooling system to maintain water temperature; *see* **Note 12**).
7. Spin chromatin for 10 min at 20,000 × *g* in a precooled centrifuge at 4 °C. A large pellet of debris indicates that the sonication or cell lysis was inefficient.
8. A NanoDrop can be used at this point to get a rough idea of the chromatin yield. This is especially useful for trying to equalize input between two samples: excess can be saved at −80 °C after flash freezing in liquid nitrogen.
9. Check chromatin size by boiling a 10 μl sample for 20–30 min with 40 μl elution buffer and 2 μl of 5 M NaCl. The positive ions in the salt will stabilize the negatively charged DNA backbone. Purify the DNA using a PCR purification kit and run it on a 1 % agarose gel with EtBr. This is only an estimate. Note that it is hard to reverse the cross-links completely by boiling and chromatin will run high. If you choose to do this, we suggest doing only one check, as the chromatin has to sit and wait and may be subject to degradation. Keep chromatin on ice during the chromatin check. If the chromatin check reveals significant material above 500 bp on the agarose gel (Fig. 2), perform further cycles of sonication before proceeding.
10. Transfer equally divided supernatant (approximately 100 μl per IP) to a fresh screw-capped 2.0 ml DNase-free microfuge tube.

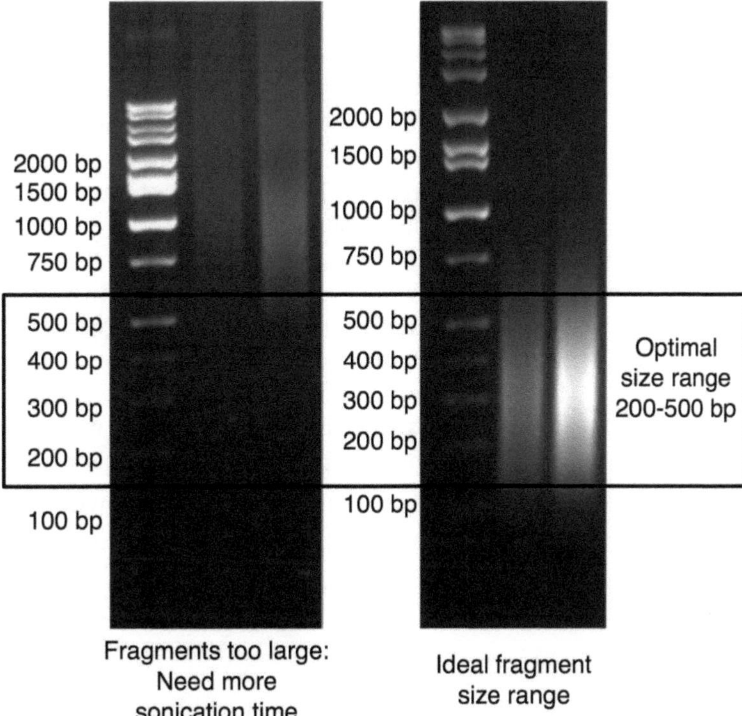

Fig. 2 Agarose gel electrophoresis demonstrating appropriate chromatin size for ChIP. This chromatin check is performed prior to IP (*see* **step 3.2, item 9**) and can be used to fine-tune the sonication time. The *left panel* shows chromatin fragments that are too large; additional sonication cycles should be performed before continuing. The *right panel* shows chromatin fragments of the optimal size range. An additional, more accurate chromatin size check can be performed after the IP, using leftover supernatant from the IP (*see* **step 3.3, item 6**)

3.3 Immunoprecipitation

1. Reserve at least 500 ng of each experimental chromatin sample for input. Store at 4 °C until **step 3.4, item 7** (reversing the cross-links of the samples).
2. Take sonicated chromatin samples which have been divided into aliquots in the 2.0 ml screw cap tubes, and add 4 volumes of IP dilution buffer with protease inhibitors to 1 volume chromatin.
3. In general, use about 3 μg of antibody for a Myc ChIP. Santa Cruz antibodies which are stored at 4 °C will retain their potency only a year. We use a similar amount of antibody for histone ChIP assays (*see* **Note 13**).
4. Incubate overnight on a rocker in a cold room.
5. Collect protein/antibody complexes by adding 15 μl of protein G magnetic beads (*see* **Note 14**) and incubating for 2 h at 4 °C on a rocker. Spin tubes briefly to remove any liquid clinging to the inside of the lid, then precipitate the beads by placing tubes

in a magnetic rack for 1 min. Carefully pipette off the supernatant. The DNA of interest is on the beads.

6. Reserve 50 μl of each supernatant to check the sonication efficiency by reversing the cross-links overnight at 67 °C, then purifying DNA and running on an agarose gel. See **step 3.4, item 9**. This will give a better idea of the true chromatin size than boiling.

3.4 Washing

1. Wash magnetic beads twice with IP dilution buffer (*see* **Note 3**): take tubes out of the magnetic rack and mix by pipetting. Return tubes to the rack for at least 1 min to allow beads to settle. Remove supernatant carefully with a pipette and discard. Repeat. Avoid loss of magnetic beads. Thorough washing is important to reduce background.
2. Wash magnetic beads twice with IP wash buffer 2.
3. Wash twice with the higher stringency IP wash buffer 3.
4. Elute chromatin by adding 100 μl elution buffer per ChIP sample to the magnetic beads. Shake samples on a vortexer for 30 min.
5. Place tube in a magnetic separation rack for 1 min until beads are pelleted. Transfer the supernatant to a low-retention or siliconized tube.
6. Add 5 M NaCl to yield a final concentration of 0.54 M NaCl (12 μl per 100 μl of elution buffer mix).
7. Retrieve the input sample that was stored on the previous day. Add 4 volumes of ChIP elution buffer to 1 volume input sample. Add 12 μl 5 M NaCl per 100 μl of diluted sample.
8. Incubate all samples (IPs and inputs) in a 67 °C water bath overnight to reverse formaldehyde cross-links.
9. Allow samples to cool. Add 1 μl of RNase A and incubate for 20 min at 37 °C.
10. Purify DNA with a Qiagen MinElute PCR clean up kit. Elute each sample twice: first with 10 μl, then repeat with 12 μl. About 10 μl will be used to test enrichment by qPCR, and 10–12 μl will be used for library preparation. Plan for 2 μl to be lost as hold up volume.

3.5 Assessment of Enrichment

1. If intending to perform ChIP-seq, set aside 5–10 μl of the eluted DNA product, and keep the rest for sequencing library preparation (*see* **Note 15**). Primer sets for Myc target genes and negative controls are given in Tables 3 and 4 for ChIP assays performed on mouse and human cells, respectively (*see* also Fig. 3).

Table 3
Suggested primer sets for PCR validation of Myc ChIP assays in mouse cells or tissues (Subheading 3.5)

Gene	Cell type tested	Product size (bp)	Forward	Reverse	Notes
Myc/N-Myc targets					
Apexl	Neurosphere	188	ACG AAC AAC CCA GAA CCA AG	CTA AGC CAG AGA CCC TCA CG	Works for qPCR. Annealing temp = 57–58 °C
CCD2	Neurosphere	221	GAC CCT CCT GCA GAA ACC TT	GGG AGA GCC AAA CCT AAA CC	Not tested for qPCR. Annealing temp = 55–58 °C
Slc25A19	Neurosphere	228	CAG GTC AGG GAG GAA ACT GA	GCT CCC TCT TCA TCT GCA TC	Not tested for qPCR. Annealing temp = 58 °C
Negative controls					
Mmaa	MEF	125	TTC AGC CCC AGT GCA AGG TAG	TTT CTT ATG CCC CAC ATC CAG AG	Works for qPCR
Ubbl	MEF	92	CCC ACA CAA AGC CCC TCA ATC	CAG CGA AAG CGA CAG GCT AAA	Works for qPCR
Yesl	MEF	119	GGC CCA AAC GTC AGC TTT CAT	CTC CCA GGA ATG GTG GCT CTC	Works for qPCR

Annealing temperature for each primer set is 60 °C unless otherwise noted

2. Reaction setup for end-point PCR with agarose gel electrophoresis:

   1.0 μl DNA eluate (undiluted) or input sample (1:50 and 1:200 dilutions).

   2.0 μl 10× reaction buffer.

   1.5 μl 25 mM $MgCl_2$.

   1.5 μl 2 mM dNTPs.

   1.5 μl 10 μM forward and reverse primer mix.

   4.0 μl 5 M betaine or combinatorial enhancer solution (CES, *see* **Note 16**).

   0.2 μl Taq polymerase.

   8.3 μl ddH2O.

   20 μl total reaction volume.

Table 4
Suggested primer sets for PCR validation of Myc ChIP assays in human cells or tissues (Subheading 3.5)

Gene	Cell type tested	Product size (bp)	Forward	Reverse	Notes
Myc/N-Myc targets					
APEX1	Tet21N, hESC, embryoid bodies	121	GGC GGG ACC TGG TGC GGG GA	ACC GCG TCA CCC ACC GAA GCA	Works for qPCR. Also positive control for H3K9ac and H3K4me3
JUB	Tet21N	180	AAG CCC ACA GAG AGA GGT GGA AG	CCA GAA GGT GGT GCC TTT TTA TTG	Works for qPCR
POU5F1	hESC	216	GTG GCA TCC GTG AGT CTT TTG AG	GGC CCC ATA ATC TAC CTG CCT TT	Works for qPCR
RPL4	Tet21N, hESC	82	GCT TCC CGG CGC GTC CTG TGC	GGT GTG GAA CTG GGA TGT GCG GCG	Works for qPCR. Also positive control for H3K9ac and H3K4me3
RPL35	Tet21N hESC, embryoid bodies	122	TGC GCG CCA GGA TGC ACG GA	AGG CGG CCG GAT CAC AGC CCT	Works for qPCR. Also positive control for H3K9ac and H3K4me3
Negative controls					
NDUFA3	Tet21N	206	TCA AGA ATG CCT GGG ACA AG	ATG GGG ATA GAT AAG GGG ATG	Not tested for qPCR
ATP50	Tet21N	232	TAC TGC CGC AGA GTT TGA TCT	CCC TTC CTG GCA TCT TAG GTA	Not tested for qPCR
TSEN34	Tet21N	204	CCA AAG AGG ATG AGA CCA GTG	TGG TGG GGA GTA CAG AGA AGA	Not tested for qPCR

Annealing temperature for each primer set is 60 °C

Run for 35 cycles:

1×	3 min at 94 °C
35×	30 s at 94 °C 30 s at annealing temperature (60 °C unless otherwise noted) 30 s at 72 °C
1×	2 min at 72 °C

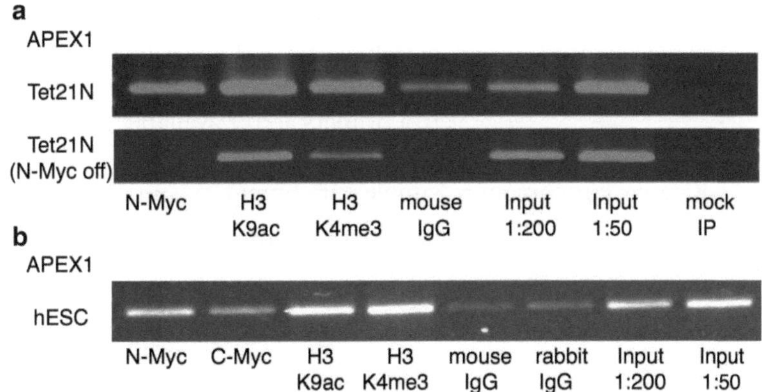

Fig. 3 Representative results showing end-point PCR validation of ChIP assays in human cells using APEX1 primers given in Table 4 (*see* **step 3.5, item 2**). (**a**) ChIP in Tet21N neuroblastoma cells. *Top panel*, cells that overexpress N-Myc; *bottom panel*, cells after 3 days of tetracycline treatment which blocks N-Myc expression. (**b**) ChIP for both c-Myc and N-Myc in hESCs (two 6-well plates of cells per IP). Antibodies: N-Myc (Santa Cruz, sc-53993), c-Myc (Abcam, ab56), H3K9ac (Abcam, ab12179), and H3K4me3 (Millipore, 04–745)

Compare enrichment of IP sample to 1:200 and 1:50 dilutions of the input sample. A good enrichment of the target will show a signal over the 1:200 dilution and ideally equal to or greater than the 1:50 dilution.

3. Reaction setup for qPCR: include a reaction for the IgG control, particularly if negative control regions are uncertain. Dilute ChIP DNA 1:5 and input DNA 1:50 (or dilute all samples to 2 ng/μl).

   1.0 μl diluted ChIP or input DNA.

   7.5 μl 2× Absolute Blue SYBR reaction mix containing Taq polymerase.

   2 μl 3 μM forward and reverse primer mix.

   4.5 μl ddH_2O.

   15 μl total reaction volume.

   Run on a qPCR machine with the following cycling conditions:

1×	15 min at 95 °C
45×	30 s at 95 °C 30 s at annealing temperature (60 °C unless otherwise noted) Melt curve from 70 to 90 °C

4. Evaluate qPCR enrichment relative to the input for each primer set:

$$\text{Sample enrichment} = 2^\wedge(\text{Cp input} - \text{Cp IP sample}).$$

Then compute the relative enrichment by dividing the enrichment for the positive regions by the enrichment for a negative control region, or divide by IgG if good negative control primers aren't available (*see* **Note 17**):

$$\text{Relative enrichment} = \text{Sample enrichment}\left(\text{positive region}\right) / \text{Sample enrichment}\left(\text{negative region}\right),$$

or

$$\text{Relative enrichment} = \text{Sample enrichment}\left(\text{positive region}\right) / \text{IgG enrichment}\left(\text{same positive region}\right).$$

4 Notes

1. Autoclave all solutions before use to rid them of possible bacterial contamination. When using solutions, divide needed amounts into aliquots in conical tubes, then add protease inhibitors and Igepal if needed. Avoid going into the stock solutions more than necessary. This will help avoid contamination of wash and elution buffers and prevent false positive results.
2. Formaldehyde solution should be replaced at the manufacturer's expiration date or when precipitate appears in the bottle.
3. Preservation of proteins may improve when protease inhibitors are added to wash buffers as well as lysis buffers. However, ChIP can also be performed successfully without using protease inhibitors in the wash buffers.
4. IP dilution and wash buffers will gel after a few days. These can be liquefied for use by heating in a water bath at 37 °C, then cooling to room temperature.
5. For cost savings, lyophilized Staph A cells (Pansorbin, Calbiochem) may be used for collection of antibody-protein complexes [11]. However, if the ultimate goal is to proceed to ChIP-seq, these cells may produce undesirable sequencing background. Thus, we recommend the use of protein A or G magnetic beads instead (Cell Signaling Technology). In addition, magnetic beads are easier to wash.
6. You can save money by taking a DIY (do-it-yourself) approach via purchasing neodymium magnets of the desired shape (Amazon.com or other suppliers) and taping them to suitable plastic racks rather than buying a commercially available magnetic rack.
7. If downstream processing of final product into libraries requires a concentrated eluate of 20 μl or less, then use the MinElute kit for the final purification of DNA in **step 3.4, item 10**. If a

concentrated product is not necessary, then any commercial PCR purification kit will suffice. We prefer the Invitrogen PureLink PCR purification kit.

8. Use of trypsin to detach adherent cells may disrupt the protein complexes on the chromatin and so should be avoided. Cross-link adherent cells directly on the plate, on a rocker in the fume hood. Tissues and suspension cells can be transferred to a 15 ml tube for cross-linking.
9. More cross-linking time has been suggested by others for increasing successful pulldown of factors not directly bound to DNA, but in our hands, cross-linking for more than 12 min will greatly increase background. Also, more cross-linking will require more sonication later.
10. The Diagenode Bioruptor recommendations include specific tubes: polystyrene 15 ml tubes from Falcon or 1.5 ml locking tubes from Eppendorf. The 15 ml conical tubes hold a maximum sonication volume of 1 ml, and the 1.5 ml tubes hold a maximum sonication volume of 300 μl. Do not exceed the recommended volume. However, sonicating in less than the recommended volume will not work well either. We recommend using the 1.5 ml tubes for ChIP and ChIP-seq sonication. Use the same type of tube consistently to reduce variation. People often autoclave tubes that are already sterile and DNase/RNase free. We find this weakens the plastic and can cause tubes to crack during sonication.
11. It is essential to test sonication time with a similar quantity of the specific type of cells to what will be used for actual experiments. A more concentrated volume of cells will require longer time. Sonication times must be optimized for each cell type, as from our experience the chromatin from different cell types can exhibit substantially different responses to sonication. If different quantities of cells are used for optimization, the resulting chromatin size will be different.
12. If more than one 15 min round of sonication is required, let cells rest on ice between rounds for at least 5 min. Remove the warm water from the sonication receptacle and replace with prechilled water (or use a cooling system to maintain water temperature). Add some ice if necessary, but be sure to stir it until the ice melts because any ice remaining in the water may decrease the sonication efficiency.
13. Some published protocols recommend the use of a very large quantity of antibody—up to 12 μg. This does not improve the precipitation recovery in our experience, yet it is very expensive.
14. A secondary bridging antibody is generally not needed when using protein G beads, as protein G binds most species' antibodies well (except antibodies raised in dog, cat, chicken, and

pig). Be careful to check the charts available for species and antibody isotype affinity for protein A and G. There are magnetic beads available with either or both protein A and protein G. When using an antibody that does not bind either protein A or G strongly, use a bridging antibody that binds both the antibody and protein G.

15. Results can be evaluated by end-point PCR or qPCR. Results may be evaluated with end-point PCR and agarose gel if you feel the precipitation was robust and the targets are excellent. If not, use qPCR which is much more sensitive. Detection by agarose gel is adequate if you don't need to compare two groups that may differ by a relatively small amount. If comparing ChIP assays for different groups by end-point PCR with agarose gels, take care that chromatin levels are equilibrated as well as possible (using estimates from the NanoDrop in **step 3.2, item 8**). From our experience qPCR is generally far superior to end-point PCR for evaluating ChIP assays.
16. CES is combinatorial enhancer solution, a cocktail of PCR additives, which will help reduce the secondary structure of genomic DNA template in end-point PCR. As yet, we have not tested CES with qPCR. CES contains 2.7 M betaine, 6.7 mM DTT, 6.7 % DMSO, 55 μg/ml BSA [12].
17. Controls for ChIP with qPCR: results are often expressed as a percentage of input. IgG may be used as a negative control as well, in the absence of negative control primers. Results can be expressed as enrichment over IgG, or enrichment over a negative control region. The latter option may be considered more accurate, but it will be difficult to find good negative control regions without prior knowledge of which genes are bound by the factors of interest. If you are interested in testing a gene as a positive Myc target, but you don't know where to begin, try to design primers in the promoter surrounding the Myc "E-box" sequence CANNTG: CACGTG, CATGTG, and CACATG.

Acknowledgment

This work was supported by CIRM grant RN2-00922 and NIH grant R01GM100782-01. We thank the Farnham laboratory for help with functional genomics protocols and in particular Henriette O'Geen for assistance with ChIP-seq.

References

1. Guccione E, Martinato F, Finocchiaro G, Luzi L, Tizzoni L, Dall' Olio V et al (2006) Myc-binding-site recognition in the human genome is determined by chromatin context. Nat Cell Biol 8:764–770
2. Staller P, Peukert K, Kiermaier A, Seoane J, Lukas J, Karsunky H et al (2001) Repression of p15INK4b expression by Myc through association with Miz-1. Nat Cell Biol 3: 392–399
3. Varlakhanova N, Cotterman R, Bradnam K, Korf I, Knoepfler PS (2011) Myc and Miz-1 have coordinate genomic functions including targeting Hox genes in human embryonic stem cells. Epigenetics Chromatin 4:20
4. Bieda M, Xu X, Singer MA, Green R, Farnham PJ (2006) Unbiased location analysis of E2F1-binding sites suggests a widespread role for E2F1 in the human genome. Genome Res 16:595–605
5. Knoepfler PS, Zhang X, Cheng PF, Gafken PR, McMahon SB, Eisenman RN (2006) Myc influences global chromatin structure. EMBO J 25:2723
6. Zippo A, De Robertis A, Serafini R, Oliviero S (2007) PIM1-dependent phosphorylation of histone H3 at serine 10 is required for MYC-dependent transcriptional activation and oncogenic transformation. Nat Cell Biol 9: 932–944
7. McMahon SB, Wood MA, Cole MD (2000) The essential cofactor TRRAP recruits the histone acetyltransferase hGCN5 to c-Myc. Mol Cell Biol 20:556–562
8. Frank SR, Parisi T, Taubert S, Fernandez P, Fuchs M, Chan H-M et al (2003) MYC recruits the TIP60 histone acetyltransferase complex to chromatin. EMBO Rep 4:575–580
9. Cotterman R, Jin VX, Krig SR, Lemen JM, Wey A, Farnham PJ et al (2008) N-Myc regulates a widespread euchromatic program in the human genome partially independent of its role as a classical transcription factor. Cancer Res 68:9654–9662
10. O'Geen H, Echipare L, Farnham PJ (2011) Using ChIP-Seq technology to generate high-resolution profiles of histone modifications. Methods Mol Biol 791:265–286
11. O'Geen H, Nicolet CM, Blahnik K, Green R, Farnham PJ (2006) Comparison of sample preparation methods for ChIP-chip assays. Biotechniques 41:577–580
12. Ralser M, Querfurth R, Warnatz HJ, Lehrach H, Yaspo ML, Krobitsch S (2006) An efficient and economic enhancer mix for PCR. Biochem Biophys Res Commun 347:747–751

Chapter 10

Methods to Quantify microRNAs in the Myc Gene Network for Posttranscriptional Gene Repression

Rui Song, Nicole Sponer, and Lin He

Abstract

As a global transcription factor, Myc regulates both protein-coding genes and noncoding microRNA genes. Myc-activated or repressed miRNAs are involved in various pathways to affect tumorigenesis, mediate apoptosis, proliferation, angiogenesis, metastasis, and metabolism downstream of Myc. Functional characterization of miRNAs in the Myc network requires the accurate detection and quantification of miRNA expression levels. Here, we describe two widely used methodologies to determine miRNA expression, including miRNA real-time PCR and miRNA northern analysis.

Key words Myc, miRNA expression, Real-time PCR, Northern analysis

1 Introduction

Myc encodes a basic helix-loop-helix transcription factor that acts as global gene regulator [1]. The aberrant activation of the *Myc* gene has been frequently observed in many human tumor types. *Myc* activation constitutes the molecular basis of many aberrant cellular processes during malignant transformation, including uncontrolled cell proliferation and cell growth, evasion of programmed cell death, enhanced angiogenesis, increased motility, cancer metabolism, and metastasis [2, 3].

As a potent transcription factor, Myc acts as a global gene regulator that activates or represses numerous target genes in the large network of major oncogenes and tumor suppressors. To date, majority of studies on Myc have focused on the Myc-regulated protein-coding genes and their diverse biological functions downstream of Myc. Yet recent studies have revealed the functional importance of noncoding RNAs, particularly microRNAs (miRNAs), in the gene network regulated by Myc. miRNAs are a class of small, noncoding RNAs that mediate posttranscriptional gene silencing of many mRNAs [4, 5]. In addition to its direct transcriptional

Laura Soucek and Nicole M. Sodir (eds.), *The Myc Gene: Methods and Protocols*, Methods in Molecular Biology, vol. 1012, DOI 10.1007/978-1-62703-429-6_10, © Springer Science+Business Media, LLC 2013

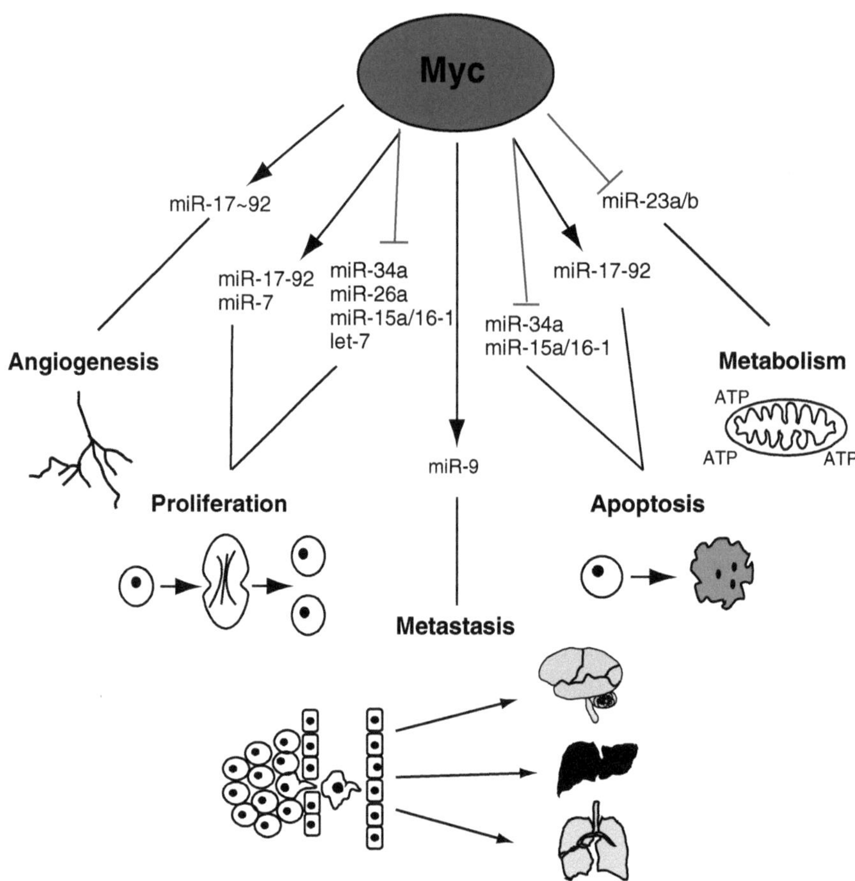

Fig. 1 Representatives Myc-activated or repressed miRNAs in various cellular processes during tumor development, including angiogenesis, cell proliferation, metastasis, apoptosis and metabolism

regulation of protein-coding genes, the ability of Myc to regulate many miRNAs—thus to indirectly regulate protein coding gene expression posttranscriptionally—further expands its capacity as a global gene regulator. Hence, miRNAs regulated by Myc constitute an important molecular mechanism underlying its pleiotropic functions during diverse developmental and pathological processes.

The polycistronic *mir-17-92* is the first miRNA identified as the target of Myc [6]. Myc can directly bind to the E-box sequence located at the *mir-17-92* promoter, increasing the level of both pri-miR-17-92 and, subsequently, its mature miRNA components [6]. Enforced *mir-17-92* expression accelerates *c-Myc*-induced B-lymphomagenesis due to impairment of *c-Myc*-induced apoptosis [7]. As a miRNA oncogene downstream of Myc, *mir-17-92* mediates pleiotropic biological functions, promoting cell survival, cell proliferation, and angiogenesis in a cell type- and context-dependent manner [8–11] (Fig. 1). Following the initial discovery of *mir-17-92* as a key mediator of Myc function, a number of miRNAs have been identified as major Myc targets that mediate its diverse developmental and pathological functions. For example, in response to the activation of the EGFR signaling, Myc activates the transcription of *miR-7*, which subsequently enhances cell proliferation and tumor formation [12] (Fig. 1). Additionally, both Myc and N-Myc upregulate *miR-9* in human mammary epithelial cells, which targets the epithelial cell adhesion molecule E-cadherin to promote cell motility and tumor invasion [13] (Fig. 1).

In addition to transcriptional activation, Myc can also regulate the transcriptional repression of miRNA genes. miRNA profiling studies using both human and mouse B-cell lymphoma cells identified the widespread repression of miRNA expression upon *Myc* activation, including *miR-15a/16-1*, *let-7*, *miR-34a*, *miR-26a*, and *miR-23a/b* [14]. Myc-repressed *miR-15a/16-1*, *let-7*, *miR-34a*, and *miR-26a*, when overexpressed, promote apoptosis and proliferation arrest by repressing antiapoptotic genes and cell cycle genes, respetively in a cell type-dependent manner [15–22] (Fig. 1). Myc-repressed miRNAs, *miR-23a*, and *miR-23b* also mediate cancer metabolism. For instance, glutaminase (GLS) is a direct target of *miR-23*. The transcriptional repression of *miR-23a/b* by Myc leads to increased expression of GLS, which ultimately upregulates glutamine metabolism in cancer [23] (Fig. 1).

Studies of miRNAs in the Myc network require sensitive and reliable methods to detect and quantify the expression of miRNAs. Here, we describe two methods to fulfill this purpose: miRNA real-time PCR and miRNA northern blot. Both methods are optimized based on previous reports [24, 25], providing a powerful and quantitative methodology for determining miRNA expression levels in the complex Myc signaling network.

2 Materials

2.1 miRNA Real-Time PCR

1. RNA isolation: TRIzol Reagent (Life Technologies), chloroform, isopropanol, 75 % ethanol, and nuclease-free water.
2. Poly(A) tailing of miRNAs: Poly(A) Polymerase Tailing Kit (Epicentre), RNaseOUT (Life Technologies), and nuclease-free water.
3. cDNA synthesis of poly(A) tailed miRNAs: small RNA reverse transcription primer (CGAATTCTAGAGCTCGAGGCAGG CGACATGGCTGGCTAGTTAAGCTTGGTACCGAGC TCGGATCCACTAGTCCTTTTTTTTTTTTTTTTTT TTTTTTTVN), 10 mM dNTP, nuclease-free water, RNaseOUT (Life Technologies), and SuperScript III reverse transcriptase (Life Technologies).
4. Real-time PCR: miRNA-specific forward primer, universal reverse primer (CGAATTCTAGAGCTCGAGGC), TaqMan probe (6FAM-

CTCGGATCCACTAGTC-MGBNFQ, Life Technologies), Taq Man Universal Master Mix without UNG AmpErase (Life Technologies), nuclease-free water. miRNA-specific forward primer is designed based on the sequences of target miRNAs, and this design should obey the general guidelines of real time PCR primer design. The primer sequences can be the exact sequences of target miRNAs, or few nucleotides shorter at 5′ or 3′ end to make qualified primers. U6 snRNA (GCAAATTCGTGAAGCGTTCC) is used as the endogenous control for real-time PCR.

5. NanoDrop 1000 Spectrophotometer (Thermo Scientific).
6. 96 well clear low-profile PCR microplate, ultra clear pressure-sensitive sealing film (Axygen), 7. 7900 HT Fast real-time PCR system (Applied Biosystems).

2.2 miRNA Northern Blot

1. Sequa Gel, 7.5 M urea gel system (National Diagnostics).
2. Dye loading buffer: 95 % formamide, 0.025 % bromophenol blue, 0.025 % xylene cyanol FF, 5 mM EDTA, stored in aliquots at −20 °C.
3. 10× TBE: 1 M Tris base, 1 M boric acid, 0.02 M EDTA, stored for up to 6 months at room temperature.
4. Decade marker system (Life Technologies): Mix 1 μl decade marker RNA (100 ng), 6 μl nuclease-free water, 1 μl 10× Kinase Reaction Buffer, 1 μl [gamma-^{32}P] ATP >3,000 Ci/mmol, and 1 μl T4 Polynucleotide Kinase (New England Biolabs). Incubate this mixture at 37 °C for 1 h. After the incubation, add 8 μl nuclease-free water and 2 μl 10× cleavage reagent, and then incubate at room temperature for 5 min. Subsequently, add 20 μl dye loading buffer. Heat the mixture at 95 °C for 5 min, before storing the labeled marker at −20 °C.
5. Running buffer and transfer buffer: 0.5× TBE in nuclease-free water.
6. Hybond N + nylon membrane (GE Healthcare).
7. Whatman 3 MM chromatography paper (Whatman).
8. 100× Denhardt's solution: 2 % Polyvinylpyrrolidone, 2 % Bovine serum albumin, 2 % Ficoll 400, 2.5 mM; store at −20 °C for long-term storage.
9. 20× SSC: 3 M NaCl, 0.3 M Trisodium citrate; adjust the pH to 7.0 before autoclave, and store at room temperature.
10. Pre-hybridization/hybridization mix: 6× SSC buffer, 5× denhardt's solution and 0.1 % SDS in RNAse free water; prepare fresh before each use.
11. Washing buffer: 2× SSC, 0.1 % SDS in nuclease-free water.
12. HL-2000 HybriLinker including HB-1000 hybridization oven and UV cross-linker (UVP).

13. Typhoon Trio variable mode imager (Amersham Biosciences) and storage phosphor screen (Molecular Dynamics).
14. Microspin G-50 columns (GE Healthcare).
15. T4 Polynucleotide Kinase (New England Biolabs).
16. [gamma-^{32}P] ATP, 6,000 Ci/mmol, 10 mCi/ml (Perkin Elmer).

3 Methods

3.1 miRNA Real-Time PCR

3.1.1 RNA Isolation

1. Add 1 ml TRIzol Reagent to ~50 mg tissue or ~5×10^6 harvested cells.
2. Homogenize tissue sample using tissue grinder or lyse cell sample by pipetting the cells up and down several times (*see* **Note 1**).
3. Incubate the homogenized sample for 5 min at room temperature.
4. Add 0.2 ml of chloroform. Shake tube vigorously by hand for 15 seconds.
5. Incubate for 2 min at room temperature.
6. Centrifuge the sample at 12,000 × *g* for 15 min at 4 °C (*see* **Note 2**).
7. Transfer the aqueous phase into a fresh Eppendorf tube. Avoid drawing any of the interphase or organic layer into the pipette.
8. Add 0.5 ml of 100 % isopropanol to the aqueous phase and mix well.
9. Incubate for 10 min at room temperature.
10. Centrifuge at 12,000 × *g* for 10 min at 4 °C (*see* **Note 3**).
11. Discard the supernatant from the tube, leaving the RNA pellet in the tube.
12. Wash the RNA pellet with 1 ml of 75 % ethanol.
13. Centrifuge at 7,500 × *g* for 5 min at 4 °C. Discard the supernatant.
14. Air dry the RNA pellet for 5–10 min (*see* **Note 4**).
15. Resuspend the RNA pellet in nuclease-free water by pipetting up and down several times.
16. Measure the concentration of the dissolved RNA using a NanoDrop.

3.1.2 Poly(A) Tailing of miRNAs

1. Combine the following components: 2 μg RNA, 1 μl Poly(A) Polymerase 10× Reaction Buffer, 1 μl 10 mM ATP, 0.5 μl RNaseOUT (40 U/μl), 0.25 μl Poly(A) Polymerase (4 U/μl). Add nuclease-free water to total 10 μl.

2. Incubate at 37 °C for 30 min.
3. Stop the reaction by immediate freezing at −20 °C (*see* **Note 5**).

3.1.3 cDNA Synthesis of Poly(A) Tailed miRNAs

1. Add 1 μl small RNA reverse transcription primer (1 μg/μl), 2 μl 10 mM dNTP, and 10 μl nuclease-free water to 10 μl of poly(A) tailed small RNAs (*see* **Note 6**).
2. Heat the mixture to 65 °C for 5 min, followed by incubation on ice for at least 1 min.
3. Collect the content of the tube by brief centrifugation, and add 7 μl 5× First-Strand Buffer, 2 μl 0.1 M DTT, 2 μl RNaseOUT (40 U/μl), and 1 μl SuperScript III reverse transcriptase (200 U/μl).
4. Mix by pipetting gently. Incubate the mixture at 50 °C for 60 min.
5. Inactivate the reaction by 15 min of heating at 70 °C.
6. Add 65 μl nuclease-free water to bring final volume to 100 μl. Store the cDNA at −20 °C.

3.1.4 Real-Time PCR

1. Dilute the stock cDNA 20 times to the working concentration using nuclease-free water (*see* **Note 7**).
2. Combine the following components into one reaction: 1 μl cDNA at the working concentration, 0.5 μl miRNA-specific forward primer (10 μM), 0.5 μl universal reverse primer (10 μM), 0.4 μl TaqMan probe (10 μM), 7.6 μl nuclease-free water, and 10 μl TaqMan Universal Master Mix without UNG AmpErase (*see* **Note 8**).
3. Transfer the reaction mixture to each well of an optical 96 well PCR microplate. Seal the plate with an ultra clear sealing film and centrifuge briefly to position reaction mix at the bottom of the well.
4. Run the following program on a real-time PCR system:

   95 °C, 10 min (enzyme activation);

   95 °C, 15 s (denaturation) and 60 °C, 1 min (annealing/extension) for 40 cycles (*see* **Note 9**).
5. Analyze the results according to instrument user guide.

3.2 miRNA Northern Blot

1. Resuspend RNA samples (~10–15 μg of total RNA) in 10 μl of dye loading buffer, and heat it at 95 °C for 5 min.
2. Load the RNA samples and the radioactively labeled decade marker onto a high-percentage (10–15 %) denaturing polyacrylamide-urea gel. Run the gel ~ 25 W until the bromophenol blue is near the bottom of the gel (*see* **Note 10**).

3. Disassemble the gel and place it in transfer buffer for 5 min. Cut six sheets of Whatman 3 MM paper and Hybond N + nylon membrane into the same size of the gel, and wet them with transfer buffer.
4. Place three Whatman 3 MM papers on anode plate of Semi-Dry Transfer Cell (Bio-Rad), and then put pre-wet Hybond N + nylon membrane on top of the papers. Lay the gel on top of membrane and add another three Whatman 3 MM papers. Remove any air bubbles by rolling a pipette on the gel sandwich. Finally, add a few drops of transfer buffer before putting the safety lid on to keep the sandwich wet.
5. Transfer at constant 800 mA for 1 h with the maximum voltage of 25 V.
6. Disassemble the sandwich and put the membrane on a Whatman 3 MM paper.
7. Expose the membrane, with the RNA surface facing up, to UV light cross-link for 1 min (254 nm, 1,200 mJ auto-cross-linking setting). Dry the membrane overnight between two Whatman 3 MM papers at 4 °C (*see* **Note 11**).
8. Put the membrane in glass hybridization bottle with 5 ml pre-hybridization mix. Pre-hybridize in a pre-hybridization mix with rotation for at least 2 h at 42 °C in a hybridization oven.
9. End label a radioactive northern probe using the complementary DNA oligo to the mature miRNA sequence. To do so, 100 ng DNA oligo is diluted in 3.25 μl nuclease-free water. Add 1 μl 10× Reaction Buffer, 0.75 μl T4 Polynucleotide Kinase, and 5 μl 6,000 Ci/mmol [gamma-^{32}P] ATP to the reaction mix, and incubate at 37 °C for 1 h. Add 60 μl nuclease-free water to the reaction, and clean the probe with G-50 spin column. Denature the probe at 95 °C for 5 min.
10. Add the radiolabeled probe in fresh hybridization mix and hybridize at 42 °C overnight (*see* **Note 12**).
11. Dispose the hybridization mix, and wash the membrane at 42 °C for 20 min in washing buffer. Repeat this washing step one more time (*see* **Note 13**).
12. Cover the membrane with plastic wrap and expose it to a phosphor imager screen for phosphorimaging. Scan the screen at 50 μm high sensitivity on Typhoon scanner after exposure for overnight or longer (*see* **Note 14**).

4 Notes

1. It is critical to homogenize the samples as quickly as possible in TRIzol. Make sure that the solution is clear and uniformly viscous.
2. The mixture separates into an upper aqueous phase, an interphase, and a lower phenol-chloroform phase. RNA remains only in the aqueous phase.
3. RNA will form a white pellet on the bottom of the tube after centrifugation.
4. Do not overdry the RNA pellet because the pellet can lose its solubility.
5. This reaction can be stopped by phenol/chloroform extraction and salt/alcohol precipitation.
6. It is recommended to use the primer purified by RNase free HPLC.
7. The dilution factor depends on the abundance of the target small RNAs in the tissue or cell samples.
8. TaqMan probe is light sensitive. Therefore, the tubes containing TaqMan probe must be wrapped with aluminum foil.
9. 3-step cycling can be used alternatively, anneal at optimal temperature for 30 s followed by extension at 60 °C for 30 s. The Tm of primers determines the optimal annealing temperature.
10. It is important to thoroughly rinse the wells before loading the samples in order to have sharp and clear bands. Remaining urea will result in fuzzy bands on the blot.
11. Do not UV cross-link the membrane more than once. Multiple cross-linkings will reduce the hybridization efficiency of the probe. EDC chemical cross-linking can be used as an alternative method [24].
12. Hybridization temperature may vary. If cross hybridization happened, higher temperatures should be used instead of 42 °C. It is also recommended to increase the hybridization time for low abundant miRNAs.
13. Higher stringent washing buffer should be used in case of high background. On the other hand, lower stringent washing buffer must be used if the signal is weak.
14. Membrane can be stripped with 0.2 % SDS in nuclease-free water. Strip at 80 °C for 1 h. Do not strip one membrane for more than 3 times. When stripping, keep the membrane wet, otherwise probe will be permanently bound to the membrane.

Acknowledgment

L.H. acknowledges the support of an RO1 and an R21 grant from NCI (R01 CA139067, 1R21CA175560-01); a joint grant from National Science Foundation (NSF, CBET-1015026); a new faculty award from CIRM (RN2-00923-1), a research grant from TRDRP (21RT-0133) and a research scholar award from American Cancer Society (ACS, 123339-RSG-12-265-01-RMC).

References

1. Meyer N, Penn LZ (2008) Reflecting on 25 years with MYC. Nat Rev Cancer 8:976–990
2. Nilsson JA, Cleveland JL (2003) Myc pathways provoking cell suicide and cancer. Oncogene 22:9007–9021
3. Dang CV (2012) MYC on the path to cancer. Cell 149:22–35
4. Bartel DP (2004) MicroRNAs: genomics, biogenesis, mechanism, and function. Cell 116: 281–297
5. He L, Hannon GJ (2004) MicroRNAs: small RNAs with a big role in gene regulation. Nat Rev Genet 5:522–531
6. O'Donnell KA, Wentzel EA, Zeller KI, Dang CV, Mendell JT (2005) c-Myc-regulated microRNAs modulate E2F1 expression. Nature 435:839–843
7. He L, Thomson JM, Hemann MT, Hernando-Monge E, Mu D, Goodson S, Powers S, Cordon-Cardo C, Lowe SW, Hannon GJ, Hammond SM (2005) A microRNA polycistron as a potential human oncogene. Nature 435:828–833
8. Olive V, Bennett MJ, Walker JC, Ma C, Jiang I, Cordon-Cardo C, Li QJ, Lowe SW, Hannon GJ, He L (2009) miR-19 is a key oncogenic component of mir-17-92. Genes Dev 23:2839–2849
9. Mu P, Han YC, Betel D, Yao E, Squatrito M, Ogrodowski P, de Stanchina E, D'Andrea A, Sander C, Ventura A (2009) Genetic dissection of the miR-17 92 cluster of microRNAs in Myc-induced B-cell lymphomas. Genes Dev 23:2806–2811
10. Hong L, Lai M, Chen M, Xie C, Liao R, Kang YJ, Xiao C, Hu WY, Han J, Sun P (2010) The miR-17-92 cluster of microRNAs confers tumorigenicity by inhibiting oncogene-induced senescence. Cancer Res 70:8547–8557
11. Dews M, Homayouni A, Yu D, Murphy D, Sevignani C, Wentzel E, Furth EE, Lee WM, Enders GH, Mendell JT, Thomas-Tikhonenko A (2006) Augmentation of tumor angiogenesis by a Myc-activated microRNA cluster. Nat Genet 38:1060–1065
12. Chou YT, Lin HH, Lien YC, Wang YH, Hong CF, Kao YR, Lin SC, Chang YC, Lin SY, Chen SJ, Chen HC, Yeh SD, Wu CW (2010) EGFR promotes lung tumorigenesis by activating miR-7 through a Ras/ERK/Myc pathway that targets the Ets2 transcriptional repressor ERF. Cancer Res 70:8822–8831
13. Ma L, Young J, Prabhala H, Pan E, Mestdagh P, Muth D, Teruya-Feldstein J, Reinhardt F, Onder TT, Valastyan S, Westermann F, Speleman F, Vandesompele J, Weinberg RA (2010) miR-9, a MYC/MYCN-activated microRNA, regulates E-cadherin and cancer metastasis. Nat cell biol 12:247–256
14. Chang TC, Yu D, Lee YS, Wentzel EA, Arking DE, West KM, Dang CV, Thomas-Tikhonenko A, Mendell JT (2008) Widespread microRNA repression by Myc contributes to tumorigenesis. Nat Genet 40:43–50
15. He L, He X, Lim LP, de Stanchina E, Xuan Z, Liang Y, Xue W, Zender L, Magnus J, Ridzon D, Jackson AL, Linsley PS, Chen C, Lowe SW, Cleary MA, Hannon GJ (2007) A microRNA component of the p53 tumour suppressor network. Nature 447:1130–1134
16. Cimmino A, Calin GA, Fabbri M, Iorio MV, Ferracin M, Shimizu M, Wojcik SE, Aqeilan RI, Zupo S, Dono M, Rassenti L, Alder H, Volinia S, Liu CG, Kipps TJ, Negrini M, Croce CM (2005) miR-15 and miR-16 induce apoptosis by targeting BCL2. Proc Natl Acad Sci U S A 102:13944–13949
17. Calin GA, Cimmino A, Fabbri M, Ferracin M, Wojcik SE, Shimizu M, Taccioli C, Zanesi N, Garzon R, Aqeilan RI, Alder H, Volinia S, Rassenti L, Liu X, Liu CG, Kipps TJ, Negrini M, Croce CM (2008) MiR-15a and miR-16-1 cluster functions in human leukemia. Proc Natl Acad Sci U S A 105:5166–5171
18. Raver-Shapira N, Marciano E, Meiri E, Spector Y, Rosenfeld N, Moskovits N, Bentwich Z, Oren M (2007) Transcriptional activation of miR-34a contributes to p53-mediated apoptosis. Mol cell 26:731–743

19. Chang TC, Wentzel EA, Kent OA, Ramachandran K, Mullendore M, Lee KH, Feldmann G, Yamakuchi M, Ferlito M, Lowenstein CJ, Arking DE, Beer MA, Maitra A, Mendell JT (2007) Transactivation of miR-34a by p53 broadly influences gene expression and promotes apoptosis. Mol cell 26: 745–752
20. Tazawa H, Tsuchiya N, Izumiya M, Nakagama H (2007) Tumor-suppressive miR-34a induces senescence-like growth arrest through modulation of the E2F pathway in human colon cancer cells. Proc Natl Acad Sci U S A 104:15472–15477
21. Johnson CD, Esquela-Kerscher A, Stefani G, Byrom M, Kelnar K, Ovcharenko D, Wilson M, Wang X, Shelton J, Shingara J, Chin L, Brown D, Slack FJ (2007) The let-7 microRNA represses cell proliferation pathways in human cells. Cancer Res 67:7713–7722
22. Kota J, Chivukula RR, O'Donnell KA, Wentzel EA, Montgomery CL, Hwang HW, Chang TC, Vivekanandan P, Torbenson M, Clark KR, Mendell JR, Mendell JT (2009) Therapeutic microRNA delivery suppresses tumorigenesis in a murine liver cancer model. Cell 137: 1005–1017
23. Gao P, Tchernyshyov I, Chang TC, Lee YS, Kita K, Ochi T, Zeller KI, De Marzo AM, Van Eyk JE, Mendell JT, Dang CV (2009) c-Myc suppression of miR-23a/b enhances mitochondrial glutaminase expression and glutamine metabolism. Nature 458:762–765
24. Pall GS, Hamilton AJ (2008) Improved northern blot method for enhanced detection of small RNA. Nat Protoc 3:1077–1084
25. Ro S, Park C, Jin J, Sanders KM, Yan W (2006) A PCR-based method for detection and quantification of small RNAs. Biochem Biophys Res Commun 351:756–763

Chapter 11

Genome-Wide Analysis of c-MYC-Regulated mRNAs and miRNAs, and c-MYC DNA Binding by Next-Generation Sequencing

Rene Jackstadt, Antje Menssen, and Heiko Hermeking

Abstract

The c-*MYC* oncogene is activated in ~50 % of all tumors, and its product, the c-MYC transcription factor, regulates numerous processes, which contribute to tumor initiation and progression. Therefore, the genome-wide characterization of c-MYC targets and their role in different tumor entities is a recurrent theme in cancer research. Recently, next-generation sequencing (NGS) has become a powerful tool to analyze mRNA and miRNA expression, as well as DNA binding of proteins in a genome-wide manner with an extremely high resolution and coverage. Since the c-MYC transcription factor regulates mRNA and miRNA expression by binding to specific DNA elements in the vicinity of promoters, NGS can be used to generate integrated representations of c-MYC-mediated regulations of gene transcription and chromatin modifications. Here, we provide protocols and examples of NGS-based analyses of c-MYC-regulated mRNA and miRNA expression, as well as of DNA binding by c-MYC. Furthermore, the validation of single c-MYC targets identified by NGS is described. Taken together, these approaches allow an accelerated and comprehensive analysis of c-MYC function in numerous cellular contexts which will further illuminate the role of this important oncogene.

Key words c-MYC/MYC, Chromatin immunoprecipitation (ChIP), mRNA, microRNA (miRNA), Next-generation sequencing (NGS), Transcriptional regulation, Target genes

1 Introduction

The protein product of the proto-oncogene *c-MYC* is a transcription factor, which is involved in a broad range of cellular processes, such as cell cycle progression, cell growth, cell motility, metabolism, metastasis, pluripotency, and vascularization, which have been implicated in the initiation and progression of tumors [1, 2]. In tumors c-MYC expression can be induced via several mechanisms: chromosomal translocations [3], amplification [4], loss of

HiSeq (1), Illumina (2), MiSeq (3), and TruSeq (4) are registered trademarks of Illumina Inc., San Diego, CA.

Laura Soucek and Nicole M. Sodir (eds.), *The Myc Gene: Methods and Protocols*, Methods in Molecular Biology, vol. 1012, DOI 10.1007/978-1-62703-429-6_11,

autoregulation [5], increased translation of *c-MYC* mRNA [5, 6], stabilizing mutations [7, 8], and activating mutations in upstream regulators such as the Wnt signaling in colorectal cancer [9–11]. In most cases these alterations lead to constitutive expression of the c-MYC protein, which is normally under strict control, as its expression is restricted to mitogenic responses. As a fail-safe mechanism against aberrant proliferation, deregulated c-MYC expression leads to the activation of the transcription factor p53, either by replicative stress or induced DNA damage in human cells and by the ARF pathway in murine cells, resulting in apoptosis and/or senescence [12–15]. In addition, c-MYC has been shown to suppress an intrinsic senescence program, which depends on the induction of *htert*, *CDK2*, and the *WRN* genes by c-MYC (reviewed in ref. 16). We have recently shown that activation of the NAD-dependent deacetylase SIRT1 by direct induction of the *NAMPT* (*Nicotinamide phosphoribosyltransferase*) gene by c-MYC contributes to the suppression of senescence [17]. Activated SIRT1 inactivates pro-apoptotic and senescence-mediating substrates, such as p53, by deacetylation.

c-MYC belongs to the basic helix–loop–helix leucine zipper (bHLH-LZ) transcription factor family. The C-terminus of c-MYC harbors a HLH-LZ motif, which mediates the heterodimerization with the bHLH-LZ transcription factor MAX. The basic regions of the c-MYC/MAX heterodimer mediate the DNA binding to the consensus E-box sequence CA(C/T)GTG [18]. The N-terminus of c-MYC encompasses two evolutionary conserved regions, MycBox I and II, which act as transactivation domains. Upon DNA binding the c-MYC/MAX heterodimer recruits cofactors, which mediate multiple effects of c-MYC on gene expression in a context-dependent manner [19]. So far chromatin modifications, histone phosphorylation, transcriptional silencing, promoter clearance [20], and transcriptional elongation [21] have been described as consequences of c-MYC action at promoters. Besides with c-MYC, MAX may heterodimerize with MAD, Mxi1, and Mnt, which thereby antagonize c-MYC function [22, 23]. Whereas direct E-box binding of MYC/MAX heterodimers generally mediates the induction of c-MYC target gene transcription, repression by c-MYC is often dependent on MYC/MAX/Miz1 complexes binding to initiator (*Inr*) elements [24–27]. This element consists of 17 bp pyrimidine-rich motifs, which mediate transcriptional initiation from TATA-less promoters [28]. Several hundred c-MYC-regulated genes were identified based on microarray screens and other genome-wide techniques, such as SAGE and ChIP arrays, reviewed in [29]. However, the design of the initial genome-wide analyses did not allow a distinction between direct and indirect c-MYC targets. The latter may for example be mediated by the c-MYC regulation of other transcription factors such as E2F1. In order to comprehensively identify direct c-MYC target genes, the

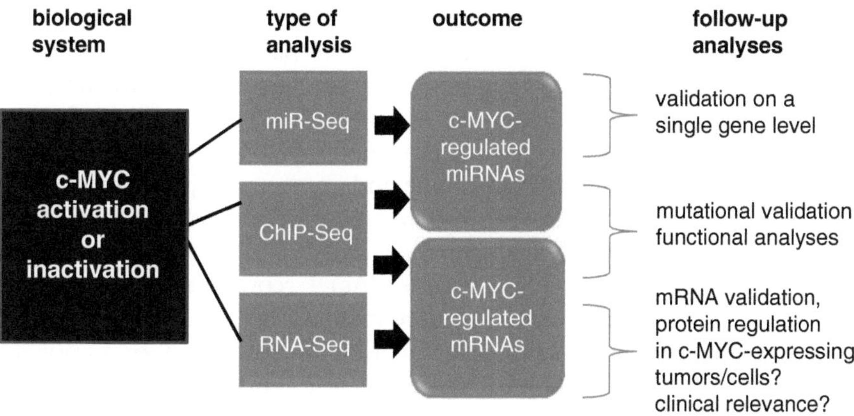

Fig. 1 Experimental outline of NGS-based analyses of c-MYC targets. After the activation or inactivation of c-MYC in different biological systems, the expression of c-MYC-regulated mRNAs and miRNAs, and the genome binding are analyzed in a genome-wide manner by next-generation sequencing. The identified c-MYC targets are analyzed in multiple assays to confirm direct regulation by c-MYC, putative cell system/type dependence of the regulation, and concomitant protein expression changes, before functional analyses may reveal their relevance for c-MYC-regulated cellular processes

expression analyses of mRNAs and miRNAs should be accompanied by the determination of genome-wide c-MYC-DNA binding via chromatin immunoprecipitation (ChIP) (Fig. 1). The ChIP technique allows studying c-MYC binding in the cellular context of chromatin. ChIP has replaced several techniques, such as the electrophoretic mobility shift assay (EMSA) and luciferase reporter assays, which were used to detect and map DNA binding by transcription factors. ChIP followed by genome-wide analyses of enriched DNA fragments by next-generation sequencing (ChIP-Seq) provides a comprehensive, genome-wide picture of c-MYC-regulated target genes. Several genome-wide studies identified genes, which show binding of c-MYC in the vicinity of the promoter region and concomitant regulation of the corresponding mRNA [29, 30]. Depending on the cell type and endogenous c-MYC expression levels, these studies demonstrated that 7.8–10.4 % of those genes, which display c-MYC occupancy had previously been reported to be regulated by c-MYC and that c-MYC binds to 10–15 % of all cellular genes [29, 31–39]. On the basis of NGS analyses, several studies demonstrated that c-MYC can act locally as well as globally on chromatin, by binding to intra- and intergenic regions via several thousand E-boxes (reviewed in refs. 18, 40–42). Presumably, global effects of c-MYC on chromatin states also contribute to the regulation of processes and states, such as stem cell self-renewal and pluripotency.

In order to compare mRNA/miRNA expression and ChIP-Seq results with previously published genome-wide analyses, two comprehensive databases can be used. The MYC homepage (http://www.myccancergene.org/site/about.asp) contains an extensive list of genes shown to be responsive to c-MYC [43], and the UCSC genome browser (http://genome.ucsc.edu/) also includes the results of several ChIP-Seq studies of c-MYC DNA binding performed in different cell lines. The latter NGS datasets can be analyzed online and compared with patterns of chromatin modification and binding of other transcription factors in a genome-wide manner.

microRNAs (miRNAs) represent a class of small ~21-nucleotide-long, noncoding RNAs and have been implicated in cancer and may function as oncogenes or tumor suppressor genes [44]. miRNAs regulate the expression of target mRNAs by inhibiting their translation and/or mediating mRNA degradation. Several miRNA-encoding genes were shown to be regulated by c-MYC [1, 45–50]. The regulation of miRNAs by c-MYC represents another important mode by which c-MYC regulates the expression of target genes and influences numerous biological processes. The integration of genome-wide miRNA and mRNA expression data with binding profiles of c-MYC enables scientists to gain a deeper insight into the regulatory networks by which c-MYC exerts its oncogenic functions.

Besides the direct binding in the vicinity of the promoter of a c-MYC-regulated gene, either encoding a protein, a miRNA, or other noncoding RNAs, some additional criteria should be fulfilled for a bona fide c-MYC target gene [51]. It should be demonstrated that expression of the target gene mRNA is induced after activation of a conditional c-MycER (estrogen receptor) fusion protein by addition of 4-hydroxy-tamoxifen (4-OHT) [52] in the presence of the *de novo* protein synthesis inhibitor cycloheximide. Thereby gene regulations mediated indirectly through c-MYC-regulated factors, such as by miRNAs or other transcription factors, can be ruled out [53]. Since the expression systems used for ectopic expression of c-MYC often induce supranormal levels of c-MYC expression, they may generate artificial gene regulations. Therefore, such studies should be complemented by loss-of-function approaches. For example, inducible expression of c-MYC-specific shRNAs or transfection of siRNAs can be used to specifically downregulate the expression and activity of endogenous c-MYC. Alternatively, direct c-MYC targets can be confirmed by comparing RAT1 cell lines with wild-type and deleted c-*MYC* alleles upon stimulation of quiescent fibroblasts with mitogens [17, 54]. The requirement of the new target gene for c-MYC-mediated biological functions is interrogated usually by experimental inactivation of the respective candidate gene.

NGS-based approaches have several advantages when compared to the microarray-based approaches. First, due to the superior dynamic range of NGS sequencing, a typical analysis is likely to represent transcripts, which are expressed at very low levels, with a few reads, up to highly expressed genes represented by millions of reads. Second, the number of transcripts, which can be detected, is not limited to previously known mRNAs, as in the case of microarrays. Currently, only unbiased genome-wide NGS analyses covering all expressed exons allow the discovery of rare or novel transcripts and alternative splicing.

Several NGS sequencing platforms are currently available [55, 56]. The most commonly used ones are presumably the HiSeq[1] (Illumina[2]), 454 (Roche), SOLiD (Applied Biosystems), and Ion Torrent (Life Technologies) systems. For further discussion of these platforms, see [56, 57]. In this protocol collection, we will focus on approaches that employ the HiSeq platform offered by Illumina, which is one of the most frequently used NGS platforms. To facilitate the imaging of single DNA sequences, the templates are amplified with primers that are covalently linked to a glass slide. Thereby, approximately 100–200 million template clusters are generated, which are then sequenced using a universal primer complementary to the free ends of the templates.

By now, most of the larger research institutions provide core facilities, which use the HiSeq 2000 or similar devices from Illumina. These allow obtaining ~180 million reads per run. Since one sample is sufficiently covered by 10–30 millions reads, single runs are often used to multiplex several probes by using primers with specific barcodes for each sample. More recently a smaller NGS sequencing device was introduced by Illumina, called MiSeq[3], which fits the requirements of single labs, as it has an output of ~15 million reads per run and is affordable by smaller budgets.

In the following sections detailed protocols for the analysis of c-MYC-mediated gene regulation by NGS are provided.

2 Materials

2.1 RNA-Seq

Materials not listed in this section are supplied with the Illumina TruSeq[4] RNA Sample Prep Kit:

1. 1.5 ml RNase/DNase-free non-sticky tubes (Ambion).
2. 10 μl multichannel pipettes.
3. 10 μl barrier pipette tips.
4. 10 μl single channel pipettes.
5. 100 % ethanol.
6. 200 μl multichannel pipettes.
7. 200 μl barrier pipette tips.
8. 200 μl single channel pipettes.
9. RNA extraction kit.
10. Agilent Technologies 2100 Bioanalyzer.

11. Agencourt AMPure XP 60 ml kit (Beckman Coulter Genomics).
12. RNase/DNase zapper (to decontaminate surfaces).
13. Freshly prepared 80 % ethanol.
14. Microseal "B" adhesive seals (Bio-Rad).
15. RNase/DNase-free disposable multichannel reagent reservoirs.
16. RNase/DNase-free strip tubes and caps.
17. SuperScript II Reverse Transcriptase (Invitrogen).
18. 10 mM Tris–HCl, pH 8.5 with 0.1 % Tween 20, or QIAGEN EB Buffer (general lab supplier or QIAGEN).
19. Ultrapure water.

2.2 miRNA-Seq

Materials not listed in this section are supplied in the Illumina Small RNA Sample Prep Kit:

1. 1 μg total RNA in 5 μl nuclease-free water.
2. 3 M NaOAc, pH 5.2.
3. 5 μm filter tube.
4. 5× Novex TBE buffer.
5. 6 % Novex TBE PAGE gel, 1.0 mm, 10 wells.
6. 200 μl, clean, nuclease-free PCR tubes.
7. Clean scalpels.
8. DNA 1000 chip Agilent.
9. DNA loading dye.
10. 100 % ethanol, −15 °C to −25 °C.
11. 70 % ethanol, room temperature.
12. Gel breaker tubes.
13. High sensitivity DNA chip.
14. SuperScript II Reverse Transcriptase with 100 mM DTT and 5× first strand.
15. T4 RNA Ligase 2.
16. 10 mg/ml ultrapure ethidium bromide.
17. Agilent Technologies 2100 Bioanalyzer.

2.3 ChIP-Seq

2.3.1 Materials Required for ChIP

1. 1 M glycine in ddH_2O.
2. Triton dilution buffer: 100 mM Tris–Cl pH 8.6, 100 mM NaCl, 5 mM EDTA pH 8.0, 0.2 % NaN3, 5 % Triton X-100.
3. SDS buffer: 50 mM Tris, pH 8.1, 0.5 % SDS, 100 mM NaCl, 5 mM EDTA.
4. Immunoprecipitation (IP) buffer: Mix 1 part of Triton X-100 dilution buffer and 2 parts SDS buffer.

5. LiCl/detergent wash: 0.5 % (w/v) deoxycholic acid (sodium salt), 1 mM EDTA, 250 mM LiCl, 0.5 % (v/v) NP-40, 10 mM Tris–Cl pH 8.0, 0.2 % NaN_3.
6. Buffer 500: 0.1 % (w/v) deoxycholic acid, 1 mM EDTA, 50 mM HEPES, pH 7.5, 500 mM NaCl, 1 % (v/v) Triton X-100, 0.2 % NaN_3.
7. Blocked beads: 50 % slurry protein A-Sepharose (Amersham; for polyclonal antibodies) or protein G-Sepharose (Sigma; for monoclonal antibodies).
8. 0.5 mg/ml fatty acid-free BSA (Sigma).
9. 0.2 mg/ml salmon sperm DNA (Promega; subjected to 8 sonication cycles).
10. 37 % formaldehyde stock.
11. Mixed micelle buffer: 150 mM NaCl, 20 mM Tris–Cl, pH 8.1, 5 mM EDTA, pH 8.0, 5.2 % w/v sucrose, 0.02 % NaN_3, 1 % Triton X-100, 0.5 % SDS.
12. Elution buffer: 10 mM EDTA, 1 % (w/v) SDS, 50 mM Tris–HCl, pH 8.0.
13. Protease inhibitors: Complete mini with EDTA (Roche).
14. 1× TBS: 1.5 M NaCl, 0.1 M Tris–HCl, pH 7.4 at 4 °C.
15. 1× PBS: 13.7 mM NaCl, 2.7 mM KCl, 80.9 mM Na_2HPO_4, 1.5 mM KH_2PO_4, pH 7.4.

2.3.2 Materials Required for ChIP-Seq Library Generation

Materials not listed in this section are supplied in the Illumina ChIP-Seq Sample Prep Kit:

1. QIAquick PCR Purification Kit (QIAGEN).
2. MinElute PCR Purification Kit (QIAGEN).
3. QIAquick Gel Extraction Kit (QIAGEN).
4. Ethidium bromide.
5. 1× TAE buffer: 40 mM Tris–acetate; 1 mM EDTA pH 8.0.
6. 2 % agarose gel in TAE.
7. 100 bp DNA ladder.
8. DNA loading buffer.
9. Fast SYBR Green Master Mix (Ambion).
10. LightCycler 480 (Roche).

3 Methods

3.1 RNA-Seq

3.1.1 mRNA Library Preparation

(According to the Manufacturer's Protocol, Illumina Part # 15008136 Rev. A)

Quality and Quantity Control of Total RNA Input

This protocol is optimized for 0.1–4 μg of total RNA (*see* **Note 1**). In-line controls are available for this protocol to confirm every single step in the final sequencing and support potential troubleshooting. The use of high quality RNA samples is essential for optimal NGS results. RNA can be obtained using different kits (*see* **Note 2**) and RNA quality control using an Agilent Bioanalyzer or similar devices should be performed. The RNA Integrity Number (RIN) value should be equal or greater than eight (*see* **Note 3**). In addition, the regulation of previously described c-MYC target genes, which are expected to be regulated in the employed biological system, should be confirmed by RT-qPCR analysis of the same RNA samples also used for generating the NGS libraries.

mRNA Purification and Fragmentation

RNA Bead Plate Preparation, Washing, and Elution

1. Dilute the total RNA (0.1–4 μg per sample) with nuclease-free ultrapure water to a final volume of 50 μl in a 96-well 0.3 ml PCR plate with the RNA Bead Plate (RBP) barcode label.
2. Vortex the thawed RNA Purification Beads tube vigorously to completely resuspend the oligo-dT beads.
3. Add 50 μl of RNA Purification Beads to each well of the RBP plate using a multichannel pipette to bind the poly-A RNA to the oligo-dT magnetic beads. Gently pipette the entire volume up and down 6 times to mix thoroughly. Change the tips after each column.
4. Seal the RBP plate with a Microseal "B" adhesive seal.
5. Place the RBP plate on the magnetic stand (provided with the kit) at room temperature for 5 min to separate the poly-A RNA bound beads from the solution.
6. Remove the adhesive seal from the RBP plate.
7. While leaving the plate on the magnet, remove and discard all of the supernatant from each well of the RBP plate using a multichannel pipette. Take care not to disturb the beads. Change the tips after each column.
8. Remove the RBP plate from the magnetic stand.
9. Wash the beads by adding 200 μl of Bead Washing Buffer in each well of the RBP plate using a multichannel pipette to remove unbound RNA. Gently pipette the entire volume up and down 6 times to mix thoroughly. Change the tips after each column.
10. Place the RBP plate on the magnetic stand at room temperature for 5 min.
11. Briefly centrifuge the thawed elution buffer to 600 × *g* for 5 s.
12. Remove and discard all of the supernatant from each well of the RBP plate using a multichannel pipette. Take care not to disturb the beads. Change the tips after each column.

13. Remove the RBP plate from the magnetic stand.
14. Add 50 μl of elution buffer in each well of the RBP plate using a multichannel pipette. Gently pipette the entire volume up and down 6 times to mix thoroughly. Change the tips after each column.
15. Seal the RBP plate with a Microseal "B" adhesive seal.
16. Store the elution buffer tube at 4 °C.
17. Place the sealed RBP plate on the preprogrammed thermal cycler. Close the lid and select mRNA elution 1 (80 °C for 2 min, 25 °C hold) to elute the mRNA from the beads. Remove the RBP plate from the thermal cycler when it reaches 25 °C.
18. Place the RBP plate on the bench at room temperature and remove the adhesive seal from the plate.

Generation/Incubation of RNA Fragmentation Plate (RFP)

1. Add 50 μl of Bead Binding Buffer to each well of the RBP plate using a multichannel pipette to allow the RNA to rebind to the beads. Gently pipette the entire volume up and down 6 times to mix thoroughly. Change the tips after each column.
2. Incubate the RBP plate at room temperature for 5 min and store the Bead Binding Buffer tube at 2–8 °C.
3. Place the RBP plate on the magnetic stand at room temperature for 5 min.
4. Remove and discard all of the supernatant from each well of the RBP plate using a multichannel pipette. Take care not to disturb the beads. Change the tips after each column.
5. Remove the RBP plate from the magnetic stand.
6. Wash the beads by adding 200 μl of Bead Washing Buffer in each well of the RBP plate using a multichannel pipette. Gently pipette the entire volume up and down 6 times to mix thoroughly. Change the tips after each column.
7. Store the Bead Washing Buffer tube at 2–8 °C.
8. Place the RBP plate on the magnetic stand at room temperature for 5 min.
9. Remove and discard all of the supernatant from each well of the RBP plate using a multichannel pipette. Take care not to disturb the beads. Change the tips after each column.
10. Remove the RBP plate from the magnetic stand.
11. Add 19.5 μl of Elute, Prime, Fragment Mix to each well of the RBP plate using a multichannel pipette. Gently pipette the entire volume up and down 6 times to mix thoroughly. Change the tips after each column. The Elute, Prime, Fragment Mix contains random hexamers for RT priming and serves as the 1st strand cDNA synthesis reaction buffer.

12. Seal the RBP plate with a Microseal "B" adhesive seal.
13. Store the Elute, Prime, Fragment Mix tube at −15 °C to −25 °C.
14. Place the sealed RBP plate on the pre-programmed thermal cycler. Close the lid and run (94 °C for 8 min, 4 °C hold) to elute, fragment, and prime the RNA.
15. Remove the RBP plate from the thermal cycler when it reaches 4 °C and centrifuge briefly.
16. Proceed immediately to synthesize first-strand cDNA.

First-Strand cDNA Synthesis

1. Place the RBP plate on the magnetic stand at room temperature for 5 min. Do not remove the plate from the magnetic stand.
2. Remove the adhesive seal from the RBP plate.
3. Transfer 17 μl of the supernatant (fragmented and primed mRNA) from each well of the RBP plate to the corresponding well of the new 0.3 ml PCR plate labeled with the cDNA plate (CDP) barcode. Some liquid may remain in each well.
4. Briefly centrifuge the thawed First Strand Master Mix to 600 × *g* for 5 s.
5. Add 50 μl SuperScript II to the First Strand Master Mix tube (ratio: 1 μl SuperScript II for each 7 μl First Strand Master Mix). Mix gently, but thoroughly, and centrifuge briefly. Label the First Strand Master Mix tube to indicate that the SuperScript II has been added.
6. Add 8 μl of First Strand Master Mix and SuperScript II mix to each well of the CDP plate using a multichannel pipette. Gently pipette the entire volume up and down 6 times to mix thoroughly. Change the tips after each column.
7. Seal the CDP plate with a Microseal "B" adhesive seal and centrifuge briefly.
8. Return the First Strand Master Mix tube back to −15 °C to −25 °C storage immediately after use.
9. Incubate the CDP plate on the thermal cycler, with the lid closed, using the following program:
   (a) 25 °C for 10 min.
   (b) 42 °C for 50 min.
   (c) 70 °C for 15 min.
   (d) Hold at 4 °C.
10. When the thermal cycler reaches 4 °C, remove the CDP plate from the thermal cycler and proceed immediately to synthesize second-strand cDNA.

Second-Strand Synthesis

1. Briefly centrifuge the thawed Second Strand Master Mix to 600×*g* for 5 s.
2. Remove the adhesive seal from the CDP plate.
3. Add 25 μl of thawed Second Strand Master Mix to each well of the CDP plate using a multichannel pipette. Gently pipette the entire volume up and down 6 times to mix thoroughly. Change the tips after each column.
4. Seal the CDP plate with a Microseal "B" adhesive seal.
5. Incubate the CDP plate on the preheated thermal cycler, with the lid closed, at 16 °C for 1 h.
6. Remove the CDP plate from the thermal cycler, remove the adhesive seal, and let the plate stand to bring it to room temperature.

cDNA Clean Up

1. Vortex the AMPure XP beads until they are well dispersed, then add 90 μl of well-mixed AMPure XP beads to each well of the CDP plate containing 50 μl of ds cDNA. Gently pipette the entire volume up and down 10 times to mix thoroughly.
2. Incubate the CDP plate at room temperature for 15 min.
3. Place the CDP plate on the magnetic stand at room temperature, for at least 5 min to ensure that all of the beads are bound to the side of the wells.
4. Remove and discard 135 μl of the supernatant from each well of the CDP plate using a multichannel pipette. Some liquid may remain in each well. Take care not to disturb the beads. Change the tips after each column.
5. With the CDP plate remaining on the magnetic stand, add 200 μl of freshly prepared 80 % EtOH to each well without disturbing the beads.
6. Incubate the CDP plate at room temperature for 30 s, then remove and discard all of the supernatant from each well using a multichannel pipette. Take care not to disturb the beads. Change the tips after each column.
7. Repeat **steps 5** and **6** once for a total of two 80 % EtOH washes.
8. Let the plate stand at room temperature for 15 min to dry and then remove the CDP plate from the magnetic stand.
9. Briefly centrifuge the thawed, room temperature resuspension buffer at 600×*g* for 5 s.
10. Add 52.5 μl resuspension buffer to each well of the CDP plate using a multichannel pipette. Gently pipette the entire volume up and down 10 times to mix thoroughly.
11. Incubate the CDP plate at room temperature for 2 min.

12. Place the CDP plate on the magnetic stand at room temperature for 5 min.
13. Transfer 50 μl of the supernatant (ds cDNA) from the CDP plate to the new 0.3 ml PCR plate labeled with the IMP barcode. Some liquid may remain in each well.

End Repair

1. Add 40 μl of End Repair Mix to each well of the Insert Modification plate (IMP) containing the ds cDNA using a multichannel pipette. Gently pipette the entire volume up and down 10 times to mix thoroughly.
2. Seal the IMP plate with a Microseal "B" adhesive seal.
3. Incubate the IMP plate on the preheated thermal cycler, with the lid closed, at 30 °C for 30 min.
4. Remove the IMP plate from the thermal cycler.
5. Remove the adhesive seal from the IMP plate.
6. Vortex the AMPure XP beads until they are well dispersed, then add 160 μl of well-mixed AMPure XP beads to each well of the IMP plate containing 100 μl of End Repair Mix. Gently pipette up and down 10 times to mix thoroughly.
7. Incubate the IMP plate at room temperature for 15 min.
8. Place the IMP plate on the magnetic stand at room temperature for at least 5 min, until the liquid appears clear.
9. Using a 200 μl multichannel pipette set to 127.5 μl, remove and discard 127.5 μl of the supernatant from each well of the IMP plate. Take care not to disturb the beads. Change the tips after each column.
10. Repeat **step 9** once. Some liquid may remain in each well.
11. With the IMP plate on the magnetic stand, add 200 μl of freshly prepared 80 % EtOH to each well without disturbing the beads.
12. Incubate the IMP plate at room temperature for at least 30 s, then remove and discard all of the supernatant from each well. Take care not to disturb the beads. Change the tips after each column.
13. Repeat **steps 11** and **12** once for a total of two 80 % EtOH washes.
14. Let the IMP plate stand at room temperature for 15 min to dry and then remove the plate from the magnetic stand.
15. Resuspend the dried pellet in 17.5 μl resuspension buffer. Gently pipette the entire volume up and down 10 times to mix thoroughly.
16. Incubate the IMP plate at room temperature for 2 min.

17. Place the IMP plate on the magnetic stand at room temperature for at least 5 min, until the liquid appears clear.
18. Transfer 15 μl of the clear supernatant from each well of the IMP plate to the corresponding well of the new 0.3 ml PCR plate labeled with the ALP barcode. Some liquid may remain in each well.

3′-End Adenylation

1. Add 2.5 μl of resuspension buffer to each well of the Adapter Ligation Plate (ALP) using a multichannel pipette.
2. Adjust the multichannel pipette to 30 μl and gently pipette the entire volume up and down 10 times to mix thoroughly.
3. Add 12.5 μl of thawed A-Tailing Mix to each well of the ALP plate using a multichannel pipette. Change the tips after each column.
4. Seal the ALP plate with a Microseal "B" adhesive seal.
5. Incubate the ALP plate on the preheated thermal cycler, with the lid closed, at 37 °C for 30 min.
6. Immediately remove the ALP plate from the thermal cycler, then proceed immediately to ligate Adapters.

Adapter Ligation

1. Briefly centrifuge the thawed RNA Adapter Index tubes (AR001–AR012 depending on the RNA Adapter Indexes being used), ligase control (if using ligase control), and Stop Ligase Mix tubes at 600 × *g* for 5 s.
2. Immediately before use, remove the DNA Ligase Mix tube from −15 °C to −25 °C storage.
3. Remove the adhesive seal from the ALP plate.
4. Add 2.5 μl of DNA Ligase Mix to each well of the ALP plate.
5. Return the DNA Ligase Mix tube back to −15 °C to −25 °C storage immediately after use.
6. Add 2.5 μl of each thawed RNA Adapter Index (AR001–AR012 depending on the RNA Adapter Indexes being used) to each well of the ALP plate using a multichannel pipette.
7. Adjust the multichannel pipette to 40 μl and gently pipette the entire volume up and down 10 times to mix thoroughly.
8. Seal the ALP plate with a Microseal "B" adhesive seal.
9. Incubate the ALP plate on the preheated thermal cycler, with the lid closed, at 30 °C for 10 min.
10. Remove the ALP plate from the thermal cycler.
11. Remove the adhesive seal from the ALP plate.
12. Add 5 μl of Stop Ligase Mix to each well of the ALP plate to inactivate the ligation mix using a multichannel pipette. Gently pipette the entire volume up and down 10 times to mix thoroughly.

13. Vortex the AMPure XP beads until they are well dispersed, then add 42 μl of mixed AMPure XP beads to each well of the ALP plate using a multichannel pipette. Gently pipette the entire volume up and down 10 times to mix thoroughly.
14. Incubate the ALP plate at room temperature for 15 min.
15. Place the ALP plate on the magnetic stand at room temperature for at least 5 min, until the liquid appears clear.
16. Remove and discard 79.5 μl of the supernatant from each well of the ALP plate using a multichannel pipette. Some liquid may remain in each well. Take care not to disturb the beads. Change the tips after each column.
17. With the ALP plate remaining on the magnetic stand, add 200 μl of freshly prepared 80 % EtOH to each well without disturbing the beads.
18. Incubate the ALP plate at room temperature for at least 30 s, then remove and discard all of the supernatant from each well. Take care not to disturb the beads. Change the tips after each column.
19. Repeat **steps 17** and **18** once for a total of two 80 % EtOH washes.
20. Let the ALP plate stand at room temperature for 15 min to dry and then remove the plate from the magnetic stand.
21. Resuspend the dried pellet in each well with 52.5 μl resuspension buffer. Gently pipette the entire volume up and down 10 times to mix thoroughly.
22. Incubate the ALP plate at room temperature for 2 min.
23. Place the ALP plate on the magnetic stand at room temperature for at least 5 min, until the liquid appears clear.
24. Transfer 50 μl of the clear supernatant from each well of the ALP plate to the corresponding well of the new 0.3 ml PCR plate labeled with the Clean up ALP plate (CAP) barcode. Some liquid may remain in each well. Change the tips after each column.
25. Vortex the AMPure XP beads until they are well dispersed, then add 50 μl of mixed AMPure XP beads to each well of the CAP plate for a second clean up using a multichannel pipette. Gently pipette the entire volume up and down 10 times to mix thoroughly.
26. Incubate the CAP plate at room temperature for 15 min.
27. Place the CAP plate on the magnetic stand at room temperature for 5 min or until the liquid appears clear.
28. Remove and discard 95 μl of the supernatant from each well of the CAP plate, using a multichannel pipette. Some liquid may

remain in each well. Take care not to disturb the beads. Change the tips after each column.

29. With the CAP plate remaining on the magnetic stand, add 200 μl of freshly prepared 80 % EtOH to each well without disturbing the beads.
30. Incubate the CAP plate at room temperature for at least 30 s, then remove and discard all of the supernatant from each well. Take care not to disturb the beads. Change the tips after each column.
31. Repeat **steps 29** and **30** once for a total of two 80 % EtOH washes.
32. Let the CAP plate stand at room temperature for 15 min to dry and then remove the plate from the magnetic stand.
33. Resuspend the dried pellet in each well with 22.5 μl resuspension buffer. Gently pipette the entire volume up and down 10 times to mix thoroughly.
34. Incubate the CAP plate at room temperature for 2 min.
35. Place the CAP plate on the magnetic stand at room temperature for at least 5 min, until the liquid appears clear.
36. Transfer 20 μl of the clear supernatant from each well of the CAP plate to the corresponding well of the new 0.3 ml PCR plate labeled with the PCR barcode. Some liquid may remain in each well. Change the tips after each column.

DNA Fragment Enrichment

1. Add 5 μl of thawed PCR Primer Cocktail to each well of the PCR plate using a multichannel pipette. Change the tips after each column.
2. Add 25 μl of thawed PCR Master Mix to each well of the PCR plate using a multichannel pipette. Change the tips after each column. Adjust the single channel or multichannel pipette to 40 μl and gently pipette the entire volume up and down 10 times to mix thoroughly.
3. Amplify the PCR plate in the preprogrammed thermal cycler, with the lid closed, using the PCR program:
   (a) 98 °C for 30 s.
   (b) 15 cycles of:
      - 98 °C for 10 s.
      - 60 °C for 30 s.
      - 72 °C for 30 s.
      - 72 °C for 5 min.

   (c) Hold at 4 °C.
4. Vortex the AMPure XP beads until they are well dispersed, and then add 50 μl of the mixed AMPure XP beads to each well of

the PCR plate containing 50 μl of the PCR amplified library using a multichannel pipette. Gently pipette the entire volume up and down 10 times to mix thoroughly.

5. Incubate the PCR plate at room temperature for 15 min.
6. Place the PCR plate on the magnetic stand at room temperature for at least 5 min, until the liquid appears clear.
7. Remove and discard 95 μl of the supernatant from each well of the PCR plate, using a multichannel pipette. Some liquid may remain in each well. Take care not to disturb the beads.
8. With the PCR plate remaining on the magnetic stand, add 200 μl of freshly prepared 80 % EtOH to each well without disturbing the beads.
9. Incubate the PCR plate at room temperature for at least 30 s, then remove and discard all of the supernatant from each well. Take care not to disturb the beads.
10. Repeat **steps 8** and **9** once for a total of two 80 % EtOH washes.
11. Let the PCR plate stand at room temperature for 15 min to dry and then remove the plate from the magnetic stand.
12. Resuspend the dried pellet in each well with 32.5 μl resuspension buffer using a multichannel pipette. Gently pipette the entire volume up and down 10 times to mix thoroughly.
13. Incubate the PCR plate at room temperature for 2 min.
14. Place the PCR plate on the magnetic stand at room temperature for at least 5 min, until the liquid appears clear.
15. Transfer 30 μl of the clear supernatant from each well of the PCR plate to the corresponding well of the new 0.3 ml PCR plate labeled with the Target sample plate 1 (TSP1) barcode. Some liquid may remain in each well.

Library Quality Control

To control the integrity of the generated library, add 1 μl of the library to the Agilent Bioanalyzer. The final library should have a peak at 260 bp for single-read libraries (*see* Fig. 2).

3.1.2 mRNA Library Sequencing

When the cDNA library material has the required quality and quantity as determined in the previous point, it may be subjected to sequencing, e.g., with the HighSeq 2000 device from Illumina. A sequencing depth of ~20 million reads and a read length from 35 to 50 bp is sufficient for a representative mRNA analysis. The processing of the resulting FASTQ files is discussed in the Bioinformatics paragraph and in [57]. In case there is no possibility to sequence the generated library in house, the sequencing may be outsourced to sequencing services providers.

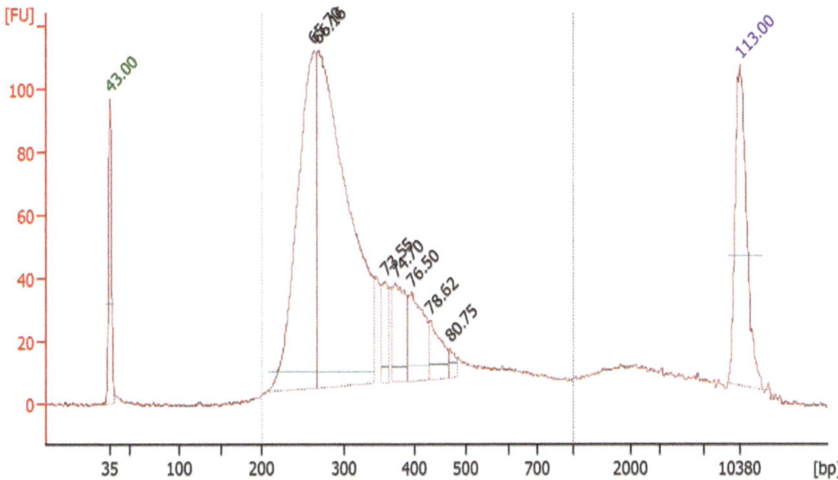

Fig. 2 Electropherogram of an mRNA library subjected to analysis with the DNA Bioanalyzer. Peaks at 35 and 10,380 bp correspond to size standards. *FU* fluorescence units

3.2 miRNA-Seq

The following paragraphs describe the protocol for generating cDNA libraries representing miRNAs and other small RNAs starting from total RNA preparations.

3.2.1 miRNA Library Preparation

(According to the Manufacturer's protocol, Illumina Part # 15004197 Rev. A)

This protocol is suitable for 1–10 μg of total RNA. The starting material, total RNA, can be isolated by a number of techniques. Here we have used the Roche High Pure RNA Isolation Kit (*see* **Note 4**). Before NGS, the differential expression of miRNAs known to be regulated by c-MYC should be confirmed in the samples, which will be subjected to library generation. In order to determine the expression of mature miRNAs stem-loop reverse transcription (RT) followed by a TaqMan PCR or LNA-based PCR (Exiqon) should be used [58].

3′-Adapter Ligation

1. Set up the ligation reaction in a sterile, nuclease-free 200 μl PCR tube on ice using the following:
   (a) RNA 3′ Adapter (RA3) (1 μl).
   (b) Total RNA in nuclease-free water (5 μl).
   Total volume (6 μl).
2. Gently pipette the entire volume up and down 6–8 times to mix thoroughly, then centrifuge briefly.
3. Incubate the tube on the preheated thermal cycler at 70 °C for 2 min and then immediately place the tube on ice.
4. Preheat the thermal cycler to 28 °C.

5. Prepare the following mix in a separate, sterile, nuclease-free 200 μl PCR tube on ice. Multiply each reagent volume by the number of samples being prepared (*see* **Note 5**):
   (a) 5× HM Ligation Buffer (HML) (2 μl).
   (b) RNase inhibitor (1 μl).
   (c) T4 RNA Ligase 2, truncated (1 μl).
   Total volume (4 μl).
6. Gently pipette the entire volume up and down 6–8 times to mix thoroughly, then centrifuge briefly.
7. Add 4 μl of the mix to the reaction tube from **step 1** and gently pipette the entire volume up and down 6–8 times to mix thoroughly. The total volume of the reaction should be 10 μl.
8. Incubate the tube on the preheated thermal cycler at 28 °C for 1 h.
9. With the reaction tube remaining on the thermal cycler, add 1 μl Stop Solution (STP) and gently pipette the entire volume up and down 6–8 times to mix thoroughly. Continue to incubate the reaction tube on the thermal cycler at 28 °C for 15 min, and then place the tube on ice.

5′-Adapter Ligation

1. Preheat the thermal cycler to 70 °C.
2. Aliquot 1.1 × *N* μl of the RNA 5′ Adapter (RA5) into a separate, nuclease-free 200 μl PCR tube, with *N* being equal to the number of samples being processed for the current experiment.
3. Incubate the adapter on the preheated thermal cycler at 70 °C for 2 min and then immediately place the tube on ice.
4. Preheat the thermal cycler to 28 °C.
5. Add 1.1 × *N* μl of 10 mM ATP to the aliquoted RNA 5′ Adapter tube, with *N* equal to the number of samples being processed for the current experiment. Gently pipette the entire volume up and down 6–8 times to mix thoroughly.
6. Add 1.1 × *N* μl of T4 RNA Ligase to the aliquoted RNA 5′ Adapter tube, with *N* equal to the number of samples being processed for the current experiment. Gently pipette the entire volume up and down 6–8 times to mix thoroughly.
7. Add 3 μl of the mix from the aliquoted RNA 5′ Adapter tube to the reaction from **step 9** of "3′-adapter ligation." Gently pipette the entire volume up and down 6–8 times to mix thoroughly. The total volume of the reaction should now be 14 μl.
8. Incubate the reaction tube on the preheated thermal cycler at 28 °C for 1 h and then place the tube on ice.

Sample Electrophoresis and RNA Gel Extraction

Repeat the sample electrophoresis and RNA gel extraction by using a 10 % TBE-urea PAGE Gel to separate the RNA.

Reverse Transcription and Amplification

Template Preparation

1. Pipette following reaction mix:
   (a) 5′ and 3′ Adapter-ligated RNA (6 μl).
   (b) RNA RT Primer (RTP) (1 μl)
   Total volume (7 μl).
2. Gently pipette the entire volume up and down 6–8 times to mix thoroughly, then centrifuge briefly.
3. Incubate the tube on the pre-heated thermal cycler at 70 °C for 2 min and then immediately place the tube on ice.
4. Pre-heat the thermal cycler to 50 °C.
5. Prepare the following mix in a separate, sterile, nuclease-free, 200 μl PCR tube placed on ice. Multiply each reagent volume by the number of samples being prepared. Make 10 % extra reagent if you are preparing multiple samples.
6. Pipette following reaction mix:
   (a) 5× first-strand buffer (2 μl).
   (b) 12.5 mM dNTP mix (0.5 μl).
   (c) 100 mM DTT (1 μl).
   (d) RNase inhibitor (1 μl).
   (e) SuperScript II Reverse Transcriptase (1 μl).
   Total volume (5.5 μl).
7. Gently pipette the entire volume up and down 6–8 times to mix thoroughly, then centrifuge briefly.
8. Add 5.5 μl of the mix to the reaction tube from **step 3**. Gently pipette the entire volume up and down 6–8 times to mix thoroughly, then centrifuge briefly. The total volume should now be 12.5 μl.
9. Incubate the tube in the preheated thermal cycler at 50 °C for 1 h and then place the tube on ice.

PCR

1. Prepare the following master mix reaction:
   (a) Ultrapure water (22.5 μl).
   (b) 5× Phusion HF Buffer (10 μl).
   (c) RNA PCR Primer (RP1) (2 μl).
   (d) RNA PCR Primer Index (RPI*X*) (2 μl).
   (e) 25 mM dNTP mix (0.5 μl).
   (f) Phusion DNA Polymerase (0.5 μl).
   Total volume (37.5 μl)
   For each reaction, only one of the 48 RNA PCR Primer Indices is used during the PCR step.

2. Gently pipette the entire volume up and down 6–8 times to mix thoroughly, then centrifuge briefly, and then place the tube on ice.
3. Add 37.5 µl of PCR master mix to the reaction tube from **step 8** of the “reverse transcription” protocol.
4. Gently pipette the entire volume up and down 6–8 times to mix thoroughly, then centrifuge briefly, and place the tube on ice. The total volume should now be 50 µl.
5. Amplify the tube in the thermal cycler using the following PCR cycling conditions:
   (a) 30 s at 98 °C.
   (b) 11 cycles of:
      - 10 s at 98 °C.
      - 30 s at 60 °C.
      - 15 s at 72 °C.
   (c) 10 min at 72 °C.
   (d) Hold at 4 °C.
6. Run each sample on a high sensitivity Bioanalyzer DNA chip according to the manufacturer's instructions.

miRNA Library Purification

At this point of the protocol, individual libraries with unique indices may be pooled and gel purified together. Combine equal volumes of the library or molar amounts and then load the samples on the gel according to the instructions below. Do not load more than 30 µl of sample per well.

1. Determine the volume of 1× TBE buffer needed. Dilute the 5× TBE buffer to 1× for use in the electrophoresis.
2. Assemble the gel electrophoresis apparatus according to the manufacturer's instructions.
3. Mix 2 µl of Custom Ladder with 2 µl of DNA loading dye.
4. Mix 1 µl of High-Resolution Ladder with 1 µl of DNA loading dye.
5. Mix the total volume containing the amplified cDNA (typically 48–50 µl) with 10 µl of DNA loading dye.
6. Load 2 µl of mixed Custom Ladder and loading dye in two wells on the 6 % PAGE Gel.
7. Load 2 µl of High-Resolution Ladder and loading dye in a different well.
8. Load two wells with 25 µl each of mixed amplified cDNA construct and loading dye on the 6 % PAGE Gel. A total volume of 50 µl should be loaded on the gel.

9. Run the gel for 60 min at 145 V or until the blue front dye exits the gel. Proceed immediately to the next step.
10. Remove the gel from the apparatus.

Recovery of miRNA Library from Gel

1. Open the cassette according to the manufacturer's instructions and stain the gel with ethidium bromide (0.5 μg/ml in water) in a clean container for 2–3 min.
2. Place the gel breaker tube into a sterile, round-bottom, nuclease-free, 2 ml microcentrifuge tube.
3. View the gel on a Dark Reader transilluminator or a UV transilluminator at a wavelength not harming the DNA (i.e., greater than 320 nm), Fig. 3.
4. Using a clean scalpel, cut out the bands corresponding to the adapter-ligated constructs derived from the 22 nt and 30 nt small RNA fragments. miRNAs often vary in length (so-called iso-miRs). The tighter the band selection, the tighter the size distribution of the final miRNA representation. (*See* Fig. 3; miRNA with adaptors should have a size of ~100 bp.)
5. Place the isolated band of interest into the 0.5 ml gel breaker tube from **step 2**.
6. Centrifuge the stacked tubes at 16,000 × *g* in a microcentrifuge for 2 min at room temperature to passage the gel through the holes into the 2 ml tube. Ensure that the gel has been completely transferred into the bottom tube.
7. Add 300 μl of gel elution buffer to the gel pieces in the 2 ml tube.

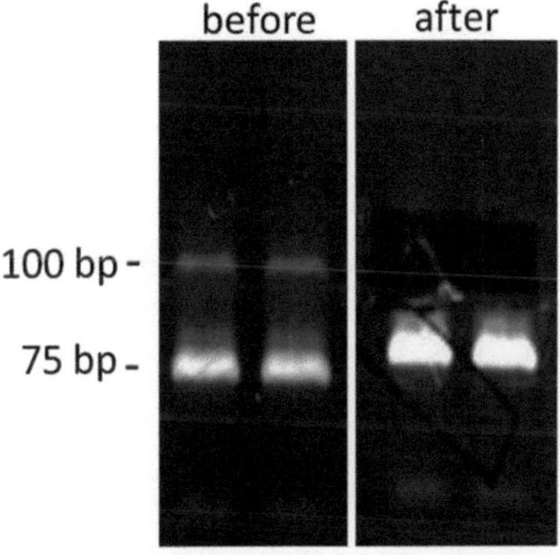

Fig. 3 Gel purification of small RNA libraries. A small RNA library separated on a 6 % PAGE gel before and after cutting the region containing the miRNA library out of the gel (gel pictures kindly provided by Anne Dück, University of Regensburg)

8. Elute the DNA by rotating or shaking the tube at room temperature for at least 2 h. The tube can be rotated overnight.
9. Transfer the eluate and the gel pieces to the top of a 5 μm filter.
10. Centrifuge the filter for 2 min at 16,000 × *g*.
11. Add 2 μl of glycogen, 30 μl of 3 M NaOAc, 2 μl of 0.1× Pellet Paint (optional), and 975 μl of prechilled 100 % ethanol (−15 °C to −25 °C).
12. Immediately centrifuge in a benchtop microcentrifuge at 16,000 × *g* for 20 min at 4 °C.
13. Remove and discard the supernatant, leaving the pellet intact.
14. Wash the pellet with 500 μl of room temperature 70 % ethanol.
15. Centrifuge at 16,000 × *g* at room temperature for 2 min.
16. Remove and discard the supernatant, leaving the pellet intact.
17. Dry the pellet by placing the tube, lid open, in a 37 °C heat block for 5–10 min or until dry.
18. Resuspend the pellet in 10 μl resuspension buffer.

3.2.2 miRNA Library Validation

Use the Agilent Technologies 2100 Bioanalyzer chip for DNA to validate your generated library (Fig. 4).

3.2.3 miRNA Library Sequencing

If the library material has the required quality and quantity, the material can be subjected to NGS sequencing. A sequencing depth of around 20 million reads and a sequencing read length of

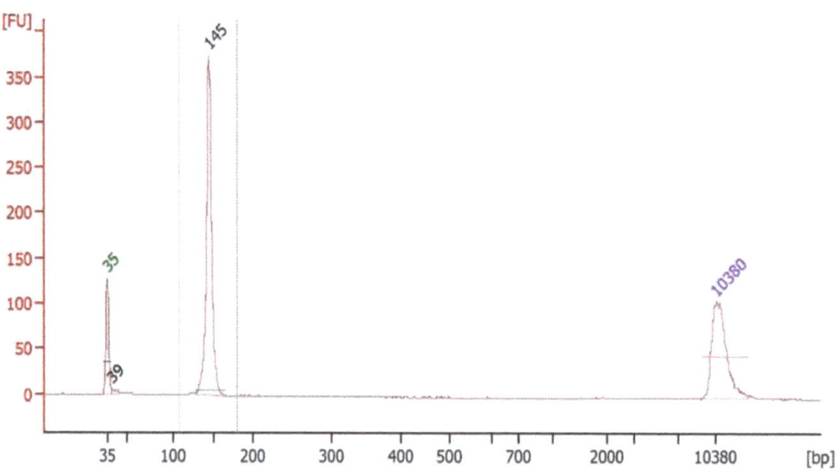

Fig. 4 Analysis of a mature miRNA library using a DNA Bioanalyzer. Peaks at 35 and 10,380 bp represent the size standard. *FU* fluorescence units

35–50 bp are sufficient to obtain a representative result. The processing of the resulting FASTQ files is discussed in the bioinformatics paragraph.

3.3 ChIP-Seq

This section describes the analysis of the genome-wide pattern of c-MYC DNA binding using a combination of chromatin immunoprecipitation and NGS. In order to immunoprecipitate c-MYC bound to DNA, several different strategies can be applied. One approach is to analyze the binding of endogenous c-MYC after subjecting cells to treatments, which either induce or decrease the expression of endogenous c-MYC, e.g., serum starvation and restimulation to increase c-MYC expression [35, 53, 59] or treatment with TGF-β to decrease it [60]. For the ChIP analysis of endogenous human c-MYC protein, the antibody N-262 (sc-764 from Santa Cruz) is commonly used [34, 35, 53, 59, 61]. As a control, unspecific polyclonal isotype control, corresponding to the c-MYC-specific antibody, should be used (*see* **Note 6**). When performing ChIP analysis after ectopic expression of tagged MYC proteins, the levels of c-MYC expression should be closely monitored since nonphysiological expression levels may result in aberrant DNA binding [29] (*see* **Note** 7). If you use a different antibody or a new lot of antibody, make sure that these allow obtaining the expected results in a conventional immunoprecipitation or in a ChIP assay [63]. Furthermore, genome-wide ChIP analyses allow determining the effect of c-MYC activation on the global organization of chromatin, e.g., by mapping the distribution of histone marks, representing repressive or active chromatin [40, 64] (*see* **Note 8**). In the following sections the isolation of c-MYC-associated DNA, the validation of c-MYC binding to known binding sites, and the generation of ChIP-DNA libraries, which can be subjected to NGS sequencing, are described.

3.3.1 ChIP

Experimental Design of ChIP for Subsequent NGS Analysis

For the initial antibody testing, three 15 cm diameter culture plates with subconfluent cell-layers per condition are enough to obtain a sufficient amount of ChIP-DNA. For the final library generation, five 15 cm diameter culture plates per condition with a cell confluency of 70 % are necessary for each state. Ideally, a ChIP analysis is accompanied by several different types of control analyses in order to rule out artifacts generated by the procedure itself. As controls, the DNA present in the IP input, mock IP, and/or IgG IP may be subject to NGS analysis (*see* **Note 9**). If ectopic expression of a tagged c-MYC protein is used, DNA obtained from the empty control vector cell line after IP with the tag-specific antibody should be subjected to NGS sequencing. The different types of control experiments and other ChIP analysis related issues are reviewed in [62].

Pre-blocking of Sepharose Beads

The pre-blocking prevents unspecific binding of DNA or proteins to the beads. For this, two different stocks of pre-blocking solutions should be prepared: one stock for preclearing and 60 µl . The other for the IP, which will be used at different times during the protocol.

1. Calculate the total bead volume for preclearing and IP. For three plates of adherent cells with ~70 % confluency (~5×10^6 cells). 30 µl beads (bed volume, add 20 % to compensate for pipetting errors) should be used for preclearing, and 60 µl beads (bed volume, add 20 % to compensate for pipetting errors) for the immunoprecipitation.
2. Start with the beads for preclearing. Wash beads three times with 1 ml PBS to get rid of ethanol, which is present in the bead storage buffer.
3. Add 2.4 µl BSA fatty acid free (50 µg/µl) and 4.8 µl (10 µg/µl) salmon sperm DNA (subjected to 8 sonication cycles) per 30 µl protein A/G beads (50 % slurry in PBS).
4. Pre-block total bead volume at least 3 h at 4 °C on a rotating wheel (beads are necessary for the immunoprecipitation and the pre-blocking step).
5. Wash beads 4 times with 1 ml PBS and once with 1 ml IP buffer after pre-blocking, by centrifugation at $400 \times g$ for 1 min.

Cross-linking and Sonication

1. During pre-blocking of beads, dissolve complete mini (CM) protease inhibitors in SDS buffer (1 tablet/5 ml) (always prepare fresh). Do not put on ice since SDS will precipitate and keep it at room temperature. Also prepare IP buffer containing CM (1 tablet/5 ml), keep on ice until usage.
2. For cross-linking, remove medium from the cells and keep it in a beaker (at room temperature).
3. Immediately add 270.3 µl from 37 % stock formaldehyde (*see* **Notes 10** and **11**) per 10 ml medium from the cells (= 1 % formaldehyde), mix and quickly add the 1 % formaldehyde containing medium back onto the cells.
4. Perform cross-linking times at RT without shaking. The optimal incubation times vary between cell lines/types (e.g., HeLa, 10 min; HDF, 5 min; MCF-7, 5 min; DLD-1, 5 min; H1299, 8 min). If processing many plates, take 3–5 at a time (*see* **Note 12**).
5. To stop the cross-linking, add 1.5 ml 1 M glycine/10 ml cross-linking medium (CM = 0.125 M glycine), mix and incubate for 2 min, and then discard.
6. Wash twice with 1× TBS (4 °C cold) using 20 ml/15 cm plate.
7. Harvest cells from one 15 cm diameter plate in 500 µl SDS/CM buffer, collect lysate and repeat harvesting step with additional 500 µl SDS buffer, combine the 2 fractions, and keep at room temperature in order to avoid precipitation of SDS.

8. Centrifugation (400 × *g* for 5 min at 4 °C) in a 15 ml Falcon.
9. Resuspend the pellet in 2 ml IP buffer. Independent of the initial cell number, use a 15 ml Falcon and a volume between >500 μl and up to 2 ml to guarantee efficient sonication.
10. For sonication, use standard sonication devices (e.g., HTU SONI 130 (G. Heinemann); SONOPULS (Bandelin)) or automated devices as the Bioruptor (Diagenode). For NGS of ChIP-DNA, it is necessary to generate DNA fragments with an average size of 300 bp or below. As shown in Fig. 5, the number of sonication cycles has to be optimized to obtain fragments of the required size. With the SONOPULS device, we use an amplitude of 80 % with the following sonication cycles: 20 s sonication pulse and 50 s pause. The samples should be kept on an ethanol ice mixture to avoid denaturation of the proteins during the sonication procedure, which generates heat.
11. Distribute lysates to 2 ml reaction tubes. Centrifuge samples 15 min at 16,000 × *g* and 4 °C, and transfer supernatant into fresh 1.5 ml reaction tube for preclearing.

Immunoprecipitation

1. Wash pre-blocked beads (for preclearing) 4 times with 1 ml PBS and once with 1 ml IP buffer. Then resuspend the beads in IP buffer.
2. Preclear the lysates for 1 h at 4 °C by adding 30 μl pre-blocked beads per condition. Preclearing can be done on the entire sample volumes before subdividing into individual IPs to maintain consistency in processing as long as possible.

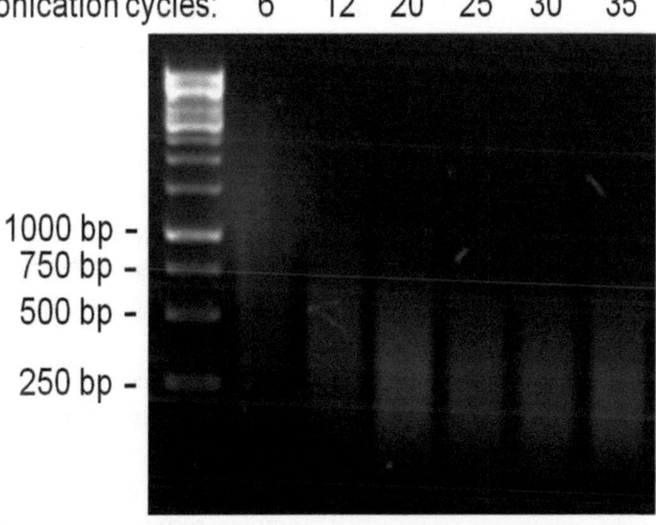

Fig. 5 Reduction of DNA fragment size by increasing the number of sonication cycles. 15 μl of sonicated input DNA from the indicated sonication cycles were separated on a 1.2 % TAE agarose gel, 1 kbp DNA marker in the *first lane*

3. Centrifuge at 400 × *g* for 5 min at 4 °C to precipitate the beads and unspecific associated material and transfer the supernatant into fresh 1.5 ml reaction tubes. Keep about 100 μl as "Input" sample and store at −20 °C until further analysis. Leave sticky protein aggregates at the walls of the tube.
4. Distribute the exact same amount of lysate to the individual IP tubes (*see* **Note 13**).
5. For immunoprecipitation rotate samples overnight at 4 °C with 2–10 μg antibody/sample. For ChIP analysis of c-MYC, the N-262 (SC-764, Santa Cruz) rabbit polyclonal antibody has been widely used. 2 μg antibody per dish (15 cm) is recommended.
6. During the immunoprecipitation, pre-block the washed protein A/G beads for 3 h as described above (for IP).
7. Wash the pre-blocked beads 4 times with 1 ml PBS and once with 1 ml IP buffer. To recover the antibody-bound protein, add 60 μl beads (bed volume)/3 plates. Rotate 2 h at 4 °C.
8. Centrifuge 1 min 400 × *g* at room temperature and discard the supernatant (use vacuum pump with pipette tip on top of a glass Pasteur pipette).
9. Washing procedure: Add 1 ml mixed micelle buffer; rotate 5 min at room temperature, centrifuge 1 min at 1000 × *g*, and remove the supernatant.
10. Repeat with 1 ml buffer 500.
11. Repeat with 1 ml LiCl/detergent wash solution.
12. Repeat with 1 ml TE pH 7.5.
13. Remove remaining supernatant carefully.
14. Add 100 μl freshly prepared elution buffer and incubate 10 min at 65 °C in a water bath. The use of a water bath is important. Flick the tube several times in order to resuspend the beads.
15. Centrifuge at maximum speed and transfer the eluate to a fresh 1.5 ml reaction tube.
16. Again add 100 μl elution buffer to the beads and repeat incubate on for 10 min at 65 °C in a water bath.
17. Centrifuge at full speed and combine both eluates.

Cross-link Removal and Purification

1. Incubate the samples at 65 °C in a water bath for a minimum of 6 h up to overnight (also include the tube with the input DNA).
2. Add 5 μl or 10 μl of a 10 mg/ml RNase solution to "input" or "IP" samples, respectively. Incubate 30 min at 37 °C.
3. Before qPCR, purify the DNA (also the input sample) using the QIAGEN PCR purification kit according to the instructions

of the manufacturer. Incubate the columns after adding the elution buffer 5 min at room temperature, and elute twice with the same 40–50 μl into the same 1.5 ml reaction tube (to obtain a higher DNA amount in the samples, necessary for sequencing library production). Quality of sheared DNA should be controlled by separating the input DNA on a 1.2 % agarose gel. Usually 15 μl of input DNA solution is sufficient to visualize DNA.

4. Store DNA solution at −20 °C until use.

Quality and Quantity Control of Immunoprecipitated DNA

In order to determine quality and quantity of the immunoprecipitated DNA before starting to generate the ChIP-Seq library, aliquots of the ChIP-DNA should be subjected to standard qPCR analyses to determine the specificity of binding to regions known to display occupancy with the respective factors. Instructions for performing ChIP-qPCR analyses can be found in paragraph 3 of Subheading 3.5. An appropriate way to analyze the size distribution of the DNA is to use a DNA Bioanalyzer (Fig. 6). If size and quantity of the enriched DNA is appropriate, it can be further processed using the ChIP-Seq Sample Prep Kit (Illumina) to obtain the libraries for sequencing.

3.3.2 ChIP-Seq Library

(According to the Manufacturer's protocol; Illumina Part # 11257047)

ChIP-Seq Library Generation

End Repair

1. Dilute Klenow DNA polymerase 1:5 with water to a final concentration of 1 U/μl.

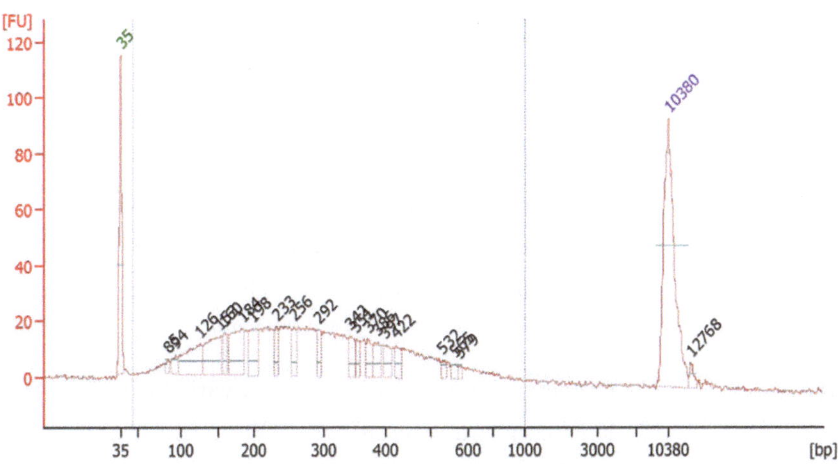

Fig. 6 ChIP-DNA analysis with a DNA Bioanalyzer prior to library generation. In this case the average fragment size is around 233 bp. It should ideally be between 150 and 300 bp. Peaks at 35 and 10,380 bp represent the size standard. *FU* fluorescence units

2. Prepare the following reaction mix:
   (a) 10 ng ChIP-enriched DNA (in 30 μl).
   (b) Water (10 μl).
   (c) T4 DNA ligase buffer with 10 mM ATP (5 μl).
   (d) dNTP mix (2 μl).
   (e) T4 DNA polymerase (1 μl).
   (f) Klenow DNA polymerase (1 μl).
   (g) T4 polynucleotide kinase (1 μl).

   The total volume should be 50 μl.
3. Incubate in the thermo-cycler for 30 min at 20 °C.
4. Purify the products by following the instructions of the QIAquick PCR Purification Kit, and use 34 μl elution buffer.

Addition of "A" Bases to the 3′ End of the DNA Fragments

1. Prepare the following reaction mix:
   (a) DNA sample (34 μl).
   (b) Klenow buffer (5 μl).
   (c) dATP (10 μl).
   (d) Klenow exo (3′ to 5′ exo minus) (1 μl).

   The total volume should be 50 μl.
2. Incubate for 30 min at 37 °C.

   Follow the instructions in the MinElute PCR Purification Kit, use 10 μl of elution buffer.

Adapter to DNA Fragment Ligation

1. Dilute the Adapter oligo mix 1:10 with water to adjust for the smaller quantity of DNA.
2. Prepare the following reaction mix:
   (a) DNA sample (10 μl).
   (b) DNA ligase buffer (15 μl).
   (c) Diluted adapter oligo mix (1 μl).
   (d) DNA ligase (4 μl).

   The total volume should be 30 μl.
3. Incubate for 15 min at room temperature.

   Purify the products by following the instructions in the MinElute PCR Purification Kit, and use 10 μl of elution buffer.

Size Selection of the Library

1. Prepare a 50 ml, 2 % agarose gel with 1x TAE buffer.
2. Final concentration of EtBr 400 ng/ml.
3. Load 500 ng of 100 bp DNA ladder on the gel.
4. Add 3 μl of loading buffer to 10 μl of the DNA from the purified ligation reaction.

5. Load the entire sample on the gel, leaving at least one empty lane between ladder and sample.
6. Run gel at 120 V for 60 min.
7. Visualize on a >320 nm wavelength UV table to avoid damaging the DNA fragments. The 100 bp ladder will allow determining the correct position to cut the library DNA out of the gel. Excise a region of the gel with DNA in the 200 ± 25 bp range a clean scalpel. Document the gel before and after the slice is excised.
8. Cut a slice of the same size from an empty well on the same gel and take this sample through gel purification and PCR. No visible PCR product should be present.
9. Use the QIAGEN Gel Extraction Kit to purify the DNA from the agarose slices and elute DNA in 36 μl.

Enrichment of Adapter-Modified DNA Fragments by PCR

1. Prepare the following PCR reaction mix:
   (a) DNA (36 μl).
   (b) 5× Phusion buffer (10 μl).
   (c) dNTP mix (1.5 μl).
   (d) PCR primer 1.1 (1 μl).
   (e) PCR primer 2.1 (1 μl).
   (f) Phusion polymerase (0.5 μl).
   The total volume should be 50 μl.
2. Amplify using the following PCR protocol:
   (a) 30 s at 98 °C.
   (b) 18 cycles of:
       - 10 s at 98 °C.
       - 30 s at 65 °C.
       - 30 s at 72 °C.
       - 5 min at 72 °C.
   (c) Hold at 4 °C
3. Follow the instructions in the MinElute PCR Purification Kit, and use 15 μl of elution buffer.

ChIP-Seq Library Validation

The amount of starting material is very low (10 ng), and after 18 cycles of PCR, the yield may still be too low to detect on a regular gel, even though it is enough for cluster generation. It is recommended to perform a more sensitive quality control analysis of the sample library using an Agilent Bioanalyzer (Fig. 7) (*see* **Note 14**).

3.3.3 Library Sequencing

If the library has the required quality and quantity, which is the case when the maximum peak at 300–400 bp is more pronounced

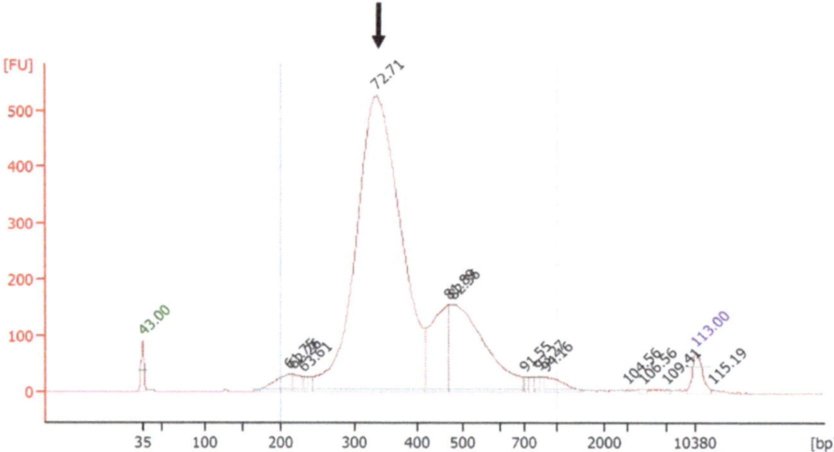

Fig. 7 Analysis of ChIP-enriched DNA after library generation using a DNA Bioanalyzer. The peak representing the library is indicated by an *arrow*. *Peaks* at 35 bp and 10,380 represent the size standard. *FU* fluorescence units

than a side product of the library generation having a size of ~500 bp, the material can be subjected to sequencing with devices from Illumina, e.g., the HighSeq 2000. A sequencing depth of around 20 million reads and sequencing read length from 35 to 50 bp is sufficient to obtain representative results. The processing of the resulting FASTQ files is discussed in the Bioinformatics paragraph and in [62, 65].

3.4 Bioinformatics Analyses of NGS Results

Once you obtained the NGS result files, these will have to be converted and subjected to multiple bioinformatics analyses. Ideally, results of the three different types of analyses suggested here will be integrated in order to generate a global picture of c-MYC-mediated regulation of mRNA, miRNA, and miRNA target expression. An outline of such a bioinformatic analysis is provided in Fig. 8.

The ChIP-, miRNA- and mRNA-Seq reads have to be mapped to the reference genome database. For human genome data, this is currently the hg19 release. Several programs for mapping reads to the genome are available, including ELAND, SOAP31, MAQ32, and RMAP33. Information about these programs can be found at the Illumina homepage (http://www.illumina.com/) or at SEQanswers (http://seqanswers.com/).

3.4.1 ChIP-Seq Bioinformatics

After the alignment to a reference genome, ChIP-Seq data are subjected to peak calling. For this several tools are available, which are discussed in [66], e.g., Find Peaks [67]. After peak calling the results may be viewed using the UCSC genome browser, and the distribution of c-MYC DNA binding in any genomic region can be compared to previously published ChIP-Seq analyses (Fig. 9). Additionally, a

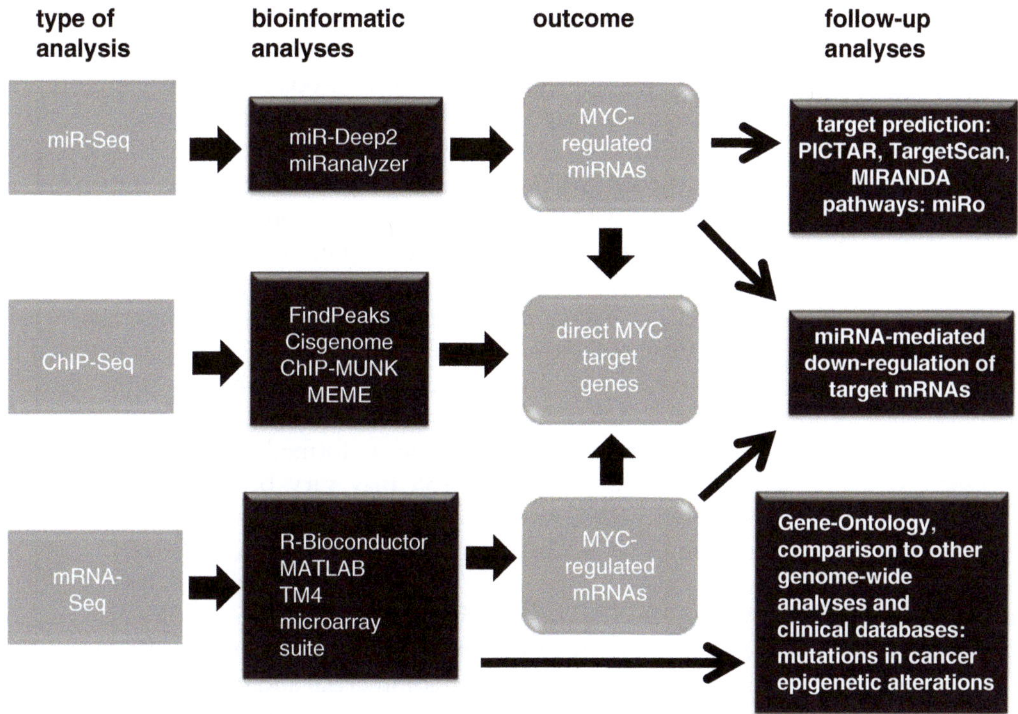

Fig. 8 Bioinformatics characterization of c-MYC-regulated mRNAs, miRNAs, and their targets. Summary of bioinformatics approaches for the comprehensive characterization of MYC-regulated miRNAs. As indicated, the programs and websites Bioconductor [75], CisGenome [76], ChIP-Munk [77], FindPeaks [67], Meme [78], MirDeep2 [79], miRanalyzer [80], miRo [81], and TM4 microarray suite [82] facilitate the analyses of data obtained by the experimental analyses described in the main text and in Fig. 1

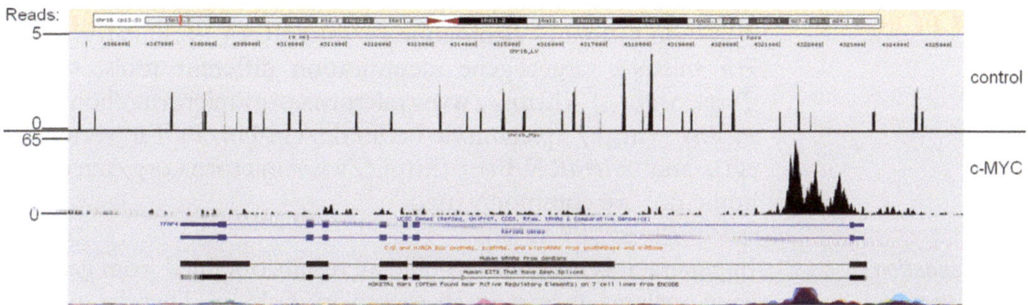

Fig. 9 c-MYC binding to the human AP4 promoter. A representative example of a bioinformatic analysis of ChIP-Seq results using the USCS genome browser. A VSV-tagged c-MYC protein was precipitated using a VSV-specific antibody after induction of a conditional *c-MYC-VSV* allele in a colorectal cancer cell line. *Upper row*: a control ChIP analysis using a VSV-specific antibody after induction of an empty vector by addition of doxycycline. Note that the *y*-axes have different scales in the c-MYC and the control IP analysis. *Lower row*: c-MYC-VSV ChIP-Seq result with the VSV-specific antibody. Enrichment in the first intron of *AP4*, in which the known c-MYC occupied E-boxes are localized (Jackstadt et al., unpublished results)

de novo motif search can be performed with the MEME software [68] to analyze whether the ChIP-Seq analysis resulted in the expected enrichment of the c-MYC E-box motif. Furthermore, the enrichment of neighboring DNA motifs, which may serve as binding sites for cooperating or antagonizing factors, may be identified using the MEME algorithm. In general it is recommended to collaborate with a bioinformatician in order to obtain adequate computational analyses and interpretation of the ChIP-Seq results. For example, it should be evaluated whether promoters that display c-MYC occupancy correspond to promoters of transcripts, which are differentially expressed after activation of c-MYC as detected by miRNA- and mRNA-Seq. Furthermore, an analysis of the distribution of E-boxes in the promoters of induced versus repressed c-MYC-regulated genes can be performed. Furthermore, the distance of E-boxes to the TSS may vary between induced versus repressed genes. An example of a bioinformatics workflow for the analysis of NGS ChIP-Seq results is also provided in [69].

3.4.2 mRNA-Seq Bioinformatics

After RNA-Seq the resulting compressed FASTQ files may be subjected to analyses by several software packages. For nonspecialists in bioinformatics, different commercial and open source programs are available [70]. For example, uploading of the FASTQ files in the Galaxy software generates a list of genes annotated to RefSeq within a few hours [71].

3.4.3 miR-Seq Bioinformatics

In order to analyze the miR-Seq results, several tools can be used. One of this is *miRDeep* algorithm, which allows analyzing deep sequencing data obtained by Illumina/HiSeq and other NGS platforms (http://www.mdc-berlin.de/de/research/research_teams/systems_biology_of_gene_regulatory_elements/projects/miRDeep/index.html). A similar tool is the miRanalyser (http://web.bioinformatics.cicbiogune.es/microRNA/miRanalyser.php). For miRNA target gene identification different tools, such as *TargetScan* (http://www.microrna.org/microrna/home.do), *PicTar* (http://pictar.mdc-berlin.de/cgi-bin/PicTar_vertebrate.cgi), and *microRNA.org* (http://www.microrna.org/microrna/home.do) are commonly used.

3.5 Validation of NGS Results

In general it is necessary to validate results obtained from genome-wide analyses at least in an exemplary fashion. Thereby, one can confirm how robust the observed regulations are and exclude potentially false-positive target gene candidates from further analyses. Especially, if further, more laborious downstream analyses are planned, results should be confirmed using alternative methods on a single gene basis. Besides the analyses of already known c-MYC targets, newly discovered targets should be confirmed using an independent method. If the results for these evaluations are positive, the genome-wide data may then be subjected to analysis tools

(Gene ontology analysis or DAVID) to discover or confirm different signatures or pathways, which may be regulated by c-MYC.

3.5.1 qPCR Validation of c-MYC-Regulated Transcripts

For RNA extraction we have used various commercial kits, which are based on the Chomchynzky RNA extraction protocol (RNAgents; Promega). For reverse transcription, the Verso cDNA Kit (Thermo Scientific) generally resulted in reliable qPCR results, also in the case of less abundant mRNAs. The following protocol uses the Fast SYBR Green Master Mix (Ambion) in combination with the LightCycler 480 (Roche) (*see* **Note 15**). Since cDNA concentrations may vary between the samples that will be compared, cDNA amounts should be adjusted after qPCR determination of housekeeping gene expression, such as *β-actin* or *ELF1α* (do not use *GAPDH* for normalizing, since it is a c-MYC target gene). Ideally all samples should be in the range of ±1 crossing point (CP) (*see* **Note 16**). Biological triplicates samples should be analyzed.

qPCR Analysis

1. Thaw the SYBR Green master mix at room temperature and mix it gently.

   Prepare a 10 μM dilution of each primer. Dilute the cDNA in a range of 1:3 to 1:10, depending on the expected abundance of the genes of interest.
2. qPCR master mix preparation (per reaction) (*see* **Note 17**):
   (a) 7.5 μl 2× SYBR Green master mix.
   (b) 1 μl diluted cDNA as starting point.
   (c) 5 μl H_2O

   15 μl total volume.
3. Mix gently without vortexing.
4. Distribute the samples in a 96 multiwell plate.
5. Add primers:
   (a) 0.75 μl forward primer (10 μM).
   (b) 0.75 μl reverse primer (10 μM).
6. The thermal-cycling condition for the LightCycler 480 are:
   (a) 20 s at 95 °C.
   (b) 40 cycles:
      - 3 s at 95 °C (denaturation).
      - 30 s at 60 °C (includes annealing and extension) (*see* **Note 18**).

The crossing point (CP) is the cycle number at which the increase in the fluorescence signal generated by the PCR product reaches an exponential phase. The CPs are used to calculate the fold change by using the second-derivative maximum method [73]. This method integrates the efficiency of the individual primer pairs, which is determined using a logarithmic dilution of cDNA to generate a linear standard curve with the CPs plotted against log of the template concentration. The primer pair efficiency can be calculated as follows: Efficiency $E = 10^{(-1/\text{slope})}$ [73]. The fold induction can be calculated as

$$\frac{(E_{\text{DNA}_x})^{\Delta\text{CP(sample-control)}}}{(E_{\text{ref}})^{\Delta\text{CP(sample-control)}}} = \text{fold induction}$$

For most primer pairs, an efficiency of ~1.9 in a CP range of 15–28 is observed. Crossing points above 28 are usually associated with lower PCR primer efficiencies. Therefore, primer pair efficiencies for CPs >28 should be experimentally deduced from qPCRs of serial template dilutions and are usually between 1.5 and 1.8. Furthermore, the primer pair specificity should also be confirmed by gel electrophoresis of the PCR product. Ideally, only one band of the expected length, which corresponds to a single peak in the melting curve, should be detected (besides a band corresponding to primer dimers). Because transcriptional inductions or repressions of c-MYC-regulated targets are often in the range of twofold, it is particularly important to analyze biological triplicates of cDNAs, which have been adjusted for equal concentration by qPCR of a housekeeping mRNA.

Primers pairs for established murine and human c-MYC targets can be found in [17, 53, 59, 61]. For the design of new qPCR primer pairs, several online tools are available that often succeed in predicting primers that generate a single specific product using cDNA as a template (and no/minimal primer dimers), e.g., Primer3. In order to avoid amplification from contaminating genomic DNA, PCR primers should ideally span introns. Furthermore, maximal PCR product generation of oligo-dT-primed cDNA can be achieved if primer pairs are directed against exons in or close to the 3′-UTR. Ideally, for use with a LightCycler device, a melting temperature of 60 °C ± 3 °C and a primer length of 20 bp with a GC content of 45–65 % are recommended. PCR products should have a length of 80–300 bp.

3.5.2 qPCR Validation of c-MYC-Regulated miRNAs

The expression of c-MYC-regulated miRNAs may be determined either at the level of the primary miRNA (pri-miRNA) transcript or of the mature miRNA. The former is analyzed essentially as described in the mRNA protocol and preferentially in combination with exon-binding primers that flank the miRNA in the pri-miRNA. This allows demonstrating the direct transcriptional regulation of the miRNA excluding direct or indirect effects of c-MYC on the processing of the respective miRNA. In some cases this may generate negative results since the expression of the miRNA, as detected by miR-Seq, may not be determined by the level of pri-miRNA expression but by regulated processing. Furthermore, many pri-miRNA/miRNA assignments are not complete and have not been experimentally validated. Additional, intragenic promoters may exist, which drive the expression of uncharacterized pri-miRNAs. In order to validate differential expression of mature miRNAs, a stem-loop PCR followed by a Taqman assay (e.g., from Applied Biosystems) or a LNA-PCR (e.g., from Exiqon) should be applied. miRNA-specific primers for miR-BASE listed miRNAs can be ordered and should be used in conjunction with the respective chemicals and enzymes provided by these companies.

3.5.3 qChIP Validation of Occupancy by c-MYC

qChIP should be used for the validation of ChIP-Seq results and is performed similarly to the qPCR analysis of mRNA. However, since promoters often contain GC-rich regions, the design of functional qChIP primers is often difficult. Primer pairs for a qChIP analysis of putative c-MYC bound regions should be positioned around the predicted E-boxes following similar rules as used for mRNA primer design. Primer pairs which can be used for negative and positive controls have been described for human and murine cell systems [53, 59, 74]. Normalization primers: correspond to, for example, *ELF1α*, a region on 16q22 or *AchR* [17, 53, 74] that are also used to adapt volumes for equal DNA input. In an initial PCR, positive and negative controls should be analyzed using 1 μl of each sample. For quantification of DNA, enrichment crossing points should be between 25 and 30. To calculate the binding of c-MYC to the DNA region analyzed, such enrichment compared to the isotype control IP is commonly expressed as percentage of input. The calculation should be performed as given below [74]:

$$2^{(\text{CPInput})-(\text{CPIP})} = \%\,\text{Input}$$

The following example shows the qChIP result prior to performing the ChIP-Seq analysis to verify the enrichment on a genomic region expected to bind to c-MYC (Fig. 10).

4 Notes

1. Lower amounts can result in inefficient ligation and low yield.
2. The High Pure RNA Isolation Kit (Roche) and the TRIzol RNA Isolation Reagents (Ambion) are suitable.
3. As an alternative method for users who do not have access to an Agilent Technologies 2100 Bioanalyzer or a similar instrument, the RNA can be separated on a formaldehyde 1 % agarose gel, and the integrity of RNA can be assessed after staining

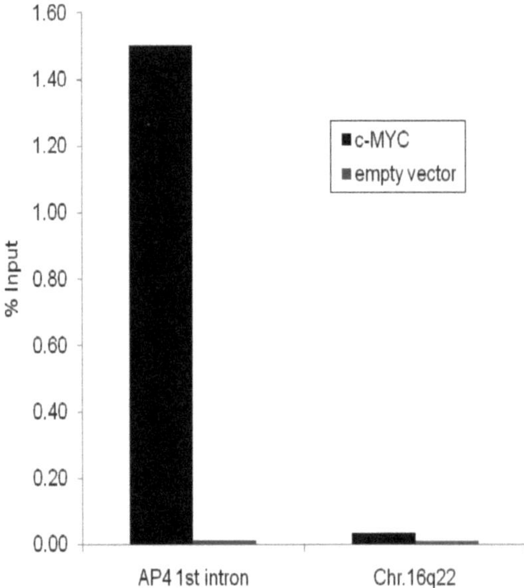

Fig. 10 qPCR analysis of c-MYC ChIP of the first intron of the *AP4* gene. The *AP4* gene is known to contain several E-boxes occupied by c-MYC [53]. ChIP analysis with a VSV-specific antibody was performed using cells treated with doxycycline either ectopically expressing a VSV-tagged c-MYC protein or harboring an empty vector. PCR of a genomic region on chromosome 16q22 devoid of E-boxes in a region 3 kbp up- or downstream served as negative control. The strong enrichment of c-MYC-bound DNA in the AP4 promoter was also seen in the ChIP-Seq analyses shown in Fig. 9 (Jackstadt et al., unpublished results)

with ethidium bromide. RNA with high quality shows a 28S rRNA and 18S rRNA band at 4.5 kbp and 1.9 kbp without any degradations.

4. Ensure that the total RNA was purified using a method that retains small RNA. The use of high-quality RNA is essential to generate reproducible results. RNA quality should be determined using the Agilent Bioanalyzer. The RNA Integrity Number (RIN) value should be eight or greater.
5. Prepare 10 % extra reagent if you have multiple samples.
6. In order to avoid false-positive signals caused by unspecific antibody interactions, the use of highly-specific antibodies directed against epitopes, which were added to the c-MYC protein, is recommended.
7. Several possibilities of performing control experiments and other ChIP analysis related issues are also reviewed in [62].
8. These related analyses are also described in another chapter of this book (*see* Chapter 9) and may be combined with the analyses described here.

9. It is recommended to use at least the IgG IP as a control.
10. Be aware of the expiration date. Expired formaldehyde doesn't work well.
11. Handle formaldehyde under a fume hood.
12. Precise timing is important in this step.
13. If possible include an analysis with an isotype matched, unspecific antibody IP for each antibody.
14. Alternative method: If you do not have access to an Agilent Technologies 2100 Bioanalyzer or similar instruments, you may try using a sensitive dsDNA measurement assay such as the Quant-iT dsDNA HS Assay Kit, for use with the Qubit fluorometer (Invitrogen). Note that this will not allow you to check the size and purity of your sample. Do not use an OD260/280 ratio for concentration measurements, since this will not distinguish dsDNA from primers and therefore cannot be used to validate the library.
15. If a different Master Mix or qPCR machine is used, the protocol needs to be adapted accordingly.
16. Some authors recommend to use up to three transcripts as normalization controls [72]. If a cDNA differs strongly in the CP, a second PCR should be performed with adapted cDNA input amounts to determine the final cDNA amount that results in roughly similar CPs as obtained for the other sample.
17. To compensate pipetting errors, prepare an excess of 10 % master mix.
18. At the end of the program, a melting curve analysis to evaluate the primer specificity should be included. Although this should have been confirmed prior to the final analysis, the individual cDNA and PCR conditions may affect the performance of the primer pair.

Acknowledgments

Work in the Hermeking lab is supported by the German–Israeli Science Foundation (GIF), the Rudolf-Bartling-Stiftung, the Deutsche Krebshilfe, the German Cancer Consortium (DKTK), and the Deutsche Forschungsgemeinschaft (DFG).

Note added in proof

While this chapters was in press the authors published a completed analysis of mRNH expression and ONH building further illustrating the approaders drescribed here [83]

References

1. Dang CV (2012) MYC on the path to cancer. Cell 149:22–35
2. Meyer N, Penn LZ (2008) Reflecting on 25 years with MYC. Nat Rev Cancer 8:976–990
3. Thorley-Lawson DA, Allday MJ (2008) The curious case of the tumour virus: 50 years of Burkitt's lymphoma. Nat Rev Microbiol 6:913–924
4. Nesbit CE, Tersak JM, Prochownik EV (1999) MYC oncogenes and human neoplastic disease. Oncogene 18:3004–3016
5. Marcu KB, Bossone SA, Patel AJ (1992) myc function and regulation. Annu Rev Biochem 61:809–860
6. Chappell SA, LeQuesne JP, Paulin FE, deSchoolmeester ML, Stoneley M, Soutar RL, Ralston SH, Helfrich MH, Willis AE (2000) A mutation in the c-myc-IRES leads to enhanced internal ribosome entry in multiple myeloma: a novel mechanism of oncogene de-regulation. Oncogene 19:4437–4440
7. Albert T, Urlbauer B, Kohlhuber F, Hammersen B, Eick D (1994) Ongoing mutations in the N-terminal domain of c-Myc affect transactivation in Burkitt's lymphoma cell lines. Oncogene 9:759–763
8. Salghetti SE, Kim SY, Tansey WP (1999) Destruction of Myc by ubiquitin-mediated proteolysis: cancer-associated and transforming mutations stabilize Myc. EMBO J 18:717–726
9. He TC, Sparks AB, Rago C, Hermeking H, Zawel L, da Costa LT, Morin PJ, Vogelstein B, Kinzler KW (1998) Identification of c-MYC as a target of the APC pathway. Science 281:1509–1512
10. van de Wetering M, Sancho E, Verweij C, de Lau W, Oving I, Hurlstone A, van der Horn K, Batlle E, Coudreuse D, Haramis AP, Tjon-Pon-Fong M, Moerer P, van den Born M, Soete G, Pals S, Eilers M, Medema R, Clevers H (2002) The beta-catenin/TCF-4 complex imposes a crypt progenitor phenotype on colorectal cancer cells. Cell 111:241–250
11. Sansom OJ, Reed KR, Hayes AJ, Ireland H, Brinkmann H, Newton IP, Batlle E, Simon-Assmann P, Clevers H, Nathke IS, Clarke AR, Winton DJ (2004) Loss of Apc in vivo immediately perturbs Wnt signaling, differentiation, and migration. Genes Dev 18:1385–1390
12. Murphy DJ, Junttila MR, Pouyet L, Karnezis A, Shchors K, Bui DA, Brown-Swigart L, Johnson L, Evan GI (2008) Distinct thresholds govern Myc's biological output in vivo. Cancer Cell 14:447–457
13. Hermeking H, Eick D (1994) Mediation of c-Myc-induced apoptosis by p53. Science 265:2091–2093
14. Vafa O, Wade M, Kern S, Beeche M, Pandita TK, Hampton GM, Wahl GM (2002) c-Myc can induce DNA damage, increase reactive oxygen species, and mitigate p53 function: a mechanism for oncogene-induced genetic instability. Mol Cell 9:1031–1044
15. Dominguez-Sola D, Ying CY, Grandori C, Ruggiero L, Chen B, Li M, Galloway DA, Gu W, Gautier J, Dalla-Favera R (2007) Non-transcriptional control of DNA replication by c-Myc. Nature 448:445–451
16. Campaner S, Doni M, Verrecchia A, Faga G, Bianchi L, Amati B (2010) Myc, Cdk2 and cellular senescence: old players, new game. Cell Cycle 9:3655–3661
17. Menssen A, Hydbring P, Kapelle K, Vervoorts J, Diebold J, Luscher B, Larsson LG, Hermeking H (2012) The c-MYC oncoprotein, the NAMPT enzyme, the SIRT1-inhibitor DBC1, and the SIRT1 deacetylase form a positive feedback loop. Proc Natl Acad Sci USA 109:E187–E196
18. Eilers M, Eisenman RN (2008) Myc's broad reach. Genes Dev 22:2755–2766
19. Jung P, Hermeking H (2009) The c-MYC-AP4-p21 cascade. Cell Cycle 8:982–989
20. Cowling VH, Cole MD (2006) Mechanism of transcriptional activation by the Myc oncoproteins. Semin Cancer Biol 16:242–252
21. Rahl PB, Lin CY, Seila AC, Flynn RA, McCuine S, Burge CB, Sharp PA, Young RA (2010) c-Myc regulates transcriptional pause release. Cell 141:432–445
22. Adhikary S, Eilers M (2005) Transcriptional regulation and transformation by Myc proteins. Nat Rev Mol Cell Biol 6:635–645
23. Ayer DE, Lawrence QA, Eisenman RN (1995) Mad-Max transcriptional repression is mediated by ternary complex formation with mammalian homologs of yeast repressor Sin3. Cell 80:767–776
24. Peukert K, Staller P, Schneider A, Carmichael G, Hanel F, Eilers M (1997) An alternative pathway for gene regulation by Myc. EMBO J 16:5672–5686
25. Herold S, Wanzel M, Beuger V, Frohme C, Beul D, Hillukkala T, Syvaoja J, Saluz HP, Haenel F, Eilers M (2002) Negative regulation of the mammalian UV response by Myc through association with Miz-1. Mol Cell 10:509–521
26. Mao DY, Watson JD, Yan PS, Barsyte-Lovejoy D, Khosravi F, Wong WW, Farnham PJ, Huang TH, Penn LZ (2003) Analysis of Myc bound loci identified by CpG island arrays shows that Max is essential for Myc-dependent repression. Curr Biol 13:882–886

27. Staller P, Peukert K, Kiermaier A, Seoane J, Lukas J, Karsunky H, Moroy T, Bartek J, Massague J, Hanel F, Eilers M (2001) Repression of p15INK4b expression by Myc through association with Miz-1. Nat Cell Biol 3:392–399
28. Smale ST, Baltimore D (1989) The "initiator" as a transcription control element. Cell 57:103–113
29. Patel JH, Loboda AP, Showe MK, Showe LC, McMahon SB (2004) Analysis of genomic targets reveals complex functions of MYC. Nat Rev Cancer 4:562–568
30. Wu CH, Sahoo D, Arvanitis C, Bradon N, Dill DL, Felsher DW (2008) Combined analysis of murine and human microarrays and ChIP analysis reveals genes associated with the ability of MYC to maintain tumorigenesis. PLoS Genet 4:e1000090
31. Zhou L, Picard D, Ra YS, Li M, Northcott PA, Hu Y, Stearns D, Hawkins C, Taylor MD, Rutka J, Der SD, Huang A (2010) Silencing of thrombospondin-1 is critical for myc-induced metastatic phenotypes in medulloblastoma. Cancer Res 70:8199–8210
32. Varlakhanova N, Cotterman R, Bradnam K, Korf I, Knoepfler PS (2011) Myc and Miz-1 have coordinate genomic functions including targeting Hox genes in human embryonic stem cells. Epigenetics Chromatin 4:20
33. Kidder BL, Yang J, Palmer S (2008) Stat3 and c-Myc genome-wide promoter occupancy in embryonic stem cells. PLoS One 3:e3932
34. Seitz V, Butzhammer P, Hirsch B, Hecht J, Gutgemann I, Ehlers A, Lenze D, Oker E, Sommerfeld A, von der Wall E, Konig C, Zinser C, Spang R, Hummel M (2011) Deep sequencing of MYC DNA-binding sites in Burkitt lymphoma. PLoS One 6:e26837
35. Perna D, Faga G, Verrecchia A, Gorski MM, Barozzi I, Narang V, Khng J, Lim KC, Sung WK, Sanges R, Stupka E, Oskarsson T, Trumpp A, Wei CL, Muller H, Amati B (2012) Genome-wide mapping of Myc binding and gene regulation in serum-stimulated fibroblasts. Oncogene 31:1695–1709
36. Ji H, Wu G, Zhan X, Nolan A, Koh C, De Marzo A, Doan HM, Fan J, Cheadle C, Fallahi M, Cleveland JL, Dang CV, Zeller KI (2011) Cell-type independent MYC target genes reveal a primordial signature involved in biomass accumulation. PLoS One 6:e26057
37. Li Z, Van Calcar S, Qu C, Cavenee WK, Zhang MQ, Ren B (2003) A global transcriptional regulatory role for c-Myc in Burkitt's lymphoma cells. Proc Natl Acad Sci USA 100:8164–8169
38. Orian A, Grewal SS, Knoepfler PS, Edgar BA, Parkhurst SM, Eisenman RN (2005) Genomic binding and transcriptional regulation by the Drosophila Myc and Mnt transcription factors. Cold Spring Harb Symp Quant Biol 70:299–307
39. Zeller KI, Zhao X, Lee CW, Chiu KP, Yao F, Yustein JT, Ooi HS, Orlov YL, Shahab A, Yong HC, Fu Y, Weng Z, Kuznetsov VA, Sung WK, Ruan Y, Dang CV, Wei CL (2006) Global mapping of c-Myc binding sites and target gene networks in human B cells. Proc Natl Acad Sci USA 103:17834–17839
40. Varlakhanova NV, Knoepfler PS (2009) Acting locally and globally: Myc's ever-expanding roles on chromatin. Cancer Res 69: 7487–7490
41. Knoepfler PS (2007) Myc goes global: new tricks for an old oncogene. Cancer Res 67:5061–5063
42. Knoepfler PS, Zhang XY, Cheng PF, Gafken PR, McMahon SB, Eisenman RN (2006) Myc influences global chromatin structure. EMBO J 25:2723–2734
43. Zeller KI, Jegga AG, Aronow BJ, O'Donnell KA, Dang CV (2003) An integrated database of genes responsive to the Myc oncogenic transcription factor: identification of direct genomic targets. Genome Biol 4:R69
44. Lujambio A, Lowe SW (2012) The microcosmos of cancer. Nature 482:347–355
45. Bui TV, Mendell JT (2010) Myc: Maestro of MicroRNAs. Genes Cancer 1:568–575
46. Frenzel A, Loven J, Henriksson MA (2010) Targeting MYC-regulated miRNAs to combat cancer. Genes Cancer 1:660–667
47. Robertus JL, Kluiver J, Weggemans C, Harms G, Reijmers RM, Swart Y, Kok K, Rosati S, Schuuring E, van Imhoff G, Pals ST, Kluin P, van den Berg A (2010) MiRNA profiling in B non-Hodgkin lymphoma: a MYC-related miRNA profile characterizes Burkitt lymphoma. Br J Haematol 149:896–899
48. Kim JW, Mori S, Nevins JR (2010) Myc-induced microRNAs integrate Myc-mediated cell proliferation and cell fate. Cancer Res 70:4820–4828
49. Mu P, Han YC, Betel D, Yao E, Squatrito M, Ogrodowski P, de Stanchina E, D'Andrea A, Sander C, Ventura A (2009) Genetic dissection of the miR-17 92 cluster of microRNAs in Myc-induced B-cell lymphomas. Genes Dev 23:2806–2811
50. Lin CH, Jackson AL, Guo J, Linsley PS, Eisenman RN (2009) Myc-regulated microRNAs attenuate embryonic stem cell differentiation. EMBO J 28:3157–3170
51. Watson JD, Oster SK, Shago M, Khosravi F, Penn LZ (2002) Identifying genes regulated in a Myc-dependent manner. J Biol Chem 277:36921–36930

52. Eilers M, Picard D, Yamamoto KR, Bishop JM (1989) Chimaeras of myc oncoprotein and steroid receptors cause hormone-dependent transformation of cells. Nature 340:66–68
53. Jung P, Menssen A, Mayr D, Hermeking H (2008) AP4 encodes a c-MYC-inducible repressor of p21. Proc Natl Acad Sci USA 105:15046–15051
54. Mateyak MK, Obaya AJ, Adachi S, Sedivy JM (1997) Phenotypes of c-Myc-deficient rat fibroblasts isolated by targeted homologous recombination. Cell Growth Differ 8:1039–1048
55. Han H, Nutiu R, Moffat J, Blencowe BJ (2011) SnapShot: high-throughput sequencing applications. Cell 146:1044, 1044 e1041–1042
56. Metzker ML (2010) Sequencing technologies—the next generation. Nat Rev Genet 11:31–46
57. Wang Z, Gerstein M, Snyder M (2009) RNA-Seq: a revolutionary tool for transcriptomics. Nat Rev Genet 10:57–63
58. Chen C, Ridzon DA, Broomer AJ, Zhou Z, Lee DH, Nguyen JT, Barbisin M, Xu NL, Mahuvakar VR, Andersen MR, Lao KQ, Livak KJ, Guegler KJ (2005) Real-time quantification of microRNAs by stem-loop RT-PCR. Nucleic Acids Res 33:e179
59. Menssen A, Hermeking H (2002) Characterization of the c-MYC-regulated transcriptome by SAGE: identification and analysis of c-MYC target genes. Proc Natl Acad Sci USA 99:6274–6279
60. Alexandrow MG, Moses HL (1995) Transforming growth factor beta and cell cycle regulation. Cancer Res 55:1452–1457
61. Menssen A, Epanchintsev A, Lodygin D, Rezaei N, Jung P, Verdoodt B, Diebold J, Hermeking H (2007) c-MYC delays prometaphase by direct transactivation of MAD2 and BubR1: identification of mechanisms underlying c-MYC-induced DNA damage and chromosomal instability. Cell Cycle 6:339–352
62. Park PJ (2009) ChIP-seq: advantages and challenges of a maturing technology. Nat Rev Genet 10:669–680
63. Fernandez PC, Frank SR, Wang L, Schroeder M, Liu S, Greene J, Cocito A, Amati B (2003) Genomic targets of the human c-Myc protein. Genes Dev 17:1115–1129
64. Martinato F, Cesaroni M, Amati B, Guccione E (2008) Analysis of Myc-induced histone modifications on target chromatin. PLoS One 3:e3650
65. Valouev A, Johnson DS, Sundquist A, Medina C, Anton E, Batzoglou S, Myers RM, Sidow A (2008) Genome-wide analysis of transcription factor binding sites based on ChIP-Seq data. Nat Methods 5:829–834
66. Zhu JY, Sun Y, Wang ZY (2012) Genome-wide identification of transcription factor-binding sites in plants using chromatin immunoprecipitation followed by microarray (ChIP-chip) or sequencing (ChIP-seq). Methods Mol Biol 876:173–188
67. Fejes AP, Robertson G, Bilenky M, Varhol R, Bainbridge M, Jones SJ (2008) FindPeaks 3.1: a tool for identifying areas of enrichment from massively parallel short-read sequencing technology. Bioinformatics 24:1729–1730
68. Bailey TL, Boden M, Buske FA, Frith M, Grant CE, Clementi L, Ren J, Li WW, Noble WS (2009) MEME SUITE: tools for motif discovery and searching. Nucleic Acids Res 37:W202–W208
69. Cantacessi C, Jex AR, Hall RS, Young ND, Campbell BE, Joachim A, Nolan MJ, Abubucker S, Sternberg PW, Ranganathan S, Mitreva M, Gasser RB (2010) A practical, bioinformatic workflow system for large data sets generated by next-generation sequencing. Nucleic Acids Res 38:e171
70. Pepke S, Wold B, Mortazavi A (2009) Computation for ChIP-seq and RNA-seq studies. Nat Methods 6:S22–S32
71. Goecks J, Nekrutenko A, Taylor J (2010) Galaxy: a comprehensive approach for supporting accessible, reproducible, and transparent computational research in the life sciences. Genome Biol 11:R86
72. Pfaffl MW (2010) The ongoing evolution of qPCR. Methods 50:215–216
73. Pfaffl MW (2001) A new mathematical model for relative quantification in real-time RT-PCR. Nucleic Acids Res 29:e45
74. Frank SR, Schroeder M, Fernandez P, Taubert S, Amati B (2001) Binding of c-Myc to chromatin mediates mitogen-induced acetylation of histone H4 and gene activation. Genes Dev 15:2069–2082
75. Reimers M, Carey VJ (2006) Bioconductor: an open source framework for bioinformatics and computational biology. Methods Enzymol 411:119–134
76. Ji H, Jiang H, Ma W, Johnson DS, Myers RM, Wong WH (2008) An integrated software system for analyzing ChIP-chip and ChIP-seq data. Nat Biotechnol 26:1293–1300
77. Kulakovskiy IV, Boeva VA, Favorov AV, Makeev VJ (2010) Deep and wide digging for binding motifs in ChIP-Seq data. Bioinformatics 26:2622–2623
78. Bailey TL (2002) Discovering novel sequence motifs with MEME. Curr Protoc Bioinformatics Chapter 2: Unit 2 4

79. Mackowiak SD (2011) Identification of novel and known miRNAs in deep-sequencing data with miRDeep2. Curr Protoc Bioinformatics Chapter 12:Unit 12 10
80. Hackenberg M, Rodriguez-Ezpeleta N, Aransay AM (2011) miRanalyzer: an update on the detection and analysis of microRNAs in high-throughput sequencing experiments. Nucleic Acids Res 39:W132–W138
81. Lagana A, Forte S, Giudice A, Arena MR, Puglisi PL, Giugno R, Pulvirenti A, Shasha D, Ferro A (2009) miRo: a miRNA knowledge base. Database (Oxford) 2009:bap008
82. Saeed AI, Bhagabati NK, Braisted JC, Liang W, Sharov V, Howe EA, Li J, Thiagarajan M, White JA, Quackenbush J (2006) TM4 microarray software suite. Methods Enzymol 411:134–193
83. Jackstadt R, Röh S, Neumann J, Jung P, Hoffmann R, Horst D, Berens C, Bornkamm GW, Kirchner T, Menssen A, Hermeking, H (2013) AP4 is a mediator of epithelial-mesenchymal transition and metastasis in colorectal cancer. J Exp Med 210 (7):1331–1350

Chapter 12

A High-Throughput siRNA Screening Platform to Identify MYC-Synthetic Lethal Genes as Candidate Therapeutic Targets

Carla Grandori

Abstract

Targeted therapeutics toward specific genes and pathways represent the future of oncological treatments. However, several commonly activated oncogenes, such as MYC, have proven difficult to target by pharmacological agents. To broaden the menu of potentially druggable therapeutic targets, we describe a method to detect genes essential for the survival of MYC overexpressing cells, which we will refer to as MYC-synthetic lethal genes (MYC-SL) (Toyoshima et al., Proc Natl Acad Sci USA 109:9545–9550, 2012). These genes represent candidate targets for drug development to be utilized for MYC-driven cancers as well as probes to further our understanding of the biology of MYC-driven tumorigenesis. The discovery platform includes the following components: (1) an isogenic cell system that enables overexpression of MYC without oncogene-induced senescence (OIS) response (Benanti and Galloway, Mol Cell Biol 24:2842–2852, 2004; Benanti et al., Mol Cancer Res 5:1181–1189, 2007); (2) arrayed siRNA libraries targeting individual genes; (3) automated laboratory equipment for dispensing of cells, siRNAs, and readout assays; and (4) bioinformatics and software for data mining and visualization. This flexible platform can be readily applied to other oncogenes or tumor suppressor genes.

Key words siRNA screens, High-throughput screening, RNA interference, Synthetic lethal genes, Therapeutic targets, Cancer, MYC, Oncogenes

1 Introduction

Inhibition of MYC is highly effective to block tumor growth as demonstrated in several transgenic mouse models [1–3]. However, development of pharmacologic inhibitors of MYC has been difficult as the MYC family of genes encode for DNA-binding proteins that are required for both normal and cancerous cell proliferation (*see* **Note 1**). Driver oncogenic mutations, such as deregulated expression of MYC, while enabling malignant phenotypes in cancer cells also confer unique vulnerabilities, which can be exploited for therapeutic purposes. MYC overexpressing cells indeed exhibit

Laura Soucek and Nicole M. Sodir (eds.), *The Myc Gene: Methods and Protocols*, Methods in Molecular Biology, vol. 1012, DOI 10.1007/978-1-62703-429-6_12,

high sensitivity to apoptotic stimuli [4, 5], increased metabolic requirements [6–8], accelerated DNA synthesis, and genomic instability [9–16]. These properties of MYC overexpressing cells represent ways to tackle MYC-driven cancers, while potentially sparing normal cells. Utilizing a functional genomics approach, we recently identified genes whose inhibition reduced cell viability only in the context of MYC overexpression, referred to as "MYC-synthetic lethal" (MYC-SL) genes. The concept to exploit synthetic lethal interactions with cancer pathways was first proposed by Hartwell and Friend [17]. Only recently, have the discovery of RNA interference and the development of siRNA/shRNA libraries targeting mammalian genomes made it possible to systematically identify synthetic lethal interactions in a broad variety of mammalian cells [18]. There are two major types of RNA interference (RNAi) screens to identify such gene targets, one referred to as the "pooled" lentiviral approach, the other, the "arrayed" high-throughput siRNA screening method. The first approach involves transduction of cells with a pool of lentiviral vectors expressing short hairpins toward broad collections of genes, including as many as 15,000–30,000 genes [19–21]. This method, which in principle enables exploration of the complete transcriptome, can achieve long-term gene knockdown and does not require lab automation, but also has several limitations, including the inherent low sensitivity of negative selection screens (*see* **Note 2**). In contrast, the arrayed or one-gene-per-well siRNA approach, which relies on the parallel measure of the effect of individual gene knockdown in microtiter plates, has increased sensitivity, and it enables a wide variety of readouts. However, knockdown is transient and this approach is limited to cells with high siRNA transfection efficiency (*see* **Note 3**). Currently, siRNA screens are best performed in an equipped high-throughput screening (HTS) facility equipped with high-precision automated liquid handling and fast-acquisition readout equipment that can accommodate processing of dozens to hundreds of microtiter plates. It is important to consider that the initial costs associated with HTS are largely offset by the knowledge gained in a very short time. For example, a typical screen can be accomplished in a week with a few hours of robotic equipment and, as every step is miniaturized, with a minimal amount of reagents. To accomplish a similar project manually would likely require many months of a full-time employee and a much larger consumption of expensive reagents and materials.

Here, we describe a platform we recently employed to identify MYC-SL genes from a collection of ~3,300 druggable genes [22]. Two other groups have recently reported the identification of MYC-SL genes in different cell systems [23, 24] and the reader should consider other cell systems and RNAi methods. Also, a detailed discussion of the nature of MYC-synthetic lethal interactions will be reported elsewhere (Cermelli, S., Jang, IS, Bernard, B., B. and Grandori C., manuscript in preparation). The accuracy of the screening platform we employed was demonstrated by the high confirmation rate of the "hits" (>90 %), which we attribute to both the genetic and epigenetic stability of the primary cells used in the screen, the robustness of the HTS method and that each gene was tested in triplicate. A schematic outline of the screening platform as well as the confirmation, prioritization, and validation strategy is outlined in Fig. 1 and detailed below.

2 Materials

2.1 Cells, Media, and Reagents

1. Human foreskin fibroblasts (HFFs) with and without c-MYC overexpression are employed as a discovery isogenic system because activated oncogenes in low passage HFFs do not trigger oncogene-induced senescence. This feature enables analysis in HFFs of transforming phenotypes and gene expression signatures following introduction of oncogenes [25, 26]. Furthermore, we have determined that MYC-related phenotypes and gene expression signatures can translate to human cancers ([27] and CG unpublished observations). HFFs are obtained as de-identified samples from otherwise discarded circumcisions performed at birth (under "nonhuman subject" IRB-approved protocol). For other isogenic cells systems that can be employed, *see* **Note 4**.
2. Cell culture media: HFFs are cultured in high glucose Dulbecco's Modified Eagle Medium with the addition of 10 % fetal bovine serum and antibiotics (*see* **Note 5**).
3. Microtiter plates: 384 well are available from several vendors and should be chosen depending on the liquid handler and readout equipment. We employed Matrix 384 flat bottom black plates (Matrix Technologies).
4. Transfection reagent: Lipofectamine RNAiMAX (Invitrogen) or DharmaFECT I (Dharmacon RNAi technology) are both suitable for HFF siRNA transfection (*see* **Note 6**).
5. Dilution reagent: Opti-MEM (Invitrogen) is used for dilution of siRNAs and transfection reagent.

2.2 Readout Assays

1. CellTiter-Glo (Promega), a luminescence cell viability assay based on ATP production.
2. Caspase-Glo assay systems (Promega) for detection of cleaved caspases if apoptosis is the desired end point.
3. Custom designed immunofluorescence-based assay utilizing specific antibodies, i.e., high content screening (*see* **Note 7**).

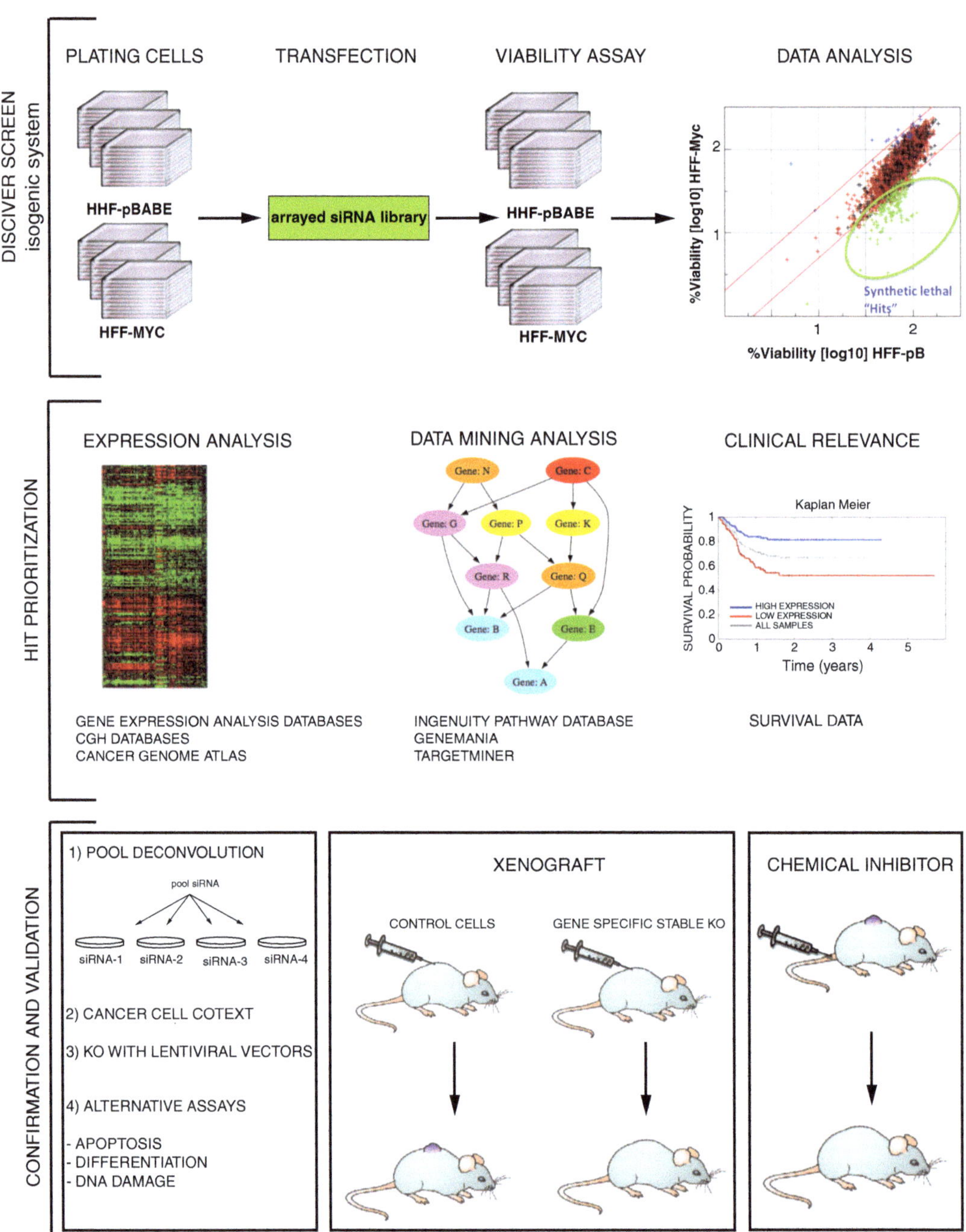

Fig. 1 Schematic of the siRNA HTS platform described. First, perform siRNA screening as a discovery step in a controlled isogenic system tailored to MYC or the oncogene in question. Second, perform statistical analysis and integrate it with genomics data to prioritize candidate genes for follow-up studies. Third, confirm and validate candidate genes with individual siRNAs in vitro and lentiviral vectors both in vitro and in vivo. If available, specific chemical compounds against selected gene targets should be used. Finally, mechanistic studies involving different assay types should also be designed to confirm and validate hits

2.3 siRNA Libraries

Arrayed siRNA libraries of short synthetic RNA duplexes (~19–25 bases in length) including genome-scale collections as well as focused gene categories and custom libraries are commercially available (Ambion, Dharmacon, and Sigma-Aldrich). Qiagen will facilitate construction of custom arrayed siRNA libraries in "FlexiPlate format". It has been our experience that siRNAs targeting individual genes are best used as pools of three to four different siRNAs. We previously employed an arrayed (384-well format) siRNA library targeting ~3,311 unique human genes with a pool of three siRNAs per gene. The library was designed to target the most druggable gene collections such as kinases, ubiquitin ligases, DNA repair, and genes with known cancer annotations (a more comprehensive version of this library is the MISSION SIGMA-ALDRICH druggable genome collection). *See* **Note 3** for arrayed shRNA libraries.

2.4 Equipment

1. Automated liquid handlers: While several different liquid handling devices are currently available, we have successfully employed both CyBi®-Well vario Pipettor and the Biomek® FXP Laboratory Automation Workstation. This type of equipment is utilized for dispensing of siRNAs in 96, 384, or 1,536 well formats. Pipet "tips" are usually employed for siRNA screening, while "pin" transfer manifold heads are available for small molecule screening but not necessary for siRNA screens.
2. Cell dispenser: The WellMate instrument (Matrix Technologies) is a high-speed, small footprint, 8-channel fluid dispenser for delivering cells/reagents efficiently to 6, 12, 24, 48, 96, and 384-well microplates. Even when coupled with the plate stacker, the unit can readily fit inside a tissue culture hood for applications where maintenance of sterility is critical. Other similar instruments are available that enable 1,536-well plate formats.
3. Plate reader: The speed of data acquisition is very important for HTS applications. The EnVision PerkinElmer Multilabel Detector is a rapid, sensitive, and versatile benchtop reader to handle fluorescence, luminescence and absorbance, fluorescence polarization, and time-resolved fluorescence detection technologies.
4. Automated microscope: The use of an automated microscope is optional if the desired readout assay involves measuring different phenotypic parameters such as DNA damage, apoptosis, or specific cellular markers on a per cell basis. Several instruments are currently available. We have employed an InCell instrument (General Electric) to quantitate nuclear γ-H2AX of siRNAs transfected cells with ~40 selected genes (*see* **Note 8**). Automated cell washing stations for carrying out plate washes of antibodies or staining steps are desirable when this type of readout is selected.

2.5 Software and Databases for Bioinformatics and Visualization

1. Microsoft Office Excel is utilized for normalization, *Z*-scores, and *P*-value calculations.
2. Miner3D or TIBCO Spotfire: for large data set visualization and analysis.
3. Ingenuity pathway analysis (IPA) for network analysis and gene annotations is accessible from academic laboratories upon subscription. GeneMANIA and TargetMine are recently available public tools. Many additional analysis resources are also available through the open-source and free Cytoscape and Bioconductor projects. In addition, the compiled Human Interaction Database [28] and the Cancer Genome Atlas database (https://tcga-data.nci.nih.gov/tcga/) are valuable resources for data mining.

3 Methods

A schematic outline of the general siRNA screening approach and steps toward the selection and prioritization of the hits is shown in Fig. 1.

3.1 siRNA Transfection Optimization

Once a cell system has been selected, it should be tested to determine its optimal siRNA transfection conditions, such as cell number, type of transfection reagent, siRNA ratio relative to transfection reagent, and readout timing. If the readout is cell viability, a functional cell toxicity assay as opposed to fluorescent-labeled oligos is best utilized to optimize screening conditions. For most cell types we found that siRNAs targeting the KIF11 gene, which encodes a kinesin motor protein, is a good positive control of transfection efficiency as it leads to cell toxicity in most cell types. A variety of recently developed negative controls are commercially available (*see* **Note 9**). The goal of the optimization process is to obtain the widest assay window, calculated via a simple statistical method referred to as *Z*-factor [29], which requires the use of a negative and a positive control. A *Z*-factor >0.5 and <1 indicates an excellent assay. We achieved a *Z*-factor for HFF-pBabe of 0.645 and for HFF-MYC of 0.726. The transfection conditions employed for screening were as follows: DharmaFECT I (1:1,000 final dilution in growth media) and siRNAs final concentration at 25 nM as outlined in Subheading 3.2.

3.2 HTS siRNA Screening Outline

1. Day 1
   Cells are seeded in 384-well tissue culture plates at 400 (HFF-pB) and 200 (HFF-MYC) cells/well in 50 μl/well of complete medium using the WellMate.
2. Day 2
   siRNA arrayed transfections are carried using a liquid handling device (Biomek FX or Cy-BIO for example) utilizing template

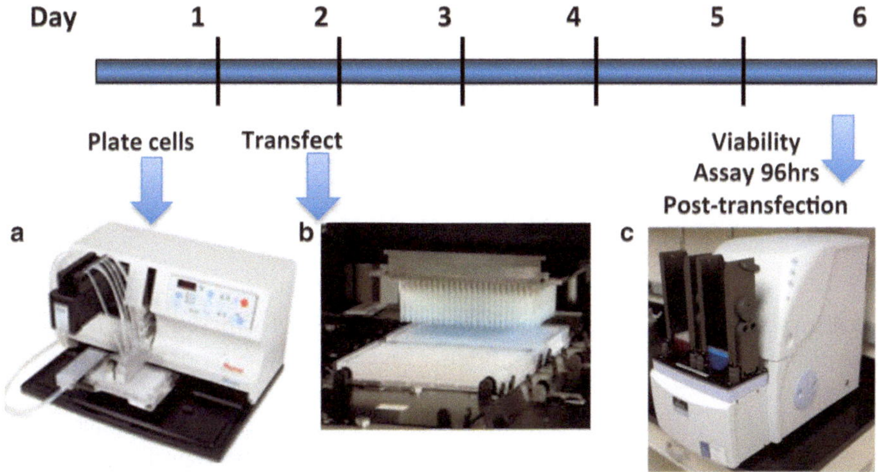

Fig. 2 Timeline and instrumentations needed to perform the high-throughput siRNA screening. (**a**) Matrix WellMate or equivalent instrument is employed for cell dispensing in 384-well plates; (**b**) a liquid handler with 384-well capabilities is needed for siRNA transfections; (**c**) a plate reader equipped with plate stacker is needed for the rapid quantification of cell viability (here shown is the EnVision, Perkin Elmer)

384-well plates containing siRNA libraries, targeting one gene/well, diluted in transfection reagent. A 2.5 μl siRNA/transfection mix aliquot from each template plate is distributed onto three cell plates to obtain three replicates. Immediately after transfection, cell plates are transferred to a 5 % CO_2 incubator at 37 °C. If using HFF the conditions are detailed below.

Mix 1	1 μl 20 μM RNA
	19 μl Opti-MEM
Mix 2	1 μl DharmaFECT I or RNAiMax
	19 μl Opti-MEM
Mix 1 and 2 and incubate at RT for 20 min	
Add 2.5 μl/well (each well has ~50 μl media)	

3. Day 5

   At 96 h (or any other time determined through optimization study) from the time of plating, cell viability is determined utilizing CellTiter-Glo or analogous viability detection kits following the instructions provided by the manufacturer (*see* **Note 7**). An EnVision multilabel plate reader or equivalent instrument equipped with plate stacker is employed for readout.

   A schematic of the timeline and example of instruments required for each step is shown in Fig. 2.

3.3 Alternative Readouts: "High Content Screening"

While we focused on assays measuring the average cell viability/well utilizing plate readers with luminescence and fluorescence capabilities, it is important to examine other readout choices. For example, automated microscopy will enable to quantify morphological changes, subcellular organelles, nuclear structures, differentiation markers by immunofluorescence, or other staining methods on a per cell basis (*see* **Note 8**). This approach is referred to as "High Content Screening."

3.4 Data Analysis and Hit Selection

Most genome-scale studies are performed in a non-replicate manner and the confidence level in the hits from primary screens is low (for review on statistical methods for HT screening *see* ref. 30). Thus, to maximize the sensitivity of the screen while increasing confidence in the hits, screens are best performed in triplicate. *Z*-scores, which are utilized to select hits, should be calculated using the median absolute deviation of the population (MAD), a method shown to increase the sensitivity of HT siRNA screens [31]. As *Z*-scores are defined by the difference between the normalized viability of an individual well and the population mean (or median), divided by the standard deviation (SD), which is increased by the presence of outliers (potential hits or detection errors) leading to increased false negatives, the utilization of the MAD is advised. For the MYC-SL described [22], the raw viability data was normalized to the negative controls on a per plate basis (*see* **Note 9**). The log2 ratio between the average-normalized viability of each siRNA in the control and MYC overexpressing cells was then calculated and employed to derive *Z*-scores as indicated above [31]. Hits were selected based on *Z*-score ≥ 2. The confirmation rate upon retesting in HFFs was close to 90 %. For additional consideration on screen's performance *see* **Note 10**.

3.5 Hits Prioritization

It is essential to consider multiple criteria when prioritizing the hits for validation experiments. If the intention is to select hits that would represent candidate drug targets, druggability is one of the criteria among others as outlined below.

1. Bioinformatics: Once statistical cuts have been made based upon inclusion of positive controls and genes known to be involved in the pathway of interest, it is imperative to mine the data with bioinformatics tools to (a) build functional networks with the tools outlined in Subheading 2, (b) derive information on the expression/mutational status of the hits in cancer databases (e.g., the Cancer Genome Atlas or more specialized databases), and (c) select hits focusing on the oncogene/cancer type in question. For example, we focused on a druggable target whose expression strongly correlated with *MYCN* amplification in neuroblastoma tumor samples based on a pediatric cancer database (Oncogenomics Section Data Center

at http://pob.abcc.ncifcrf.gov/cgi-bin/JK). Finally, we recommend carrying forward several potential hits through the next confirmation/validation tests. Weaker hits with strong implication through their gene expression patterns or their link to cancer pathways should be considered, as stable knockdown with lentiviral vectors of "weak hits" can indeed result in complete growth inhibition in long-term colony assays and, importantly, in halting tumor growth in vivo [22]. Therefore, we advise that ranking of the hits in the screen should not be the only criteria for selection. However, it is also important to pursue "strong hits" from siRNA screens even in the absence of supporting evidence, as they may lead to unforeseen insights into cancer pathways (*see* **Note 11**).

2. Hits confirmation: Pool deconvolution or independent siRNAs. The first level of screen follow-up involves a simple technical confirmation through siRNA pool deconvolution, i.e., testing individual siRNAs, or additional siRNAs in the same cell system. This step, which eliminates siRNAs with possible off-target effects [32], can be easily executed manually using a 96-well plate format. In our experience, utilization of isogenic systems minimizes scoring of siRNAs with off-target effects.
   (a) Other types of readout can be employed manually also in 96-well format, such as apoptosis via the Caspase-Glo kits (Promega). Apoptotic readouts can be normalized through a CellTiter-Glo or alamarBlue assay via parallel transfected plates. Other commercially available kits may allow simultaneous detection of apoptosis and viability. For additional assays, such as estimation of DNA damage or other phenotypic readouts, specific for the oncogene of interest, the use of automated microscopy is recommended.
3. Validation strategy:
   (a) Test siRNAs in a broad set of cancer cell lines stratified for MYC expression or in alternative isogenic systems (*see* **Note 4**) (or, if focused on other oncogenic mutation, using other criteria of stratification).
   (b) Utilize stable RNA interference to demonstrate effects in long-term assays in vitro and in vivo, with xenograft models. For construction of lentiviral vectors, it is best to employ multiple short hairpins/gene. In our study, we employed a dual vector lentiviral platform for conditional knockdown in vitro and in xenografts [33]. Note that for best preclinical assessment, tumors must be allowed to establish before conditional expression of the shRNA. Recently, all-in-one vectors employing a constitutively expressed tetracycline reverse transcriptional activator (rtTA3) and green fluorescent marker (eGFP), coupled

with an inducible cassette for expression of the short hairpin and of red fluorescent marker (rRFP), have been developed [34]. These types of vectors as well as others [35] can be used in vitro for long- and short-term growth assays and for in vivo studies (*see* **Note 12**).

(c) Confirmation with drugs, small molecules, antibodies, or other agents targeting the candidate gene products should be utilized, if available. However, it is important to point out that knockdown may not be equivalent to inhibition of activity, such as for kinase or other enzymatic or druggable domains. Thus, it is possible that different effects will be obtained with inhibitors versus gene knockdown; if this is the case, rescue experiments with enzymatically dead mutant cDNAs should be carried out to predict activity of small molecule inhibitors against candidate genes.

4 Notes

As there are several options for instrumentations and siRNA libraries, I encourage the readers to investigate what may best fit the needs of their laboratory and experimental setting.

1. MYC is here employed as a broad term to include potentially any family member, such as c-MYC, MYCN, and L-MYC.
2. The pooled lentiviral approach is a valuable tool when the aim of the screen is to identify tumor suppressor genes as there is strong positive selection for their loss of expression. This facilitates their detection [36]. Arrayed lentiviral screening platforms are other tools that eliminate the need to utilize transfection and overcome the limitations of the "pooled approach" and enabling a more durable knockdown. Such platforms are commercially available, but costly and obtaining consistent viral particles/well is difficult. However, they should be considered when cells are not suitable for siRNA transfection (*see* **Note 3**).
3. Arrayed siRNA screens are successful when the transfection efficiency is at least >70 % of the cell population as determined through feasibility and optimization studies. In general, while siRNA transfections are significantly more efficient than transfections of large DNA molecule (i.e., plasmids), non-adherent cells tend to be more difficult to transfect even with siRNAs. However, before excluding an siRNA screening approach, experimentation on a given cell type should be attempted by varying the transfection reagents, the substratum on microtiter plates (e.g., by pre-coating plates with extracellular matrix

proteins), or through "reverse transfection," i.e., utilizing plates pre-coated with siRNAs. Electroporation is currently not a method of choice for high-throughput screening.

4. Other isogenic cell systems that can be utilized to identify/ confirm MYC-SL genes are the P493-6 lymphoma cell line with conditional c-MYC overexpression as a model for MYC-driven cancers such as Burkitt's lymphoma [37]. Human mammary epithelial cells (HMECs) overexpressing either c-MYC or the conditional fusion protein MYC-ER retroviral vector are also a potential model to derive MYC-SL potentially relevant for breast cancer, as recently shown [31]. A consideration is that HMECs are a slow growing cell population better suited for lentiviral screening platforms. However, most inducible systems do exhibit some degree of leakiness that could affect the siRNA screening results. As a model of MYCN driven neuroblastoma, the Tet-21N neuroblastoma cell line with conditional expression of MYCN is a potentially valuable system [38]. Finally, the Rat-1A cell line with c-MYC gene knockout (*c-myc*$^{-/-}$) and its isogenic pair reconstituted with a c-MYC transgene could be utilized for siRNA screening [39]. It is important to keep in mind that Rat-1A *c-myc*$^{-/-}$ cells exhibit a much reduced growth rate, making the comparison with MYC overexpression difficult. The Rat-1A parental *c-myc*$^{+/+}$ normal rat fibroblasts could also be used for control.

5. For other cells, media may require optimization for growth in microtiter plates. The addition of 25 mM Hepes to the media is advisable to minimize changes in pH during the plate handling.

6. The reader is encouraged to research the best commercially available transfection reagents for the cell in questions.

7. It is possible to lower the amounts of CellTiter-Glo reagent but the linearity of the assay to cell number will have to be established with the cells in questions. A cheaper alternative cell viability assay is the fluorescence-based alamarBlue assay. The latter can also be utilized in parallel with luminescence readouts, for example, to normalize apoptosis assay results. However, the alamarBlue assay is time sensitive and as such can produce biases if readout and incubation timing across plates is not carefully controlled.

8. For other readout assays, which measure specific phenotypes associated with MYC, markers of chromatin modification, mitochondrial activity, ribosomal and protein synthesis, and DNA replication stress could represent alternative screening strategies. These types of readouts, while requiring the use of automated microscopy that can add significant cost and labor

to the screening project, also offer a broad range of assays and will enable gaining detailed mechanistic insights of MYC-associated phenotypes.

9. Negative control siRNAs are commercially available and examples are AllStars from Qiagen or UNI from SIGMA. Occasionally these siRNAs can exhibit significant toxicity against specific cell types and, if this is the case, normalization is carried out to mock, i.e., transfection reagents only.
10. Quality control of each screen should be performed. A few suggestions include: examination of the performance of positive and negative controls, the distribution of viability across the entire library and within plates, and *P*-values derived from triplicate replicas. For example, edge effects lowering nonspecifically cell viability may be detected, although the magnitude varies among cell types. For this reason, the outermost columns of each plate are not utilized. Also, it is important to place negative and positive controls throughout the plate.
11. It has been our experience that strong hits, which lead to complete growth inhibition in a 5 day assay, often correspond to genes involved in normal mitotic control or affect microtubule dynamics, such as KIF11. The knockdown of these genes also considerably affects viability of normal cells. To eliminate potentially highly toxic genes, hits with >50 % loss of viability in normal cells were eliminated even if they exhibited statistically significant differential toxicity in MYC overexpressing cells.
12. For in vivo knockdown studies utilizing xenografts, cancer cells should be injected into mice within a few days after lentiviral transductions because expression of the short hairpin is often lost upon expansion in tissue culture.

Acknowledgments

I am indebted to Maki Imakura and James Annis for sharing their expertise in high-throughput screening. I also thank Hamid Bolouri, Daniel Diolait-i, and Christopher Kemp for critical reading of this chapter and Daniel Diolait-i for figure design, Aaron Chang and Hamid Bolouri for discussion of bioinformatics tools, and Kristin Robinson in Denise Galloway's laboratory for isolating HFFs cultures.

References

1. Soucek L, Whitfield J, Martins CP, Finch AJ, Murphy DJ, Sodir NM, Karnezis AN, Swigart LB, Nasi S, Evan GI (2008) Modelling Myc inhibition as a cancer therapy. Nature 455: 679–683
2. Felsher DW, Bishop JM (1999) Reversible tumorigenesis by MYC in hematopoietic lineages. Mol Cell 4:199–207
3. Jain M, Arvanitis C, Chu K, Dewey W, Leonhardt E, Trinh M, Sundberg CD, Bishop JM, Felsher DW (2002) Sustained loss of a neoplastic phenotype by brief inactivation of MYC. Science 297:102–104
4. Evan GI, Wyllie AH, Gilbert CS, Littlewood TD, Land H, Brooks M, Waters CM, Penn LZ, Hancock DC (1992) Induction of apoptosis in fibroblasts by c-myc protein. Cell 69:119–128
5. Juin P, Hueber AO, Littlewood T, Evan G (1999) c-Myc-induced sensitization to apoptosis is mediated through cytochrome c release. Genes Dev 13:1367–1381
6. Morrish F, Isern N, Sadilek M, Jeffrey M, Hockenbery DM (2009) c-Myc activates multiple metabolic networks to generate substrates for cell-cycle entry. Oncogene 28:2485–2491
7. Dang CV (2011) Therapeutic targeting of Myc-reprogrammed cancer cell metabolism. Cold Spring Harb Symp Quant Biol 76: 369–374
8. Barna M, Pusic A, Zollo O, Costa M, Kondrashov N, Rego E, Rao PH, Ruggero D (2008) Suppression of Myc oncogenic activity by ribosomal protein haploinsufficiency. Nature 456:971–975
9. Mai S, Fluri M, Siwarski D, Huppi K (1996) Genomic instability in MycER-activated Rat1A-MycER cells. Chromosome Res 4:365–371
10. Felsher DW, Zetterberg A, Zhu J, Tlsty T, Bishop JM (2000) Overexpression of MYC causes p53-dependent G2 arrest of normal fibroblasts. Proc Natl Acad Sci USA 97: 10544–10548
11. Wade M, Wahl GM (2006) c-Myc, genome instability, and tumorigenesis: the devil is in the details. Curr Top Microbiol Immunol 302: 169–203
12. Goga A, Yang D, Tward AD, Morgan DO, Bishop JM (2007) Inhibition of CDK1 as a potential therapy for tumors over-expressing MYC. Nat Med 13:820–827
13. Dominguez-Sola D, Ying CY, Grandori C, Ruggiero L, Chen B, Li M, Galloway DA, Gu W, Gautier J, Dalla-Favera R (2007) Non-transcriptional control of DNA replication by c-Myc. Nature 448:445–451
14. Robinson K, Asawachaicharn N, Galloway DA, Grandori C (2009) c-Myc accelerates S-phase and requires WRN to avoid replication stress. PLoS One 4:e5951
15. Yang D, Liu H, Goga A, Kim S, Yuneva M, Bishop JM (2010) Therapeutic potential of a synthetic lethal interaction between the MYC proto-oncogene and inhibition of aurora-B kinase. Proc Natl Acad Sci USA 107: 13836–13841
16. Moser R, Toyoshima M, Robinson K, Gurley KE, Howie HL, Davison J, Morgan M, Kemp CJ, Grandori C (2012) MYC-driven tumorigenesis is inhibited by WRN syndrome gene deficiency. Mol Cancer Res 10:535–545
17. Hartwell LH, Szankasi P, Roberts CJ, Murray AW, Friend SH (1997) Integrating genetic approaches into the discovery of anticancer drugs. Science 278:1064–1068
18. Kaelin WG Jr (2005) The concept of synthetic lethality in the context of anticancer therapy. Nat Rev Cancer 5:689–698
19. Paddison PJ, Silva JM, Conklin DS, Schlabach M, Li M, Aruleba S, Balija V, O'Shaughnessy A, Gnoj L, Scobie K, Chang K, Westbrook T, Cleary M, Sachidanandam R, McCombie WR, Elledge SJ, Hannon GJ (2004) A resource for large-scale RNA-interference-based screens in mammals. Nature 428:427–431
20. Hu G, Kim J, Xu Q, Leng Y, Orkin SH, Elledge SJ (2009) A genome-wide RNAi screen identifies a new transcriptional module required for self-renewal. Genes Dev 23: 837–848
21. Blakely K, Ketela T, Moffat J (2011) Pooled lentiviral shRNA screening for functional genomics in mammalian cells. Methods Mol Biol 781:161–182
22. Toyoshima M, Howie HL, Imakura M, Walsh RM, Annis JE, Chang AN, Frazier J, Chau BN, Loboda A, Linsley PS, Cleary MA, Park JR, Grandori C (2012) Functional genomics identifies therapeutic targets for MYC-driven cancer. Proc Natl Acad Sci USA 109: 9545–9550
23. Kessler JD, Kahle KT, Sun T, Meerbrey KL, Schlabach MR, Schmitt EM, Skinner SO, Xu Q, Li MZ, Hartman ZC, Rao M, Yu P, Dominguez-Vidana R, Liang AC, Solimini NL, Bernardi RJ, Yu B, Hsu T, Golding I, Luo J, Osborne CK, Creighton CJ, Hilsenbeck SG, Schiff R, Shaw CA, Elledge SJ, Westbrook TF (2012) A SUMOylation-dependent transcriptional subprogram is required for Myc-driven tumorigenesis. Science 335:348–353
24. Liu L, Ulbrich J, Muller J, Wustefeld T, Aeberhard L, Kress TR, Muthalagu N, Rycak

L, Rudalska R, Moll R, Kempa S, Zender L, Eilers M, Murphy DJ (2012) Deregulated MYC expression induces dependence upon AMPK-related kinase 5. Nature 483: 608–612

25. Benanti JA, Galloway DA (2004) Normal human fibroblasts are resistant to RAS-induced senescence. Mol Cell Biol 24: 2842–2852
26. Benanti JA, Wang ML, Myers HE, Robinson KL, Grandori C, Galloway DA (2007) Epigenetic down-regulation of ARF expression is a selection step in immortalization of human fibroblasts by c-Myc. Mol Cancer Res 5:1181–1189
27. Wang ML, Walsh R, Robinson KL, Burchard J, Bartz SR, Cleary M, Galloway DA, Grandori C (2011) Gene expression signature of c-MYC-immortalized human fibroblasts reveals loss of growth inhibitory response to TGFbeta. Cell Cycle 10:2540–2548
28. Cerami EG, Gross BE, Demir E, Rodchenkov I, Babur O, Anwar N, Schultz N, Bader GD, Sander C (2011) Pathway commons, a web resource for biological pathway data. Nucleic Acids Res 39:D685–690
29. Zhang JH, Chung TD, Oldenburg KR (1999) A simple statistical parameter for use in evaluation and validation of high throughput screening assays. J Biomol Screen 4:67–73
30. Birmingham A, Selfors LM, Forster T, Wrobel D, Kennedy CJ, Shanks E, Santoyo-Lopez J, Dunican DJ, Long A, Kelleher D, Smith Q, Beijersbergen RL, Ghazal P, Shamu CE (2009) Statistical methods for analysis of high-throughput RNA interference screens. Nat Methods 6:569–575
31. Chung N, Zhang XD, Kreamer A, Locco L, Kuan PF, Bartz S, Linsley PS, Ferrer M, Strulovici B (2008) Median absolute deviation to improve hit selection for genome-scale RNAi screens. J Biomol Screen 13:149–158
32. Jackson AL, Linsley PS (2010) Recognizing and avoiding siRNA off-target effects for target identification and therapeutic application. Nat Rev Drug Discov 9:57–67
33. Chau BN, Diaz RL, Saunders MA, Cheng C, Chang AN, Warrener P, Bradshaw J, Linsley PS, Cleary MA (2009) Identification of SULF2 as a novel transcriptional target of p53 by use of integrated genomic analyses. Cancer Res 69:1368–1374
34. Meerbrey KL, Hu G, Kessler JD, Roarty K, Li MZ, Fang JE, Herschkowitz JI, Burrows AE, Ciccia A, Sun T, Schmitt EM, Bernardi RJ, Fu X, Bland CS, Cooper TA, Schiff R, Rosen JM, Westbrook TF, Elledge SJ (2011) The pINDUCER lentiviral toolkit for inducible RNA interference in vitro and in vivo. Proc Natl Acad Sci USA 108:3665–3670
35. Zuber J, McJunkin K, Fellmann C, Dow LE, Taylor MJ, Hannon GJ, Lowe SW (2010) Toolkit for evaluating genes required for proliferation and survival using tetracycline-regulated RNAi. Nat Biotechnol 29:79–83
36. Scuoppo C, Miething C, Lindqvist L, Reyes J, Ruse C, Appelmann I, Yoon S, Krasnitz A, Teruya-Feldstein J, Pappin D, Pelletier J, Lowe SW (2012) A tumour suppressor network relying on the polyamine-hypusine axis. Nature 487:244–248
37. Schuhmacher M, Staege MS, Pajic A, Polack A, Weidle UH, Bornkamm GW, Eick D, Kohlhuber F (1999) Control of cell growth by c-Myc in the absence of cell division. Curr Biol 9:1255–1258
38. Lutz W, Stohr M, Schurmann J, Wenzel A, Lohr A, Schwab M (1996) Conditional expression of N-myc in human neuroblastoma cells increases expression of alpha-prothymosin and ornithine decarboxylase and accelerates progression into S-phase early after mitogenic stimulation of quiescent cells. Oncogene 13:803–812
39. Mateyak MK, Obaya AJ, Adachi S, Sedivy JM (1997) Phenotypes of c-Myc-deficient rat fibroblasts isolated by targeted homologous recombination. Cell Growth Differ 8:1039–1048

Chapter 13

Investigating Myc-Dependent Translational Regulation in Normal and Cancer Cells

John T. Cunningham, Michael Pourdehnad, Craig R. Stumpf, and Davide Ruggero

Abstract

There is an increasing realization that a primary role for Myc in driving cellular growth and cell cycle progression relies on Myc's ability to increase the rate of protein synthesis. Myc induces myriad changes in both global and specific mRNA translation. Herein, we present three assays that allow researchers to measure changes in protein synthesis at the global level as well as alterations in the translation of specific mRNAs. Metabolic labeling of cells with ^{35}S-containing methionine and cysteine is presented as a method to measure the overall rate of global protein synthesis. The bicistronic reporter assay is employed to determine levels of cap-dependent and cap-independent translation initiation in the cell. Finally, isolation of polysome-associated mRNAs followed by next-generation sequencing, microarray or quantitative real-time PCR (qRT-PCR) analysis is utilized to detect changes in the abundance of specific mRNAs that are regulated upon Myc hyperactivation. The protocols described in this chapter can be used to understand how and to what extent Myc-dependent regulation of translation influences normal cellular functions as well as tumorigenesis.

Key words Myc, Translation, Protein synthesis, Ribosome

1 Introduction

Myc has an evolutionarily conserved role in regulating cell growth. This function of Myc is dependent on its ability to increase protein synthesis. Importantly, numerous studies have demonstrated that Myc directly regulates the expression of many components of the protein synthesis machinery. Myc transcriptional target genes include translation initiation and elongation factors, tRNA synthetases, Pol III, nucleolar assembly components, and proteins belonging to the small and large ribosomal subunits [1–3]. As such, one of the major functions of Myc is to control ribosome

John T. Cunningham, Michael Pourdehnad, and Craig R. Stumpf have contributed equally.

Laura Soucek and Nicole M. Sodir (eds.), *The Myc Gene: Methods and Protocols*, Methods in Molecular Biology, vol. 1012, DOI 10.1007/978-1-62703-429-6_13, © Springer Science+Business Media, LLC 2013

biogenesis and mRNA translation. The role of Myc-dependent regulation of protein synthesis in physiologic cellular processes or under pathological conditions is not well understood. However, the importance of alterations in translation induced by Myc has been highlighted in the setting of oncogenic Myc activity, where restoring protein synthesis to normal levels leads to suppression of cell growth and decreased Myc-driven tumorigenesis [4, 5]. Additionally, Myc oncogenic activity has been shown to deregulate the translational control of specific mRNAs during mitosis leading to genomic instability [4]. Therefore, Myc-dependent alterations in mRNA translation direct an oncogenic program, involving distinct cellular processes, which is active during multiple stages of cancer initiation and progression [6–8].

The purpose of this chapter is to describe in detail several methods used to investigate the effects of Myc on protein synthesis. The first protocol explains *^{35}S metabolic labeling*, which allows the measurement of global protein synthesis rates by quantifying the incorporation of ^{35}S-labeled methionine and cysteine into newly translated proteins (Fig. 1a). Increased uptake of ^{35}S methionine in neoplasms compared to normal tissue was first described in the 1950s [9]. Subsequent studies demonstrated that increased Myc activity led to increased protein synthesis, which was coupled to increased cell mass [4, 10–12]. Importantly, these and other studies demonstrated that Myc has an evolutionarily conserved role across multiple cell types in promoting protein synthesis, which leads to cell growth, cell division, and, when deregulated, can lead to cancer [4, 5, 13].

The second protocol describes *measuring cap-dependent and IRES-dependent translation initiation using the bicistronic reporter assay* (Fig. 1b). The best studied mechanism of translation initiation is cap-dependent, which requires the assembly of an initiation complex on the 5′ end of mRNAs that recruits the ribosome and scans along the 5′ untranslated region prior to initiation of translation at the start codon. Most mRNAs are translated by this method of initiation. IRES-dependent translation, on the other hand, is an alternative method of translation initiation in which the ribosome is recruited by an RNA structural element, known as the internal ribosome entry site (IRES). It is postulated that IRES-mediated translation is an important alternative mode of translation that can differentially regulate the translation of specific mRNAs during distinct physiological contexts (e.g., during cell cycle progression and development) or in response to certain stimuli (e.g., hypoxia and endoplasmic reticulum stress) [4, 7, 14–17]. There is no existing structure or sequence prediction method that can accurately identify IRES elements in an mRNA. Therefore, translation initiation from these elements must be validated experimentally via methods such as the bicistronic reporter assay. This approach measures the

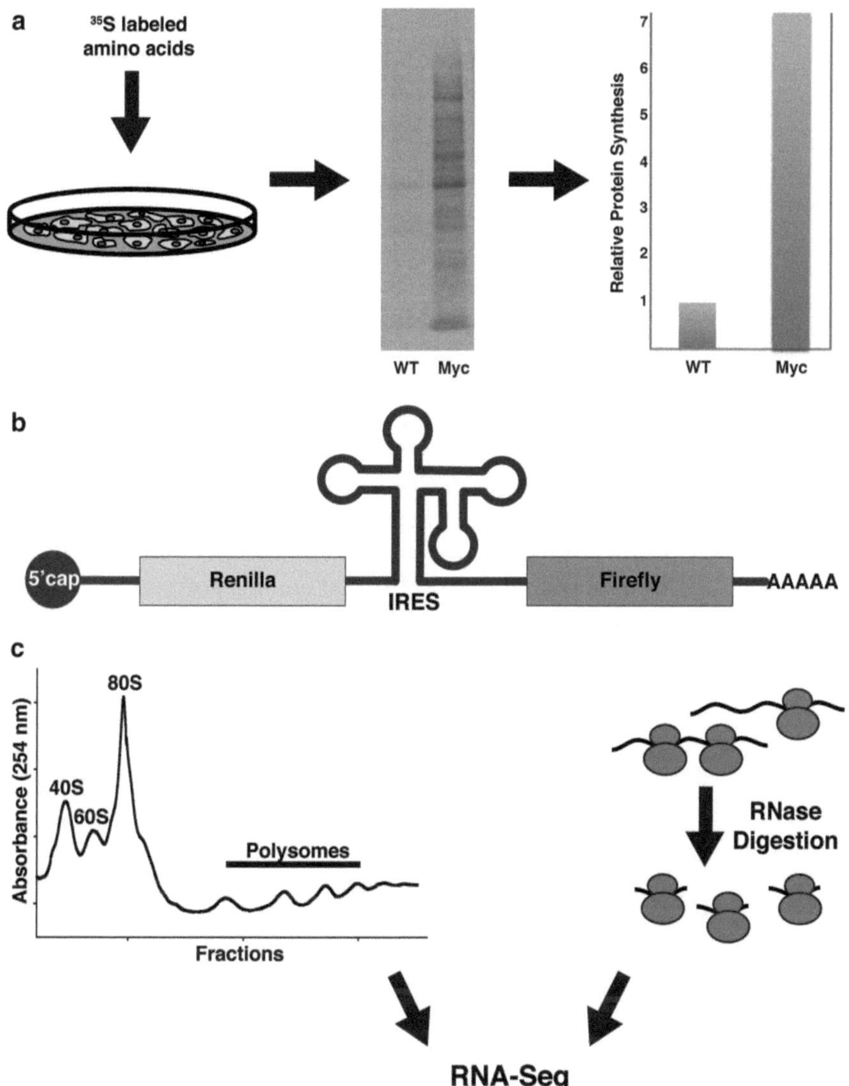

Fig. 1 Methodologies presented to dissect translational changes downstream of Myc hyperactivation. (**a**) Flowchart of ^{35}S data acquisition and processing in wild-type (WT) or Myc-overexpressing (Myc) cells. (**b**) Illustration of a bicistronic cap/IRES translation initiation reporter. In this reporter mRNA, translation of the *Renilla* luciferase open reading frame is driven by a 5′ 7-methylguanosine cap-dependent mechanism, whereas the firefly luciferase protein is expressed via translation initiation that relies on ribosome recruitment through an IRES-dependent mechanism. (**c**) Sucrose gradient fractionation of cytoplasmic RNA reveals translationally inactive (40S, 60S, 80S/monosome) and active (polysome) fractions (*left*). A schema depicting the steps involved in ribosome profiling is presented (*right*). Both procedures can be used to generate sequencing libraries for analysis of the genome-wide changes in translation upon Myc hyperactivation (*bottom*)

levels of cap-dependent and IRES-dependent translation in cells. Changes in the relative levels of cap-dependent and IRES-dependent translation have been observed during various stages of tumorigenesis [7, 18]. In this regard, several translation initiation factors are transcriptional targets of Myc, including eIF4E, and have been shown to cooperate with Myc-driven tumorigenesis [19, 20]. Additionally, Myc has been shown to deregulate the switch between cap- and IRES-dependent translation that occurs during the mitotic phase of the cell cycle [15, 16, 21]. Specifically, Myc hyperactivation leads to decreased expression of the endogenous IRES-dependent isoform of Cdk11 (p58-PITSLRE) during mitosis, which leads to genomic instability [4]. The mechanism for maintaining the normal balance between these modes of translation as well as how oncogenic signaling pathways impinge on this translational control is not well understood.

The third protocol describes *analyzing gene-specific changes in mRNA translation by isolation of polysome-associated mRNA* (Fig. 1c). While the ^{35}S methionine incorporation assay measures changes in global protein synthesis, analyzing the fraction of mRNAs associated with multiple ribosomes (polysome-associated) can be used to detect changes in the translation of specific mRNAs [22, 23]. Polysome-associated mRNAs undergo high levels of translation, while mRNAs that are not polysome-associated are less efficiently translated. After their isolation, polysome-associated mRNAs can be analyzed by qRT-PCR to determine the relative level of translation for each specific mRNA. This approach has uncovered Myc-dependent changes in the translation of specific genes. For example, B cells overexpressing Myc specifically upregulate the translation of VEGF, while VEGF mRNA levels are unchanged [22]. Two approaches utilizing next-generation RNA sequencing can be used to identify genome-wide changes in polysome-associated mRNAs. Polysome-associated mRNAs can be characterized by RNA sequencing to identify translationally regulated genes [24]. Alternatively, ribosome profiling is a novel method to analyze genome-wide translation (Fig. 1c) [25]. Recent studies have used this technology to characterize the differential translation of mRNAs downstream of an oncogenic signaling pathway [26]. The ribosome profiling protocol has been recently published and a detailed protocol specifically for analyzing polysome-associated mRNAs is provided below [27]. These technologies can be used to address a major unresolved question by identifying the translational landscape of mRNAs that are regulated by Myc.

Taken together, these protocols provide tools to better understand the effects of Myc on global protein synthesis rates, the regulation of different modes of translation, and the translational regulation of specific mRNAs. These protocols can be utilized across tissue and cell types and under various physiological and pathological conditions to unravel outstanding questions regarding the role of Myc-dependent translation regulation in normal cellular processes and cancer.

2 Materials

2.1 ^{35}S Metabolic Labeling Assay

1. Radioactivity license.
2. Tissue culture equipment (e.g., laminar flow hood, incubator, tissue culture dishes).
3. Methionine- and cysteine-free culture medium (Sigma).
4. ^{35}S-Express Protein Labeling Mix (Perkin-Elmer).
5. Dialyzed fetal bovine serum (FBS).
6. Charcoal filters.
7. Rubber policeman.
8. Gel electrophoresis equipment and buffers.
9. Autoradiography equipment.
10. Image analysis software.
11. Bradford Assay kit.
12. RIPA buffer: 50 mM Tris, pH 7.4, 150 mM NaCl, 0.1 % sodium dodecyl sulfate, 0.5 % sodium deoxycholate, 1 % Triton X-100, 10 mM β-glycerophosphate, 50 mM NaF.
13. Complete protease inhibitors (Roche).
14. 1 μM phenylmethanesulfonyl fluoride (PMSF).
15. Phosphate buffered saline (PBS).
16. Primary antibodies: tubulin, glyceraldehyde-3-phosphate dehydrogenase (GAPDH), β-actin.
17. Secondary anti-mouse and anti-rabbit antibody conjugated to Horseradish peroxidase (HRP).

2.2 Measuring Cap-Dependent and Cap-Independent Translation Initiation Using the Bicistronic Reporter Assay

1. Bicistronic plasmid containing Renilla and firefly luciferase under the control of a T7 promoter.
2. Restriction endonuclease.
3. mMessage mMachine T7 transcription kit (Ambion).
4. TransMessenger mRNA transfection reagent (Qiagen).
5. Dual-Luciferase assay kit (Promega).
6. Luminometer.

2.3 Analyzing Gene-Specific Changes in mRNA Translation by Isolation of Polysome-Associated mRNA

1. Sucrose.
2. Gradient buffer: 25 mM Tris pH 7.4, 25 mM NaCl, 5 mM $MgCl_2$.
3. Heparin: 50 mg/ml in DEPC water.
4. DTT: 1 M in DEPC water.
5. Cycloheximide: 50 mg/ml in ethanol.
6. PBS.

7. Refrigerated microcentrifuge.
8. Ultracentrifuge with SW40 (or equivalent) rotor and matching tubes.
9. Lysis buffer: 10 mM Tris, pH 8.0, 140 mM NaCl, 1.5 mM $MgCl_2$, 0.25 % NP-40, 0.1 % Triton X-100, 20 mM DTT, 0.15 mg/ml cycloheximide, 0.6 U/ml RNasin.
10. Trizol solution (Life Technologies).
11. PureLink RNA miniprep kit (Life Technologies).
12. Optional: automated gradient maker and fraction collector, microwave, trypan blue.

3 Methods

3.1 ^{35}S Metabolic Labeling Assay

1. On the day before labeling, seed cells on 6-well tissue culture dishes such that the dish is approximately 70–85 % confluent at the time of labeling (*see* **Note 1**).
2. Aspirate medium and replace with 1.5 ml of methionine- and cysteine-free medium supplemented with dialyzed FBS. Incubate in methionine- and cysteine-free medium for 30 min (*see* **Note 2**).
3. Add 3 μL (33 μCi) of Express Protein Labeling Mix directly to each well of the 6-well tissue culture dish. Return dish to tissue culture incubator and incubate for 1 h. Follow proper institutional guidelines on handling radioactive materials such as co-incubating cells with activated charcoal filters to absorb any volatile ^{35}S compounds.
4. Using institutional guidelines governing handling and disposal of radioactive materials, harvest cells in 1 ml/well ice-cold PBS using a rubber policeman and collect in Eppendorf tubes.
5. Lyse cells in RIPA lysis buffer supplemented with 1× complete protease inhibitors and 1 mM PMSF. Approximately 20–100 μL of RIPA lysis buffer per sample should be used. Incubate lysis for 15–20 min on ice. Centrifuge the lysed cells at maximum speed for 5 min and collect the supernatant.
6. Measure the protein concentration of the lysates using a Bradford Assay.
7. Run a standard 10 % SDS-PAGE gel using 10–30 μg of total protein from each of the samples per well. *See* **Note 3** for discussion of alternative approaches to detection of ^{35}S incorporation.
8. Transfer gel to PVDF membrane.
9. Remove PVDF membrane containing bound ^{35}S-labeled proteins and immediately wrap in cling wrap and expose to film in

a cassette (*see* **Note 4**). Use a Geiger counter to help ascertain the exposure time necessary for obtaining the appropriate signal intensity. Typical exposure times will range from 2 to 48 h depending on the metabolic activity of the cells used.

10. Once a suitable exposure within the linear intensity range of film has been obtained, perform western blot analysis with tubulin, GAPDH, or β-actin antibodies for internal loading control.
11. Use Image J or other imaging software to calculate the intensity of scanned images (*see* **Note 5**). Express data as a ratio of ^{35}S signal intensity/loading control signal intensity.

3.2 Measuring Cap-Dependent and Cap-Independent Translational Initiation Using the Bicistronic Reporter Assay

1. Linearize bicistronic plasmid by digesting with a restriction endonuclease downstream of the firefly luciferase gene (*see* **Note 6**).
2. Transcribe capped mRNA using the mMessage mMachine in vitro transcription kit. Confirm the transcript is the proper length by gel electrophoresis and phenol extract the RNA.
3. Transfect 2 μg mRNA per 100,000 cells following the TransMessenger protocol.
4. Incubate cells for 6–8 h to allow for translation of the luciferase genes.
5. Harvest the cells and perform the Dual-Luciferase assay according to the manufacturer's protocol.
6. Normalize the luciferase readings to the amount of reporter mRNA as measured by quantitative PCR or northern blot.

3.3 Analyzing Gene-Specific Changes in mRNA Translation by Isolation of Polysome-Associated mRNA

1. Prepare 10 % and 50 % sucrose solutions in gradient buffer. Add 0.1 mg/ml heparin and 2 mM DTT. Prepare 10–50 % linear sucrose gradient in an ultracentrifuge tube (*see* **Note 7**).
2. Treat cells with 0.1 mg/ml cycloheximide (*see* **Note 8**).
3. Wash cells in PBS containing 0.1 mg/ml cycloheximide.
4. Resuspend in lysis buffer (~ 100 μl per 10 million cells). Lyse cells at 4 °C for 30 min. Vortex cells every 10 min during the lysis (*see* **Note 9**).
5. Centrifuge lysate at 10,000 × *g* at 4 °C for 5 min.
6. Transfer cleared lysate to clean RNase-free microfuge tube. Save 1 % of the lysate to analyze total cellular RNA levels.
7. Carefully layer the lysate onto the 10–50 % sucrose gradient. Place the centrifuge tube in the appropriate bucket, ensuring that opposing buckets are balanced.
8. Centrifuge samples at 243,000 × *g* (at maximum radius) at 4 °C for 2.5 h in SW40 rotor.

9. After centrifugation, collect 12–24 fractions from each sample (*see* **Note 10**).
10. Isolate RNA from each fraction using the Trizol modification of the PureLink RNA miniprep kit protocol.
11. Analyze RNA levels of genes of interest by standard quantitative PCR or analyze global polysome-associated mRNA profiles by next-generation RNA sequencing or microarray (*see* **Notes 11** and **12**).

4 Notes

1. Several variables need to be considered at this step. ^{35}S labeling should be performed on metabolically active cells. Therefore, take care to seed cells at a density that will permit this condition to be met (i.e., ensure cells are not too confluent). Myc-overexpressing cells typically proliferate faster and may be larger than normal counterparts. In this respect, different amounts of cells may be initially plated such that the density and total number of cells is equal at the time of ^{35}S labeling. To quantify cell size prior to plating, a Coulter Z2 Particle Count and Size Analyzer or flow cytometer may be used.

   Myc-dependent increases in global protein synthesis rates are mediated through a coordinated transcriptional response involving genes encoding ribosome biogenesis factors, ribosomal proteins, initiation factors, as well as the ribosomal RNA and transfer RNAs. Therefore, if using an inducible system of Myc activation such as MycER, or Tet-inducible Myc overexpression, care should be taken to optimize the timing between activation of Myc and incubation of cells with ^{35}S-containing methionine and cysteine. Typically, using the MycER system, we routinely observe increased target gene expression within 3–6 h and increased protein synthesis rates within 24 h following 4-hydroxytamoxifen administration.

   Labeling experiments using cells freshly isolated from living tissue should be performed using the minimal amount of time necessary between harvest and ^{35}S labeling and should begin with 30 min of culture in methionine- and cysteine-free medium.
2. Methionine and cysteine starvation is performed solely to increase the uptake and incorporation of ^{35}S-labeled methionine and cysteine. The incubation time necessary to achieve proper starvation is dependent on cellular metabolism. Care should be taken not to incubate cells too long in the absence of methionine and cysteine, as this could elicit an amino acid starvation response through GCN2, which phosphorylates eIF2α and suppresses global protein synthesis. The levels of

methionine and cysteine in serum are typically several hundredfold below that of common medium formulations such as DMEM and RPMI; therefore using dialyzed FBS is not absolutely necessary.

3. To measure ^{35}S incorporation into proteins, many suitable methods exist. We have presented our preferred method, but alternative approaches may also be utilized. In addition to the method presented above, other researchers have precipitated whole cell lysates with trichloroacetic acid and used a scintillation counter to measure ^{35}S incorporation. Using this method, data is expressed as ^{35}S signal intensity per cell or per μg of protein. Alternatively, instead of transferring protein from acrylamide gel to PVDF membrane, gel may be stained with Coomassie to ensure equal loading between samples followed by gel drying and exposing to film.
4. Avoid allowing the PVDF membrane to dry out, as this will interfere with downstream western blotting applications. If the membrane happens to dry out, a quick rinse in methanol should be performed to prepare the membrane for western blotting. Use of a phosphor imaging screen as an alternative to film can significantly reduce exposure times and expand the linear range of detection.
5. In Image J (http://rsbweb.nih.gov/ij/), use the rectangle tool to outline the lanes of interest on a grayscale image of your autoradiograph. Use Ctrl+1 to select the first lane followed by dragging the outlined region to the next lane and pressing Ctrl+2 to select each additional lane. Once all lanes have been established in this manner, press Ctrl+3 to plot the intensity of the signal within the lanes. Use the wand tool to select the area underneath the peaks within each lane to quantify the overall signal intensity for each lane. It is important to note that many gels will have some degree of background signal and it may be necessary to manually draw a horizontal baseline using the straight line tool in each of the plots of the lanes to account for this prior to using the wand tool. Quantify both the ^{35}S autoradiograph and the image of the loading control western blot (e.g., tubulin, β-actin, GAPDH). Data should be expressed as ^{35}S signal intensity/loading control signal intensity for each lane.
6. In addition to this protocol for analyzing cap-dependent and IRES-dependent translation in cell culture, a mouse model harboring a widely expressed bicistronic reporter mRNA has also been developed [28].
7. Use of an automated gradient maker (e.g., Teledyne ISCO) is recommended for producing uniform gradients. Alternatively, gradients can be made by manually layering the lower

percentage sucrose solution over the higher percentage sucrose solution, capping the tube, and carefully storing the tube on its side for 1–4 h. Store gradients at 4 °C until samples are ready to load.

8. A single cell suspension must be generated from tissue or cultured cells prior to beginning the lysis protocol. Ideally start with greater than 10 million cells.
9. Lysis usually takes approximately 30–45 min depending on cell type. Check for completion of lysis by staining a small aliquot with trypan blue.
10. Optional: An automated fraction collector will increase the reproducibility of collecting fractions. Prepare fraction collector according to manufacturers specifications. For the ISCO fraction collector, dissolve 60 g sucrose to 100 ml DEPC water to push gradient into UV detector/fraction collector. Alternatively, fractions can be collected by pipetting off 0.5–1.0 ml fractions from the top of the gradient into collection tubes. Measuring and plotting the absorbance at 260 nm across the fractions will give a graph representing the presence of ribosomes among the fractions.
11. It is generally recommended to correlate changes in polysome association to changes in cellular protein levels measured by western blot analysis to confirm translational regulation.
12. Alternative: Ribosome profiling is an emerging technology that can be used as an alternative to global polysome-associated mRNA analysis by RNA-seq. Ribosome profiling characterizes the fragment of mRNA protected by the ribosome during RNase digestion.

    The advantage to using the ribosome profiling approach is the precise, codon level, positioning information along each individual mRNA. This positional data may provide unique mechanistic insights into specific regulatory elements important for the proper translation of mRNAs, such as upstream open reading frames or ribosome pause sites [29].

Acknowledgments

We would like to thank members of the Ruggero lab for their input and comments on this manuscript. Thank you to Kimhouy Tong for editing the manuscript. This work is supported by ACS #121364-PF-11-184-01-TBG (J.T.C.), NIH/NRSA F32CA162634 (C.R.S.), and NIH R01CA140456 (D.R.). D.R. is a Leukemia & Lymphoma Society Scholar.

References

1. Zeller KI, Zhao X, Lee CWH, Chiu KP, Yao F, Yustein JT, Ooi HS, Orlov YL, Shahab A, Yong HC, Fu Y, Weng Z, Kuznetsov VA, Sung W-K, Ruan Y, Dang CV, Wei C-L (2006) Global mapping of c-Myc binding sites and target gene networks in human B cells. Proc Natl Acad Sci USA 103:17834–17839
2. Fernandez PC, Frank SR, Wang L, Schroeder M, Liu S, Greene J, Cocito A, Amati B (2003) Genomic targets of the human c-Myc protein. Genes Dev 17:1115–1129
3. Gomez-Roman N, Grandori C, Eisenman RN, White RJ (2003) Direct activation of RNA polymerase III transcription by c-Myc. Nature 421:290–294
4. Barna M, Pusic A, Zollo O, Costa M, Kondrashov N, Rego E, Rao PH, Ruggero D (2008) Suppression of Myc oncogenic activity by ribosomal protein haploinsufficiency. Nature 456:971–975
5. Bywater MJ, Poortinga G, Sanij E, Hein N, Peck A, Cullinane C, Wall M, Cluse L, Drygin D, Anderes K, Huser N, Proffitt C, Bliesath J, Haddach M, Schwaebe MK, Ryckman DM, Rice WG, Schmitt C, Lowe SW, Johnstone RW, Pearson RB, McArthur GA, Hannan RD (2012) Inhibition of RNA polymerase I as a therapeutic strategy to promote cancer-specific activation of p53. Cancer Cell 22:51–65
6. Stumpf CR, Ruggero D (2011) The cancerous translation apparatus. Curr Opin Genet Dev 21(4):474–483
7. Silvera D, Formenti SC, Schneider, RJ (2010) Translational control in cancer. Nat Rev Cancer 10(4):254–266
8. Ruggero D (2009) The role of Myc-induced protein synthesis in cancer. Cancer Res 69:8839–8843
9. Bloch HS, Hitchcock CR, Kremen AJ (1951) The distribution of S35 from labeled DL-methionine in mice bearing carcinoma of the breast, neoplasms of the hematopoietic system, or liver abscesses. Cancer Res 11:313–317
10. Schuhmacher M, Staege MS, Pajic A, Polack A, Weidle UH, Bornkamm GW, Eick D, Kohlhuber F (1999) Control of cell growth by c-Myc in the absence of cell division. Cur Biol 9(21):1255–1258
11. Iritani BM, Eisenman RN (1999) c-Myc enhances protein synthesis and cell size during B lymphocyte development. Proc Natl Acad Sci USA 96:13180–13185
12. Mateyak MK, Obaya AJ, Adachi S, Sedivy JM (1997) Phenotypes of c-Myc-deficient rat fibroblasts isolated by targeted homologous recombination. Cell Growth Differ 8:1039–1048
13. Ji H, Wu G, Zhan X, Nolan A, Koh C, De Marzo A, Doan HM, Fan J, Cheadle C, Fallahi M, Cleveland JL, Dang CV, Zeller KI (2011) Cell-type independent MYC target genes reveal a primordial signature involved in biomass accumulation. PLoS One 6:e26057
14. Muranen T, Selfors LM, Worster DT, Iwanicki MP, Song L, Morales FC, Gao S, Mills GB, Brugge JS (2012) Inhibition of PI3K/mTOR leads to adaptive resistance in matrix-attached cancer cells. Cancer Cell 21:227–239
15. Qin X, Sarnow P (2004) Preferential translation of internal ribosome entry site-containing mRNAs during the mitotic cycle in mammalian cells. J Biol Chem 279:13721–13728
16. Pyronnet S, Pradayrol L, Sonenberg N (2000) A cell cycle-dependent internal ribosome entry site. Mol Cell 5:607–616
17. Hart L, Cunningham J, Datta T, Dey S, Tameire F, Lehman S, Qiu B, Zhang H, Cerniglia G, Bi M, Li Y, Gao Y, Liu H, Li C, Maity A, Thomas-Tikhonenko A, Perl A, Koong A, Fuchs S, Diehl J, Mills I, Ruggero D, Koumenis C (2012) Endoplasmic reticulum stress-mediated autophagy promotes myc-dependent transformation and tumor growth. J Clin Invest 122(12):4621–4634
18. Bellodi C, Krasnykh O, Haynes N, Theodoropoulou M, Peng G, Montanaro L, Ruggero D (2010) Loss of function of the tumor suppressor DKC1 perturbs p27 translation control and contributes to pituitary tumorigenesis. Cancer Res 70:6026–6035
19. Ruggero D, Montanaro L, Ma L, Xu W, Londei P, Cordon-Cardo C, Pandolfi PP (2004) The translation factor eIF-4E promotes tumor formation and cooperates with c-Myc in lymphomagenesis. Nat Med 10:484–486
20. Miluzio A, Beugnet A, Grosso S, Brina D, Mancino M, Campaner S, Amati B, de Marco A, Biffo S (2011) Impairment of cytoplasmic eIF6 activity restricts lymphomagenesis and tumor progression without affecting normal growth. Cancer Cell 19:765–775
21. Pyronnet S, Dostie J, Sonenberg N (2001) Suppression of cap-dependent translation in mitosis. Genes Dev 15:2083–2093
22. Yoon A, Peng G, Brandenburger Y, Zollo O, Xu W, Rego E, Ruggero D (2006) Impaired control of IRES-mediated translation in X-linked dyskeratosis congenita. Science 312:902–906
23. Mezquita P, Parghi SS, Brandvold KA, Ruddell A (2004) Myc regulates VEGF production in B cells by stimulating initiation of VEGF mRNA translation. Oncogene 24:889–901
24. Thoreen CC, Chantranupong L, Keys HR, Wang T, Gray NS, Sabatini DM (2012) A unifying model for mTORC1-mediated regulation of mRNA translation. Nature 485: 109–113

25. Ingolia NT, Ghaemmaghami S, Newman JR, Weissman JS (2009) Genome-wide analysis in vivo of translation with nucleotide resolution using ribosome profiling. Science 324: 218–223
26. Hsieh AC, Liu Y, Edlind MP, Ingolia NT, Janes MR, Sher A, Shi EY, Stumpf CR, Christensen C, Bonham MJ, Wang S, Ren P, Martin M, Jessen K, Feldman ME, Weissman JS, Shokat KM, Rommel C, Ruggero D (2012) The translational landscape of mTOR signalling steers cancer initiation and metastasis. Nature 485:55–61
27. Ingolia NT, Brar GA, Rouskin S, McGeachy AM, Weissman JS (2012) The ribosome profiling strategy for monitoring translation in vivo by deep sequencing of ribosome-protected mRNA fragments. Nat Protoc 7:1534–1550
28. Bellodi C, Kopmar N, Ruggero D (2010) Deregulation of oncogene-induced senescence and p53 translational control in X-linked dyskeratosis congenita. EMBO J 29:1865–1876
29. Ingolia NT, Lareau LF, Weissman JS (2011) Ribosome profiling of mouse embryonic stem cells reveals the complexity and dynamics of mammalian proteomes. Cell 147:789–802

Chapter 14

Studying Myc's Role in Metabolism Regulation

Anne Le and Chi V. Dang

Abstract

The *MYC* oncogene encodes a master transcription factor, Myc, which regulates genes involved in ribosome biogenesis, lipid synthesis, nucleic acid synthesis, intermediary metabolism, and cell growth and proliferation. The genomics of Myc target genes has been well-established through global mapping of Myc binding sites in a variety of different cancer cell lines. These studies highlight the importance of Myc in regulating glucose and glutamine metabolism as well as mitochondrial and ribosomal biogenesis. These genomic studies, however, only become relevant with the companion metabolic studies using a variety of methods to measure oxygen consumption, glucose uptake, or metabolic pathways based on ^{13}C-labeled glucose or glutamine uptake. These methods are described herein.

Key words MYC, Cancer metabolism, Oxygen consumption, Glucose uptake, Stable isotope resolved metabolomics (SIRM), 13C-labeled glucose or glutamine

1 Introduction

We have contributed to a richer understanding of the function of the *MYC* oncogene over the last several decades [1, 2]. Having established that the oncogenic Myc protein functions mainly as a transcription factor that dimerizes with Max, we have focused on the regulation of Myc target genes and how these genes confer Myc-mediated cellular phenotypes [3]. Studies from the last several decades indicate that Myc is a master regulator of cell growth through its direct regulation of ribosomal biogenesis coupled with regulation of cell metabolism [4, 5]. Once cell size reaches a critical point, Myc collaborates with E2F to trigger entry into S phase through direct regulation of nucleotide metabolism and DNA replication [6, 7]. Upon completion of S phase, cells in G2 phase depend on Myc to traverse through mitosis to complete their duplication [2]. The *MYC* gene is activated by growth factors, is sensitive to nutrient availability, and is diminished by nutrient deprivation and hypoxia [8]. By contrast, dysregulated MYC-induced cancer cells undergo enforced cell mass accumulation and hence

Laura Soucek and Nicole M. Sodir (eds.), *The Myc Gene: Methods and Protocols*, Methods in Molecular Biology, vol. 1012, DOI 10.1007/978-1-62703-429-6_14, © Springer Science+Business Media, LLC 2013

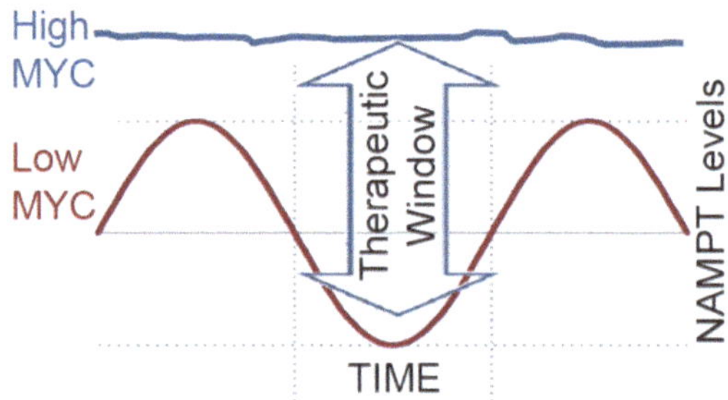

Fig. 1 Hypothetical therapeutic window of MYC-induced cancer

are addicted to nutrients, whose withdrawal triggers metabolic catastrophe and death of cells with sustained MYC expression [1]. We believe that this difference provides a therapeutic window for metabolic therapy; however, there are significant similarities between normal activated lymphocyte metabolism and the cancer metabolism of lymphoma cells [9]. As such, an additional therapeutic opportunity could exist in the window between circadian regulation of normal cell metabolism and the dysregulated, sustained non-circadian cancer metabolism (Fig. 1).

Our laboratory documented through genomic studies that Myc regulates ribosome biogenesis (cell growth) and virtually all genes involved in glycolysis, mediating the conversion of glucose to pyruvate [10]. We first identified that Myc directly regulates lactate dehydrogenase A (LDHA), which converts pyruvate to lactate, conferring aerobic glycolysis or the Warburg effect [11]. The Warburg effect, however, is unable to account for the composition of a growing cancer cell, which depends on more than just glucose for cell mass accumulation and cell division [12–14]. In this regard, we recently documented that Myc also regulates glutamine metabolism by inducing key glutamine plasma membrane transporters, such as SLC1A5 (ASCT2), and glutaminase (GLS), which converts glutamine to glutamate, for its further oxidation in the TCA cycle as α-ketoglutarate [15]. We also documented that Myc directly stimulates genes involved in mitochondrial biogenesis and enhances cellular respiration [16]. In this regard, proliferating cells use aerobic glycolysis to shunt carbon skeletons toward ribose and glycerol. They also depend on a hybrid TCA cycle comprising of both glucose and glutamine carbons to provide the necessary building blocks, such as fatty acids, amino acids and nucleotides, for the proliferating cancer cell [17, 18]. Thus, Myc promotes the acquisition of nutrients and energy for cell growth as well as regulating genes directly involved in ribosome biogenesis and cell proliferation [5].

Cell growth requires the generation of new proteins, lipid membranes, complex carbohydrates, and nucleic acids. As such, Myc appears to be involved in all these processes, by stimulating the expression of a plethora of metabolic genes prior to DNA replication [3]. Acute in vivo adenoviral-mediated overexpression of MYC in hepatocytes results in massive cellular hypertrophy with little evidence of cell proliferation, indicating that Myc is sufficient to induce ribosome biogenesis, increase cell size, and trigger G1 checkpoint, such as a p53-dependent one [19]. Biomass accumulation, however, should be accompanied by a commensurate amount of nutrients and bioenergetics, such that constitutive biomass accumulation enforced by ectopic Myc renders cells addicted to a constant supply of bioenergetic substrate. Indeed, we documented that ectopic Myc expression induces glucose-dependent apoptosis, and Yuneva et al. demonstrated that human cells overexpressing MYC are addicted to glutamine [20, 21]. These observations suggest that inhibition of enzymes involved in glucose or glutamine metabolism should be toxic to Myc-dependent cancer cells. In this regard, we provided proof-of-concept experiments that inhibition of LDHA with FX11 or glutaminase with BPTES or 968 could curb tumor progression in vivo [18, 22, 23].

Because normal lymphocytes (and presumably adult tissue stem cells) depend on Myc-mediated glycolysis and glutaminolysis for mitogenesis [9], pharmacological inhibition of metabolic enzymes could have undesirable immunosuppressive or myelotoxic effects [24]. Normal leukocytes, however, do display circadian gene expression, which is disrupted in chronic myelogenous leukemia cells [25]. In this regard, we began to determine whether ectopic Myc might disrupt cellular circadian rhythm in cancer cells, rendering them continuously susceptible to metabolic therapies while sparing normal cells at time points when normal metabolic activity is lowest. Thus, we hypothesize that the metabolic therapeutic window would be widened at certain times of the day when normal circadian-responsive tissues are at their lowest metabolic rates and hence would be spared from the toxic effects of metabolic inhibitors (Fig. 1).

To fully understand the role of Myc in metabolism, we have determined the role of Myc in mitochondrial biogenesis and established that Myc could increase oxygen consumption and glucose uptake [16, 22, 26]. Further, we have used ^{13}C-glucose or ^{13}C,^{15}N-glutamine coupled with stable isotope resolved metabolomics (SIRM) methods for tracking the fate of individual atoms from ^{13}C labeled precursors through various metabolic pathways [18, 27]. SIRM employs complementary analytical platforms, such as high resolution NMR, FT-ICR-MS, and GC-MS, to detect the magnetic or mass difference between ^{12}C and ^{13}C and other stable isotopes such as ^{15}N. NMR quantifies positional isotopomer distributions (the number and location of ^{13}C or ^{15}N atoms in a molecule), whereas MS methods determine the mass isotopologue distributions (numbers of ^{13}C or ^{15}N atoms in a molecule). In this chapter, we will provide protocols for measuring oxygen consumption, glucose uptake, and preparing isotopically labeled cellular extracts for metabolomics.

2 Materials

2.1 Measuring Oxygen Consumption

1. P493 human lymphoma B cells.
2. RPMI 1640 (Life Technologies).
3. Fetal bovine serum.
4. Penicillin-Streptomycin (P/S).
5. Oxytherm System (Hansatech Instruments Ltd).

2.2 Measuring NBD-Glucose Uptake by Flow Cytometry

1. 2-(*N*-(7-nitrobenz-2-oxa-1,3-diazol-4-yl)amino)-2-deoxyglucose (2-NBDG): 5 mg (Life Technologies).
2. Phosphate buffered saline (PBS).
3. Optional: Propidium Iodide (PI) (Life Technologies).

2.3 Labeling P493 Cells with ^{13}C Glucose or Glutamine

1. D-Glucose-$^{13}C_6$ (Sigma).
2. L-Glutamine-$^{13}C_5$,$^{15}N_2$ (Sigma).
3. RPMI without glucose, with glutamine (Life Technologies).
4. RPMI without glutamine, with glucose (Life Technologies).
5. RPMI without glutamine, without glucose (Teknova, or from Rainbow Scientific, Inc).
6. Dialyzed fetal bovine serum 10,000 MW ready to use (Life Technologies).
7. ^{13}C Glucose labeling medium: 0.2 % ^{13}C (make the stock 25 % in PBS and sterile filter), 10 % dialyzed FBS, 1 % penicillin-streptomycin. Bring up the remaining volume with RPMI with glutamine without glucose.
8. ^{13}C Glutamine labeling medium: 2 mM ^{13}C-Glutamine (make the stock 200 mM in PBS and sterile filter), 10 % dialyzed FBS (make the stock 2 M in PBS and sterile filter), 1 % Penicillin-Streptomycin (P/S). Bring up the remaining volume with RPMI with glucose without glutamine.

3 Methods

See **Note 1**.

3.1 Measuring Oxygen Consumption

Oxygen consumption is measured based on oxygen diffusing through the membrane, which traps a thin layer of electrolyte (KCl solution) on an electrode disc. The generated current bears a direct stoichiometric relation to the oxygen reduced and is converted to a digital signal.

1. Maintain P493 human lymphoma B cells at density of 10^5 cells/ml in RPMI 1640 with 10 % fetal bovine serum (FBS) and 1 % penicillin-streptomycin.
2. Collect 5×10^6 cells in 1 ml medium after a designed treatment time for measuring oxygen consumption using a Clark-type oxygen electrode (Oxytherm System).
3. Set up the Oxytherm System according to the manufacturer's instructions.
4. Pipette cells and medium into the chamber above a membrane that is permeable to oxygen.
5. Record the oxygen consumption using the provided software.

3.2 Measuring NBD-Glucose Uptake by Flow Cytometry

2-NBDG is a fluorescent glucose analog and is used to monitor glucose uptake in live cells. Its molecular formula is $C_{12}H_{14}N_4O_8$ and its molecular weight is 342.26. NBD fluorescence typically displays excitation/emission maxima of ~465/540 nm which can be detected by flow cytometry.

1. Treat the cells for a designed time (*see* **Note 2**).
2. Add fluorescent 100 μM 2-NBDG, with and without the designed treatment.
3. Return plates in the incubator for 30 min and keep them in the dark before flow cytometry analysis.
4. Remove the incubation medium and wash the cells twice with pre-cold PBS.
5. Resuspend cells in 200 μl pre-cold PBS.
6. Optional: Add PI at a final concentration of 1 μg/ml.
7. Place samples on ice and perform flow cytometry analysis within 30 min.
8. Measure FL1 and FL3 fluorescence intensity (2-NBDG and PI, respectively) of cells. Glucose uptake is assessed on the sub-population that has 2-NBDG positive and PI negative.

3.3 Labeling P493 Cells with ^{13}C Glucose or Glutamine

1. Prepare ^{13}C labeling glucose or glutamine medium.
2. Remove the incubation medium and wash the cells twice with PBS.
3. Add ^{13}C labeled glucose or glutamine to cells and let them grow for 24 h (*see* **Note 3**).
4. Use stable isotope resolved metabolomics (SIRM) methods, such as high resolution NMR, FT-ICR-MS, and GC-MS to track the fate of ^{13}C glucose or glutamine.

4 Notes

1. Store all stock solutions at -20 °C.
2. Dissolve 2-NBDG in distilled and filtered water at a concentration of 5 mM.
3. Adherent cell density in the plate should be about 80–95 % confluence in a 10 cm cell culture. Suspension cells should grow at 10^5 cells/ml.

References

1. Dang CV (2012) MYC on the path to cancer. Cell 149:22–35
2. Eilers M, Eisenman RN (2008) Myc's broad reach. Genes Dev 22:2755–2766
3. Dang CV, O'Donnell KA, Zeller KI, Nguyen T, Osthus RC, Li F (2006) The c-Myc target gene network. Semin Cancer Biol 16:253–264
4. Johnston LA, Prober DA, Edgar BA, Eisenman RN, Gallant P (1999) Drosophila myc regulates cellular growth during development. Cell 98:779–790
5. van Riggelen J, Yetil A, Felsher DW (2010) MYC as a regulator of ribosome biogenesis and protein synthesis. Nat Rev Cancer 10:301–309
6. Zeller KI, Zhao X, Lee CW, Chiu KP, Yao F, Yustein JT, Ooi HS, Orlov YL, Shahab A, Yong HC, Fu Y, Weng Z, Kuznetsov VA, Sung WK, Ruan Y, Dang CV, Wei CL (2006) Global mapping of c-Myc binding sites and target gene networks in human B cells. Proc Natl Acad Sci USA 103:17834–17839
7. Liu YC, Li F, Handler J, Huang CR, Xiang Y, Neretti N, Sedivy JM, Zeller KI, Dang CV (2008) Global regulation of nucleotide biosynthetic genes by c-Myc. PLoS One 3:e2722
8. Okuyama H, Endo H, Akashika T, Kato K, Inoue M (2010) Downregulation of c-MYC protein levels contributes to cancer cell survival under dual deficiency of oxygen and glucose. Cancer Res 70:10213–10223
9. Wang R, Dillon CP, Shi LZ, Milasta S, Carter R, Finkelstein D, McCormick LL, Fitzgerald P, Chi H, Munger J, Green DR (2011) The transcription factor Myc controls metabolic reprogramming upon T lymphocyte activation. Immunity 35(6):871–882.
10. Dang CV, Kim JW, Gao P, Yustein J (2008) The interplay between MYC and HIF in cancer. Nat Rev Cancer 8:51–56
11. Shim H, Dolde C, Lewis BC, Wu CS, Dang G, Jungmann RA, Dalla-Favera R, Dang CV (1997) c-Myc transactivation of LDH-A: implications for tumor metabolism and growth. Proc Natl Acad Sci USA 94:6658–6663
12. Vander Heiden MG, Cantley LC, Thompson CB (2009) Understanding the Warburg effect: the metabolic requirements of cell proliferation. Science 324:1029–1033
13. Dang CV (2012) Links between metabolism and cancer. Genes Dev 26:877–890
14. Koppenol WH, Bounds PL, Dang CV (2011) Otto Warburg's contributions to current concepts of cancer metabolism. Nat Rev Cancer 11:325–337
15. Gao P, Tchernyshyov I, Chang TC, Lee YS, Kita K, Ochi T, Zeller KI, De Marzo AM, Van Eyk JE, Mendell JT, Dang CV (2009) c-Myc suppression of miR-23a/b enhances mitochondrial glutaminase expression and glutamine metabolism. Nature 458:762–765
16. Li F, Wang Y, Zeller KI, Potter JJ, Wonsey DR, O'Donnell KA, Kim JW, Yustein JT, Lee LA, Dang CV (2005) Myc stimulates nuclearly encoded mitochondrial genes and mitochondrial biogenesis. Mol Cell Biol 25: 6225–6234
17. DeBerardinis RJ, Mancuso A, Daikhin E, Nissim I, Yudkoff M, Wehrli S, Thompson CB (2007) Beyond aerobic glycolysis: transformed cells can engage in glutamine metabolism that exceeds the requirement for protein and nucleotide synthesis. Proc Natl Acad Sci USA 104:19345–19350
18. Le A, Lane AN, Hamaker M, Bose S, Gouw A, Barbi J, Tsukamoto T, Rojas CJ, Slusher BS, Zhang H, Zimmerman LJ, Liebler DC, Slebos RJ, Lorkiewicz PK, Higashi RM, Fan TW, Dang CV (2012) Glucose-independent glutamine metabolism via TCA cycling for proliferation and survival in B cells. Cell Metab 15:110–121
19. Kim S, Li Q, Dang CV, Lee LA (2000) Induction of ribosomal genes and hepatocyte hypertrophy by adenovirus-mediated

expression of c-Myc in vivo. Proc Natl Acad Sci USA 97:11198–11202
20. Shim H, Chun YS, Lewis BC, Dang CV (1998) A unique glucose-dependent apoptotic pathway induced by c-Myc. Proc Natl Acad Sci USA 95:1511–1516
21. Yuneva M, Zamboni N, Oefner P, Sachidanandam R, Lazebnik Y (2007) Deficiency in glutamine but not glucose induces MYC-dependent apoptosis in human cells. J Cell Biol 178:93–105
22. Le A, Cooper CR, Gouw AM, Dinavahi R, Maitra A, Deck LM, Royer RE, Vander Jagt DL, Semenza GL, Dang CV (2010) Inhibition of lactate dehydrogenase A induces oxidative stress and inhibits tumor progression. Proc Natl Acad Sci USA 107:2037–2042
23. Wang JB, Erickson JW, Fuji R, Ramachandran S, Gao P, Dinavahi R, Wilson KF, Ambrosio AL, Dias SM, Dang CV, Cerione RA (2010) Targeting mitochondrial glutaminase activity inhibits oncogenic transformation. Cancer Cell 18:207–219
24. Gerriets VA, Rathmell JC (2012) Metabolic pathways in T cell fate and function. Trends Immunol 33:168–173
25. Yang MY, Yang WC, Lin PM, Hsu JF, Hsiao HH, Liu YC, Tsai HJ, Chang CS, Lin SF (2011) Altered expression of circadian clock genes in human chronic myeloid leukemia. J Biol Rhythms 26:136–148
26. Kim JW, Gao P, Liu YC, Semenza GL, Dang CV (2007) Hypoxia-inducible factor 1 and dysregulated c-Myc cooperatively induce vascular endothelial growth factor and metabolic switches hexokinase 2 and pyruvate dehydrogenase kinase 1. Mol Cell Biol 27:7381–7393
27. Liu W, Le A, Hancock C, Lane AN, Dang CV, Fan TW, Phang JM (2012) Reprogramming of proline and glutamine metabolism contributes to the proliferative and metabolic responses regulated by oncogenic transcription factor c-MYC. Proc Natl Acad Sci USA 109:8983–8988

Chapter 15

Generation of a Tetracycline Regulated Mouse Model of MYC-Induced T-Cell Acute Lymphoblastic Leukemia

Kavya Rakhra and Dean W. Felsher

Abstract

The tetracycline regulatory system provides a tractable strategy to interrogate the role of oncogenes in the initiation and maintenance of tumorigenesis through both spatial and temporal control of expression. This approach has several potential advantages over conventional methods to generate genetically engineered mouse models. First, continuous constitutive overexpression of an oncogene can be lethal to the host impeding further study. Second, constitutive overexpression fails to model adult onset of disease. Third, constitutive deletion does not permit, whereas conditional overexpression of an oncogene enables the study of the consequences of restoring expression of an oncogene back to endogenous levels. Fourth, the conditional activation of oncogenes enables examination of specific and/or developmental state-specific consequences.

Hence, by allowing precise control of when and where a gene is expressed, the tetracycline regulatory system provides an ideal approach for the study of putative oncogenes in both the initiation and maintenance of tumorigenesis. In this protocol, we describe the methods involved in the development of a conditional mouse model of MYC-induced T-cell acute lymphoblastic leukemia.

Key words Tetracycline inducible, Reversible expression, Doxycycline, MYC on, MYC off

1 Introduction

The development of genetically engineered mouse models (GEMM) of cancer has been invaluable to the study of oncogenes in the initiation of tumorigenesis [1, 2]. The development of the transgenic mouse enabled the interrogation of the function of specific genes. The more recent development of conditional transgenic models has enabled the study of both the adult onset of oncogene activation that would be characteristic of most types of human neoplasia, as well as the influence of the suppression of oncogene expression to study its role in the maintenance of a neoplastic phenotype.

To augment the utility of conventional GEMMs, several strategies have been employed to provide conditional transgene expression:

Laura Soucek and Nicole M. Sodir (eds.), *The Myc Gene: Methods and Protocols*, Methods in Molecular Biology, vol. 1012, DOI 10.1007/978-1-62703-429-6_15, © Springer Science+Business Media, LLC 2013

(1) the tetracycline regulatory system (Tet system) [3], (2) the Cre-Lox system [4], (3) hormone regulated expression of oncogenes via estrogen receptor or ecdysone inducible system [5], (4) hormone-based estrogen receptor (ER)-tamoxifen system [6], and (5) Lac repressor IPTG inducible system [7]. Each system has unique advantages and disadvantages. The Tet system enables reversible expression of a transgene. The Cre-Lox approach allows for irreversible activation or inactivation of endogenous gene expression. In the ER-tamoxifen system, regulation of transgene activity occurs posttranslationally, allowing for induction in a graded and rapid manner. In this chapter, we describe the generation of a GEMM of c-MYC-induced T-cell acute lymphoblastic leukemia (T-ALL) using the Tet system and discuss some of the unique advantages of this experimental system.

The Tet system was developed by the laboratory of Dr. Hermann Bujard [3] by modifying genetic elements of the bacterial tetracycline resistance operon for use in mammalian cells [3]. This genetic system can be used to conditionally induce or suppress target gene expression. In the Tet-OFF version of the system, the tetracycline transactivating protein (tTA), consisting of the bacterial Tet repressor protein fused to the Herpes Simplex Virus (HSV) VP16 transactivating domain, binds with high affinity to Tet operator (tetO) sequences and activates expression of a downstream gene. In the absence of tetracycline or a tetracycline analog such as doxycycline, there is robust expression induced by tTA binding to tetO. When doxycycline is present, tTA can no longer bind the tetO sequences and gene expression is turned off. Conversely, the Tet-ON system utilizes a reverse tetracycline transactivating protein (rtTA); tetracyclines induce the ability of rtTA to bind to tetO resulting in induction of gene expression, whereas in the absence of tetracyclines gene expression is suppressed [8].

We have used both the Tet-OFF and Tet-ON versions of the Tet system to conditionally regulate oncogene expression (Fig. 1). The Tet-ON version of the system is useful to rapidly induce gene expression. The Tet-OFF version is most useful to enable the rapid shut down of gene expression. The choice of which system to use is dictated by the length of time for which mice will be treated with tetracycline. For long-term tumor initiation studies, the Tet-OFF system allows for oncogene expression without requiring continuous administration of tetracycline, while for short-term studies, the Tet-ON system provides rapid induction of oncogene expression upon treatment with tetracycline.

We used the Tet-OFF system to generate a transgenic mouse line in which our gene of interest, the human c-MYC oncogene, can be reversibly expressed [9]. The expression of human c-MYC is induced in the hematopoietic tissues of FVB/N mice and results in a disease resembling human T-cell acute lymphoblastic leukemia (T-ALL). We discuss below how we characterized this model system [9].

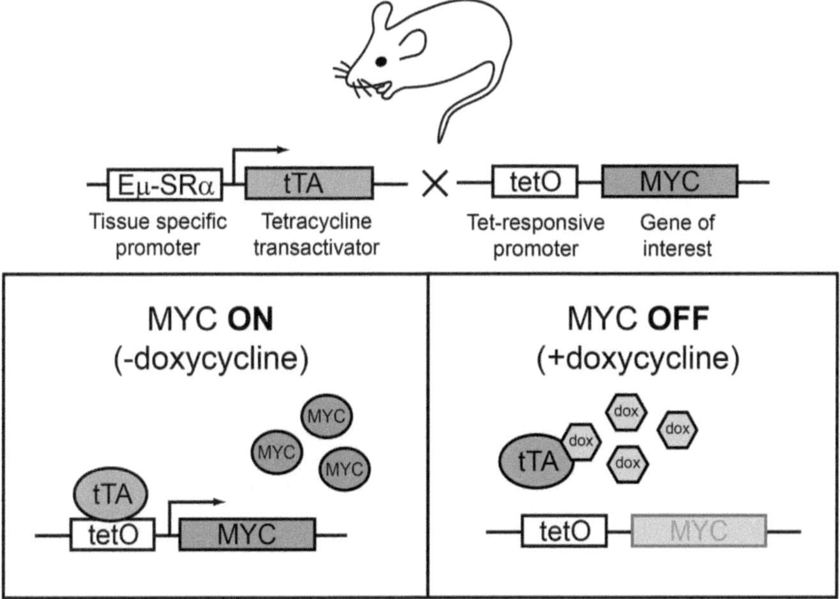

Fig. 1 The Tet-OFF System. A transgenic mouse line that expresses the tetracycline transactivator (tTA) downstream of a tissue-specific promoter (Eμ-SRα) is crossed with a mouse line carrying the MYC oncogene downstream of a tetracycline sensitive promoter (tet-O). The progeny of this cross now contains both transgenes. In the absence of doxycyline, tTA can bind to the tet-O promoter turning on MYC expression. However, when doxycycline is present, tTA binds to doxycycline and undergoes a conformational change, preventing it from binding the tet-O promoter and thus MYC expression is shut down

Our protocol focuses on the application of the Tet system to regulate the c-MYC oncogene in hematopoietic cells, but similar strategies can be used for the inducible expression of any gene of interest in other cellular lineages.

The Tet system is a bi-transgenic system and here we will describe the generation of two transgenic mouse lines. One transgenic mouse line expresses the tetracycline transactivator (tTA) gene downstream of a tissue-specific promoter, while the other carries the human c-MYC transgene downstream of a tTA/tetracycline-dependent (tetO) promoter (Fig. 1). The following protocols will be described below: (1) Generating Tet-o-MYC and EμSRα-tTA constructs, (2) In vitro validation of conditional expression of the Tet-o-MYC construct, (3) Genotypic analysis of founders and progeny, (4) Identification of founders with Conditional Gene Expression, (5) Mouse necroscopy and sample collection, (6) FACS analysis of primary tumor to identify the phenotype of cells comprising the tumor, (7) Generating cell lines from primary tumor tissue, (8) Evaluating conditional expression of Tet system in vivo, and (9) In vivo tumor transplantation studies.

2 Materials

2.1 Generation of the Tet-o-MYC and EμSRα-tTA Constructs

1. c-*Myc* pSP65 plasmid (received from M Bishop).
2. pUHD10-3 (received as a gift from H Bujard).
3. pUHD15-1 (received as a gift from H Bujard).
4. EμSRα plasmid (received as a gift from J Adams).
5. *EcoRI*, *BamH1*, *XhoI*, *HindIII*, and *NotI* enzymes.
6. T4 DNA ligase.
7. Klenow fragment of DNA polymerase.
8. 1 and 0.8 % Agarose gel.
9. Standard gel extraction kit (i.e., Qiagen QIAEX II gel extraction kit).

2.2 In Vitro Validation of Conditional Expression of the Tet-o-MYC Construct

1. 20 ng/ml Doxycycline hydrochloride in cell growth medium.
2. Mammalian cell lines with optimized expression of either tTA or rtTA, CHO-tTA, and HELA-tTA.

2.3 Genotypic Analysis of Founders and Progeny

1. 10 mg/ml Proteinase K.
2. 1× digestion buffer: mix 5 ml 1 M Tris–HCl pH 8.0, 20 ml 0.5 M Ethylenediaminetetraacetic acid (EDTA), 2 ml 5 M NaCl, 10 ml 10 % Sodium dodecyl sulfate (SDS), and 63 ml sterile distilled water.
3. 1× TE buffer: mix 10 ml of 1 M Tris–HCl pH 7.5, 2 ml 0.5 M EDTA pH 8.0, and 988 ml sterile distilled water.
4. 100 % ethanol.
5. 70 % ethanol.
6. Primer sequences to identify the Tet-o-MYC founders:

   MYC1: TAG TGA ACC GTC AGA TCG CCT G.

   MYC2: TTT GAT GAA GGT CTC GTC GTC C.
7. Primer sequences to identify the EμSRαtTA founders:

   EμSRαtTA1: AGG CCT GTA CGG AAG TGT.

   EμSRαtTA2: CTC TGC ACC TTG GTG ATC.
8. PCR reaction mixture used for genotyping:
   - Template DNA.
   - Primer Mix.
   - Taq DNA polymerase (5 U/μl).
   - dNTP mixture (10 mM).
   - QIAGEN PCR buffer (10×).

- Autoclaved distilled water.
- Details are provided in methods below.

9. DNA loading buffer.
10. 1 % Agarose gel.
11. Ethidium bromide.

2.4 Identification of Founders with Conditional Gene Expression

1. 1× Phosphate Buffer Saline (PBS).
2. 100 μg/ml filter sterilized solution of doxycycline hydrochloride in PBS.
3. Needles.
4. Syringes.
5. 100 μg/ml solution of doxycycline hydrochloride in the drinking water.

2.5 Mouse Necroscopy and Sample Collection

1. Dissection scissors.
2. Styrofoam board.
3. 70 % ethanol.
4. 10 % neutral buffered formalin.
5. Optimal Cutting Temperature (OCT) freezing medium.
6. Liquid nitrogen.

2.6 FACS Analysis of Primary Tumor to Identify the Phenotype of Cells Comprising the Tumor

1. 100 μm cell strainers.
2. 50 ml conical tubes.
3. PBS.
4. Hemocytometer.
5. FACS buffer: 1 % bovine serum albumin in PBS.
6. FACS tubes.
7. Fluorescently labeled antibody for various markers of hematopoietic cells such as CD4, CD8, C19, Mac1, Gr1.
8. Flow cytometer.

2.7 Generating Cell Lines from Primary Tumor Tissue

1. Ice cold and filter sterilized or autoclaved PBS.
2. RPMI-1640 medium.
3. RPMI-1640 medium with 20 % Fetal Bovine Serum (FBS): Mix 500 ml of RPMI 1640, 100 ml FBS, 5 ml Penicillin-Streptomycin (100×) and 0.55 ml β-mercaptoethanol.
4. Incubator at 37 °C and 5 % CO_2.
5. Freezing medium: Mix 45 ml of heat inactivated FBS and 5 ml Dimethyl sulfoxide (DMSO).

2.8 Evaluating Conditional Expression of Tet System In Vivo

1. Cell culture 6 well plates.
2. 20 ng/ml of doxycycline.

2.9 In Vivo Tumor Transplantation Studies

1. Hemocytometer.
2. Cold sterile PBS.
3. Electric shaver or depilatory cream.
4. 27 G needles.
5. Syringes.

3 Methods

3.1 Generation of the Tet-o-MYC and EμSRα-tTA Constructs

3.1.1 To Prepare the Tet-o-MYC Transgenic Construct

1. Set up a restriction digest of MYC containing pSP65 plasmid (made in Michael Bishop's laboratory) using *EcoRI*. Resolve the digested DNA on a 1 % agarose gel and purify the 1.4 kb fragment corresponding to human c-MYC cDNA using a standard gel extraction kit. Also digest pUHD10-3 with EcoRI and purify the linearized plasmid.
2. Ligate the purified 1.4 kb c-MYC fragment from pSP65 into the linearized pUHD10-3 using T4 DNA ligase, generating pUHD10-3-c-MYC (*see* Fig. 2a for plasmid map).

3.1.2 To Prepare the EμSRα-tTA Transgenic Construct

1. Set up a double digest of pUHD15-1 with *EcoRI* and *BamHI*. Resolve the digested DNA on a 1 % agarose gel and purify the 1 kb fragment containing the tTA coding sequence. Generate blunt ends by treating the tTA fragment with the Klenow fragment of DNA polymerase. Also digest the EμSRα plasmid with *EcoRV*. Resolve the digested DNA on a 1 % agarose gel and purify the linearized plasmid using a standard gel extraction kit.
2. Ligate the 1 kb blunted tTA fragment into the linearized EμSRα plasmid using T4 DNA ligase, generating EμSRα-tTA (*see* Fig. 2b for plasmid map).

3.1.3 To Prepare the Transgenic Constructs for Microinjection

1. Set up a digest of 50 μg of pUHD10-3-c-MYC plasmid with *XhoI* and *HindIII*, and a separate digest of 50 μg of EμSRα-tTA with *NotI*. Resolve the digested DNA on a 0.8 % agarose gel. From the pUHD10-3-c-MYC digest, excise the 2.37 kb fragment containing the Tet-o-MYC sequence, and from the EμSRα-tTA digest, excise the 3.7 kb fragment containing the EμSRα-tTA sequence. Purify each fragment using a standard DNA purification kit.
2. Microinjection is a specialized procedure usually carried out by a transgenic animal facility. Details of a standard microinjection protocol are outlined in [10].

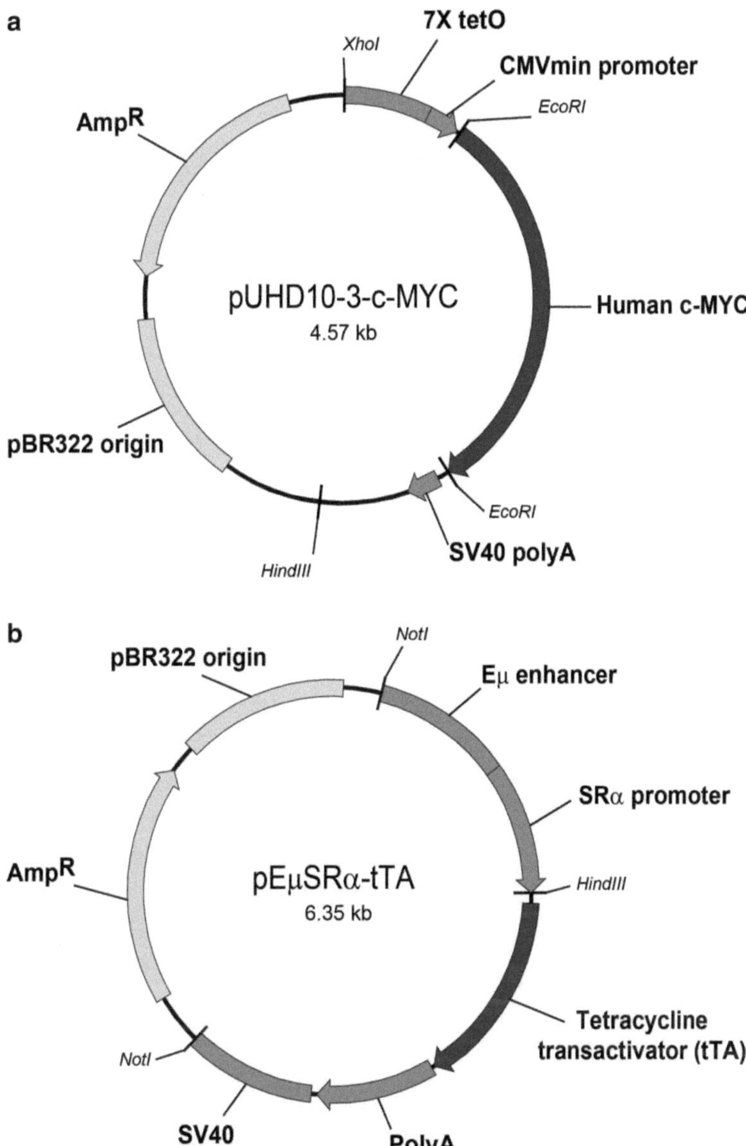

Fig. 2 Plasmids used to generate transgenic mice. (**a**) The human c-MYC cDNA is excised from the pSP65 plasmid and cloned into the pUHD-10-3 plasmid downstream of the tet-O promoter. (**b**) The tTA fragment is excised from plasmid pUHD15-1 and cloned downstream of the SR-α promoter in the Eμ-SRα plasmid

3.2 In Vitro Validation of Conditional Expression of the Tet-o-MYC Construct

Prior to the use of the constructs for the generation of transgenic mice, it is critical to ascertain that their conditional expression can be regulated by doxycycline. There are several commercially available mammalian cell lines with optimized expression of either tTA or rtTA for the purpose of testing your tet-responsive construct including CHO-tTA and HELA-tTA. In vitro validation involves transfection of a tTA/rtTA-expressing cell line with your tetO plasmid in the presence and absence of doxycycline and assessing both mRNA and protein levels of your gene using quantitative real-time PCR and western blotting, respectively. Transcriptional downregulation will be very rapid (within hours), while the downregulation of protein levels will mainly depend on the natural turnover of the protein. MYC protein has an extremely short half-life of 20–30 min [11] (*see* **Note 1**).

3.3 Genotypic Analysis of Founders and Progeny

The Tet-o-MYC and EμSRαtTA founders will be identified by a PCR-based screening assay using DNA isolated from the tails of transgenic mice as described below.

DNA Extraction from Mouse Tails:

1. When mice are between 15 and 20 days of age, use a pair of dissection scissors to cut the ends of their tails (0.1–0.5 cm) that are placed in a previously labeled 1.5 ml centrifuge tube. After the tail is cut, control the bleeding at the end of the tail by applying pressure or rubbing the tail in styptic powder (*see* **Notes 2** and **3**).
2. Add 600 μl of digestion buffer and 25 μl of Proteinase K (10 mg/ml) to each tube containing a single tail biopsy.
3. Incubate the tubes for approximately 12 h (overnight) at 55 °C with constant shaking or rotation.
4. After this incubation, spin tubes for 15 min at full speed in a centrifuge to obtain a clear supernatant.
5. Transfer 500 μl of the supernatant into a centrifuge tube containing 1 ml of absolute ethanol at room temperature.
6. Gently invert the tubes several times. Precipitation of DNA should be visible as white transparent strands.
7. Spin the tubes briefly (15 s) in the centrifuge and discard the supernatant.
8. Add 400 μl of 70 % ethanol, spin briefly again, and discard the supernatant.
9. Dry the DNA pellets by leaving the centrifuge tubes with their caps open for 10 min at room temperature.
10. Dissolve the DNA in 500 μl of TE buffer.
11. Use this DNA to perform PCR assays to check for the presence or absence of tTA and human c-MYC. Use the primers MYC1 and MYC2 (sequences are provided in subheading 2.3) to identify the Tet-o-MYC founders and the primers EμSRαtTA1 and EμSRαtTA2 (sequences are provided in subheading 2.3)

to identify the EμSRαtTA founders. The size of the amplicon is 450–500 bp and 500 bp, respectively. The PCR reaction and the PCR cycle conditions used for genotyping are described below.

PCR reaction mixture used for genotyping:

Component	Volume (μl)
Template DNA	1
Primer mix (10 μM each)	0.5 (×2)
Taq DNA polymerase (5 U/μl)	0.15
dNTP mixture (10 mM)	0.5
Qiagen PCR buffer (10×)	2.5
Autoclaved distilled water	20

PCR cycle conditions for genotyping:

Step 1: 94 °C for 5 min.

Step 2: 94 °C for 30 s (Denature).

Step 3: 57 °C for 30 s (Anneal).

Step 4: 72 °C for 30 s.

Step 5: Cycle **steps 2–5** 35 times.

Step 6: 72 °C for 5 min (Extend).

Step 7: 4 °C for 99 h (End Hold).

12. Once the PCR reaction is complete, add DNA loading buffer to the samples and run them on a 1 % agarose gel containing ethidium bromide to detect the appropriate PCR products.

Based on the results of the genotyping, the correct founders must be chosen for each transgene, Tet-o-MYC and EμSRαtTA. The founders that are chosen for breeding must express the transgenic protein and have the ability to pass on the transgene to their progeny.

3.4 Identification of Founders with Conditional Gene Expression

Once the appropriate founders have been obtained and stable breeding lines have been established, cross the Tet-o-MYC mice with the EμSRαtTA mice. To obtain progeny that possess both transgenes, it is necessary to mate male and female pairs of mice each with one of the Tet system transgenes.

When using the Tet-OFF version of the system, the Tet-regulated transgene (in this case MYC) is expressed in the absence of tetracycline or its derivatives like doxycycline. Notably, constitutive activity of a transgene in some tissues can be associated with rapid disease onset. Indeed, we have noted that constitutive expression of MYC in the liver is associated with rapid mortality due to liver cancer. Hence, it is important that mice are carefully observed at least twice a week for signs of disease such as ruffled fur, lethargy, hunched posture, and loss of appetite.

Constitutive expression of MYC in the hematopoietic compartments is associated with neoplastic disease onset as early as 8 weeks and in all mice by 12 weeks of age. Tumor formation due to MYC overexpression is apparent in the increased size of lymphoid organs such as the thymus, spleen, and lymph nodes. Enlargement of the liver was also seen due to infiltration of tumor cells. The mice succumbed to invasive tumors within 5 months of age.

To rapidly turn off MYC expression in transgenic mice, we inject intraperitoneally (i.p.) 100 μl of a 100 μg/ml filter sterilized solution of doxycycline hydrochloride in PBS. Then, doxycycline administration is maintained through oral administration of a 100 μg/ml solution of doxycycline hydrochloride in the drinking water (*see* **Notes 4** and **5**). After treatment with doxycycline, mice with tumors begin to show evidence for regression within 2 days with full recovery within 1 week. In some cases, especially in mice with very large tumor burden, the rapid regression of tumors is associated with tumor lysis syndrome and mice rapidly die.

3.5 Mouse Necropsy and Sample Collection

When mice are moribund with tumor, they can be euthanized through carbon dioxide asphyxiation and necropsies can be performed to analyze the characteristics of the MYC-induced tumorigenesis.

After euthanasia:

1. Pin all four extremities of the mouse to a styrofoam board and spray down with 70 % ethanol to avoid contamination of the murine tissues that will be harvested.
2. Use dissection scissors to cut the mouse from the pubis till the neck in a vertical line through the middle of the body.
3. Cut open the peritoneal cavity to display the organs in the thorax and abdomen.
4. Lymphomas cause enlargement of organs including spleen, lymph nodes, thymus, and liver.
5. From each organ, remove the tumors and cut them into pieces that are about 5 × 5 mm in size. Fix one piece in 10 % neutral buffered formalin overnight to make paraffin embedded blocks. Tissue sections from these blocks can be used to perform hematoxylin and eosin (H&E) staining to study tissue morphology. One of the tumor pieces should be frozen in an embedding compound such as Optimal Cutting Temperature (OCT) freezing medium and frozen at −80 °C to preserve the antigen integrity of the tissue so that it can be stained to study the expression of different proteins of interest. Pieces of tumor

about 2 × 2 mm in size are also flash frozen in liquid nitrogen and stored at −80 °C for PCR and western blotting analysis in the future. Tumors that arise spontaneously from a mouse are referred to as primary tumors for the rest of this chapter, to distinguish them from tumors that will be obtained from cell lines derived from primary tumors.

3.6 FACS Analysis of Primary Tumor to Identify the Phenotype of Cells Comprising the Tumor

1. Gently crush a piece of the tumor tissue obtained from the mouse necropsy in between the ends of 2 frosted glass slides or by using the back of a syringe plunger to pass the tumor pieces through a 100 μm cell strainer in a 50 ml conical tube.
2. Wash these cells with 30 ml of PBS.
3. Count the cells using a hemocytometer and make a cell suspension of 10^7/ml in FACS buffer (PBS with 1 % bovine serum albumin). This serves as a blocking buffer while staining the cells with FACS antibodies.
4. For each antibody being used, mix 100 μl of this cell suspension (~1×10^6 cells) with 1 μl of fluorescently labeled antibody solution in a FACS tube. For MYC-induced lymphomas, we typically stain for various markers of hematopoietic cells (e.g., CD4, CD8, C19, Macl, Grl) to identify the specific lineage and differentiation stage of the tumor.
5. Incubate for 20 min at 4 °C protected from light to prevent the fluorescent label of the FACS antibodies from photobleaching.
6. Wash the cells in FACS buffer twice and centrifuge the cells at 300 × *g* and discard the supernatant.
7. Resuspend the cell pellet in 500 μl of PBS and run the sample on a flow cytometer.

More than 80 % of the tumors we analyze are immature $CD4^+CD8^+$ T-cell lymphomas also positive for Thyl, CD3, and CD5. However, rare tumors arise from the myeloid lineage and stain positive for Grl and Macl surface markers. Also, introduction of other genetic events can be associated with a change in the phenotype of the resulting tumors suggesting that it is important to characterize multiple tumors to study the consequence of ectopic oncogene overexpression (*see* **Note 6**).

3.7 Generating Cell Lines from Primary Tumor Tissue

Here, we describe a protocol to develop cell lines from T-cell lymphomas. This protocol is adapted from *Methods in Molecular Biology, Vol. 134: T Cell Protocols, Chapter 14*. For many experimental purposes, it is useful to generate tumor-derived cell lines. We have found that these lines can be readily generated and that they continue to conditionally express the MYC oncogene as regulated by the Tet system.

1. Dissect tumor pieces from the necropsy of a Tet-o-MYC X EμSRαtTA mouse in which MYC has been turned on from birth and lymphomas have developed (this will occur at around 6–8 weeks of age).
2. Wash these tumor pieces several times with ice cold and filter sterilized or autoclaved PBS.
3. Cut the tumor piece into several small pieces (1–2 mm) and transfer them into a tissue culture plate with 5 ml of complete RPMI-1640 medium with 20 % FBS.
4. Transfer the tumor pieces and the medium into a 15 ml centrifuge tube and add complete RPMI-1640 medium to make up the volume to 13 ml. Place this tube in a stand and allow the tumor pieces to settle for 2–3 min.
5. Once tumor pieces have settled to the bottom of the tube, the free floating cells remain in suspension, which will become turbid. Remove the turbid suspension without disturbing the settled tumor tissue.
6. Add 13 ml of complete RPMI-1640 medium to the tube and gently shake it to allow the tumor pieces to settle once again and remove the cells in suspension.
7. Repeat this process (4–5 times) until the medium appears clear after shaking with the tumor pieces.
8. Remove the tumor pieces from the bottom of the tube with the remaining medium and transfer these into a 25 cm^2 sterile tissue culture flask. Rinse out the tube with 5 ml of complete RPMI-1640 medium twice and add this medium to the flask.
9. Transfer the flask into an incubator at 37 °C and 5 % CO_2 and observe the flask carefully for the next few days. A layer of adherent cells will form along with some cells in suspension as well. Replace 50 % of the medium in the flask as it turns acidic or yellow in color.
10. After 2–4 weeks, you will observe tumor cells growing in suspension as clumps. These cells can be split into fresh medium by taking 1 ml of these cells and transferring them into a new flask containing 10 ml complete RPMI-1640 medium. Once cell lines have been adapted to growth in vitro, they can be maintained using complete RPMI medium containing 10 % FBS.
11. Once in vitro cell line growth has been achieved, freeze down aliquots of cells after they have reached the late log phase of growth. Cells are frozen at a density of $6–10 \times 10^6$ cells/ml in 10 % DMSO in heat inactivated FBS and store at −80 °C. This is the first passage of the new lymphoma cell line.

3.8 Evaluating Conditional Expression of Tet System In Vivo

Once a lymphoma cell line is generated from tumor bearing Tet-o-MYC X EμSRαtTA mice, it is imperative to check whether or not these tumor cell lines express MYC and to test whether MYC expression in these cell lines is inducible to confirm that the tumor generated in the mouse was indeed due to MYC expression using the Tet system.

1. From a cell line grown in vitro, set up a seed culture of 6 ml at a density of 100,000 lymphoma cells per ml in all the wells of two 6 well plates.
2. To generate samples in triplicate, in 9 of these wells (6 from one plate and 3 from the other plate), add 20 ng/ml of doxycycline. In the remaining 3 wells, add the same volume of medium as a control. At various time points after the addition of doxycycline (12, 24 and 48 h), spin down the cells from each of 3 wells treated with doxycycline and snap freeze the pellets in liquid nitrogen before storing at −80 °C. Similarly, pellet out cells from the 3 untreated wells as samples for the 0 h time point.
3. Perform PCR and western blotting on the samples generated to check for the expression of the human *c-MYC* gene and protein to test whether its expression is inhibited by doxycycline treatment.

3.9 In Vivo Tumor Transplantation Studies

To further study the behavior of a single primary tumor in vivo, cell lines obtained from the primary tumor using the protocol described above are transplanted into mice subcutaneously. 10^7 tumor cells are injected underneath the skin of a shaved area on the back of the mouse.

1. Count in vitro passaged cell line using a hemocytometer.
2. Measure out the appropriate volume of your required cell culture and centrifuge to spin down the cells at 300 × *g* for 5 min.
3. Decant the supernatant without disturbing the cell pellet. Resuspend the cell pellet in ice cold sterile PBS. Centrifuge this at 300 × *g* for 5 min and repeat this wash twice.
4. After the last wash, resuspend the cell pellet in ice cold sterile PBS at a concentration of 10^7 cells per ml.
5. Before injecting these cells, shave a small area on the back of the mouse using an electric shaver or a depilatory cream. Load this cell suspension into a syringe with a 27 G needle and inject 100 μl into the subcutaneous space on the shaved back of the mouse (*see* **Note 7**).

The tumor will grow to a size of 1 × 1 cm within 9–14 days. Once the tumors have been established, mice can be treated with doxycycline orally or intraperitoneally as described before to test for responsiveness to doxycycline and to carry out further studies in vivo. Subcutaneous tumors can be harvested like the primary tumors described above.

The initial characterization of this transgenic model is described in [9]. This model has been used to test how tumor cells behave in vivo when MYC is turned off, how the oncogenic potential of MYC varies at different stages of development, and to titrate the amount of MYC required to initiate tumorigenesis [12]. Similar strategies can be used to reversibly express MYC in various tissues such as liver [13], bone [14], and other tissues of interest depending on the tissue-specific promoter/enhancer elements used. Similarly, other conditional mouse models can help understand the nuanced role of oncogenes in tumor initiation and discover new phenomena like oncogene addiction.

4 Notes

1. Our experience for many different constructs is that we observe significant downregulation of protein levels within 2–4 h after addition of doxycycline.
2. It is preferable to perform tail biopsies in young mice since they yield more DNA. If they are being performed in older mice, the use of anesthesia is recommended.
3. Make sure that instruments are cleaned thoroughly with ethanol and/or chlorhexidine when moving from one mouse to the next as DNA from one sample might contaminate other samples and yield erroneous PCR genotyping results.
4. As doxycycline hydrochloride is sensitive to light, amber drinking water bottles must be used to administer doxycycline orally and must be replaced every week. Alternatively, mouse chow with doxycycline is commercially available from some vendors.
5. Doxycycline stock solutions (20 mg/ml) can be prepared and stored at −20 °C for up to 6 months at this temperature. Doxycycline solutions when stored at 4 °C are stable for at least 2 weeks.
6. When the EμSRα construct is expressed upstream of c-MYC in the FVB/N strain of mice, MYC expression is restricted to T cells and T-cell lymphomas arise [15]. In contrast, in the C57BL6 background, the tumors arise in B cells.
7. To inject cells subcutaneously in the mouse, make sure to prepare an excess volume of cells to account for loss of cells in the syringe. For example, to inject 5 mice, make up at least 600 μl of cell suspension for injection.

Acknowledgments

We would like to acknowledge Peter Choi for his help with the figures in this manuscript. The work described in this chapter was supported by the following grants: Burroughs Wellcome Fund Career Award, the Damon Runyon Foundation Lilly Clinical Investigator Award, NIH RO1 grant number CA 089305, 105102, National Cancer Institute's In vivo Cellular and Molecular Imaging Center grant number CA 114747, Integrative Cancer Biology Program grant number CA 112973, NIH/NCI PO1 grant number CA034233, and the Leukemia and Lymphoma Society Translational Research grant number R6223-07.

References

1. Politi K, Pao W (2011) How genetically engineered mouse tumor models provide insights into human cancers. J Clin Oncol 29: 2273–2281
2. Sharpless NE, Depinho RA (2006) The mighty mouse: genetically engineered mouse models in cancer drug development. Nat Rev Drug Discov 5:741–754
3. Gossen M, Bujard H (1992) Tight control of gene expression in mammalian cells by tetracycline-responsive promoters. Proc Natl Acad Sci U S A 89:5547–5551
4. Utomo AR, Nikitin AY, Lee WH (1999) Temporal, spatial, and cell type-specific control of Cre-mediated DNA recombination in transgenic mice. Nat Biotechnol 17: 1091–1096
5. No D, Yao TP, Evans RM (1996) Ecdysone-inducible gene expression in mammalian cells and transgenic mice. Proc Natl Acad Sci U S A 93:3346–3351
6. Littlewood TD, Hancock DC, Danielian PS, Parker MG, Evan GI (1995) A modified oestrogen receptor ligand-binding domain as an improved switch for the regulation of heterologous proteins. Nucleic Acids Res 23:1686–1690
7. Cronin CA, Gluba W, Scrable H (2001) The lac operator-repressor system is functional in the mouse. Genes Dev 15:1506–1517
8. Zhou X, Vink M, Klaver B, Berkhout B, Das AT (2006) Optimization of the Tet-On system for regulated gene expression through viral evolution. Gene Ther 13:1382–1390
9. Felsher DW, Bishop JM (1999) Reversible tumorigenesis by MYC in hematopoietic lineages. Mol Cell 4:199–207
10. Cho A, Haruyama N, Kulkarni AB (2009) Generation of transgenic mice. Curr Protoc Cell Biol Chapter 19, Unit 19.11. doi: 10.1002/0471143030.cb1911s42
11. Hann SR, Eisenman RN (1984) Proteins encoded by the human c-myc oncogene: differential expression in neoplastic cells. Mol Cell Biol 4:2486–2497
12. Shachaf CM, Gentles AJ, Elchuri S, Sahoo D, Soen Y, Sharpe O, Perez OD, Chang M, Mitchel D, Robinson WH, Dill D, Nolan GP, Plevritis SK, Felsher DW (2008) Genomic and proteomic analysis reveals a threshold level of MYC required for tumor maintenance. Cancer Res 68:5132–5142
13. Shachaf CM, Kopelman AM, Arvanitis C, Karlsson A, Beer S, Mandl S, Bachmann MH, Borowsky AD, Ruebner B, Cardiff RD, Yang Q, Bishop JM, Contag CH, Felsher DW (2004) MYC inactivation uncovers pluripotent differentiation and tumour dormancy in hepatocellular cancer. Nature 431:1112–1117
14. Jain M, Arvanitis C, Chu K, Dewey W, Leonhardt E, Trinh M, Sundberg CD, Bishop JM, Felsher DW (2002) Sustained loss of a neoplastic phenotype by brief inactivation of MYC. Science 297:102–104
15. Kemp DJ, Harris AW, Cory S, Adams JM (1980) Expression of the immunoglobulin C mu gene in mouse T and B lymphoid and myeloid cell lines. Proc Natl Acad Sci U S A 77:2876–2880

Chapter 16

Methods to Assess Myc Function in Intestinal Homeostasis, Regeneration, and Tumorigenesis

David J. Huels, Patrizia Cammareri, Rachel A. Ridgway, Jan P. Medema, and Owen J. Sansom

Abstract

Within the intestinal epithelium, c-Myc has been characterized as a target of β-catenin-TCF signalling (He et al., Science 281:1509–1512, 1998). Given the most commonly mutated tumor suppressor gene within colorectal cancer (CRC) is the *APC* (Adenomatous Polyposis Coli) gene, a negative regulator of β-catenin-TCF signalling (Korinek et al., Science 275:1784–1787, 1997), loss of APC leads to Myc deregulation in the vast majority of CRC. This probably explains the numerous studies investigating c-Myc function within the intestinal epithelium. These have shown that c-Myc inhibition or deletion in the adult intestine results in proliferative defects (Muncan et al., Mol Cell Biol 26:8418–8426, 2006; Soucek et al., Nature 455:679–683, 2008). Importantly, intestinal enterocytes are able to survive in the absence of c-Myc which has allowed us (and others) to test the role of c-Myc in intestinal regeneration and tumorigenesis. Remarkably c-Myc deletion suppresses all the phenotypes of the *Apc* tumor suppressor gene loss and stops intestinal regeneration (Ashton et al., Dev Cell 19:259–269, 2010; Sansom et al., Oncogene 29:2585–2590, 2007). This suggests a clear therapeutic rationale for targeting c-Myc in CRC. Moreover haploinsufficiency for c-Myc in this tissue also reduces intestinal tumorigenesis (Athineos and Sansom, Oncogene 29:2585–2590, 2010; Yekkala and Baudino, Mol Cancer Res 5:1296–1303, 2007), and overexpression of c-Myc affects tissue homeostasis (Finch et al., Mol Cell Biol 29:5306–5315, 2009; Murphy et al., Cancer Cell 14:447–457, 2008).

In this chapter we will provide an overview of our current laboratory protocols to characterize c-Myc function in intestinal homeostasis, regeneration, and tumorigenesis in vivo and in vitro.

Key words Intestinal regeneration, Crypt culture, Myc, Allograft

1 Introduction

1.1 Intestinal Regeneration

It has been shown that the modulation of c-Myc expression, as a target of β-catenin-TCF signalling [1], directly affects tissue homeostasis [3, 4, 9, 10]. Moreover the inducible expression of c-Myc is absolutely required for the ability of the intestine to regenerate post damage [3–5]. Injury can be provoked within the mouse using either radiation, cytotoxic injury (e.g., cisplatin), or physical damage (e.g., DSS, dextran sulfate) [11]. Pioneering

Laura Soucek and Nicole M. Sodir (eds.), *The Myc Gene: Methods and Protocols*, Methods in Molecular Biology, vol. 1012, DOI 10.1007/978-1-62703-429-6_16, © Springer Science+Business Media, LLC 2013

studies over 30 years ago by the Potten laboratory developed the intestinal microcolony assay to study the output of this damage in terms of "clonogenic survival." In that elegant study, 72 h post a high level of DNA damage, the numbers of surviving intestinal crypts were scored. As the DNA-damaging agents kill stem cells (now known to be Lgr5—leucine-rich repeat-containing G protein-coupled receptor 5—positive) and daughter cells, if all stem cells and repopulating daughter cells are killed, the crypt will die. If killing is not absolute, the crypt will regrow following the injury, and the number of the surviving crypts is indicative of clonogenic survival. The size of the regenerating crypts allows assessment of how well the crypt has regenerated. c-Myc deletion stops all crypts from regenerating and only a small number of cystic crypts remains. We and others have shown that the immediate apoptotic response is uncoupled from the regenerative burst of crypts at 48 h. It is important to investigate a time course of apoptosis and mitosis throughout a 72 h period. Below is our adaptation of the Potten microcolony assay [12–14].

1.2 Intestinal Crypt Culture

Crypt culture or "minigut system" allows analysis of the crypt epithelial cells without the stromal component (Fig. 1). The growth factors and the Paneth cells create a niche that allows maintenance of the stem cells [15, 16]. This system allows assessment of the regrowth of intestinal crypts in vitro, stem capacity, proliferation, and differentiation. Genetic deletion of genes such as *c-Myc* can be performed in vivo prior the establishment of organoids or once the crypts have established in vitro. To do this we use mouse models that have the gene of interest (e.g., *c-Myc*) flanked by two loxP sites in their genomic DNA. The addition of tamoxifen leads to the expression of a Cre recombinase under the control of a specific promoter (e.g., Villin-CreER). The recognition of the loxP sites by the Cre recombinase leads to the deletion of the loxP site-flanked gene. This provides an excellent system to study one's gene of interest that can be further manipulated.

1.3 Adenoma Culture

The APC (Adenomatous Polyposis Coli) gene is the most commonly deregulaleted gene in CRC [2]. Downregulation of c-Myc expression can attenuate the APC-deficiency phenotype [6–8]. We use this protocol to culture tumor cells from adenomas formed from mice that carry a single constitutive *Apc* mutation, e.g., $APC^{Min/+}$ or $APC\ 1322^{T/+}$ polyps [17, 18]. The spheres in culture recapitulate the tumors in vivo including their differentiation patterns (Fig. 2).

1.4 Injection of Cells/Spheres into Mice

The subcutaneous injection of tumor cells into immunocompromised mice allows the tumorigenic and self-renewal capacity of tumor spheres to be tested. Moreover by injecting cells into the original tissue (orthotopically), this should recapitulate better the

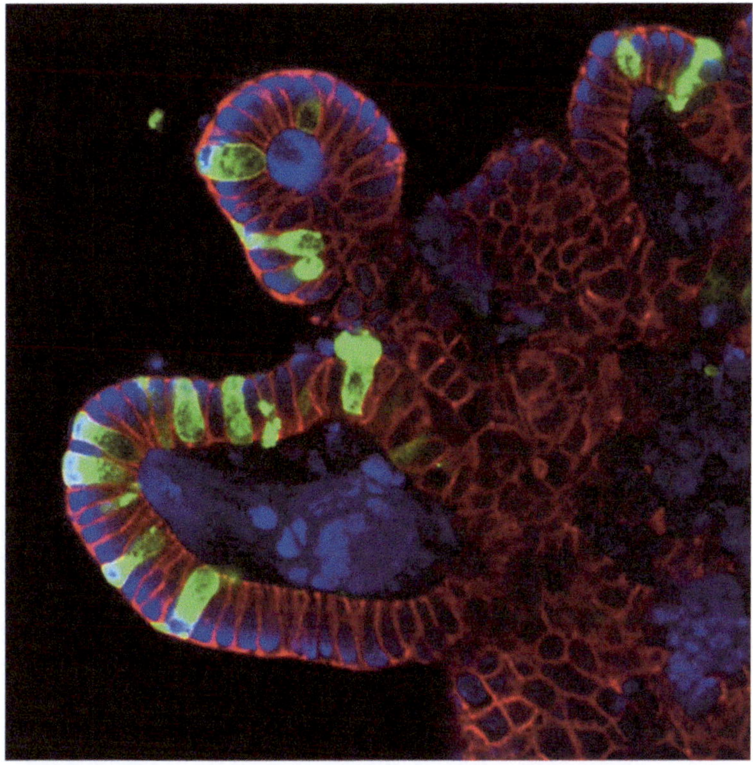

Fig. 1 Immunofluorescence staining of wild-type organoid at day 5. The Paneth cells are stained for lysozyme (*green*), the cell membrane for β-catenin (*red*), and the nucleus with DAPI (*blue*)

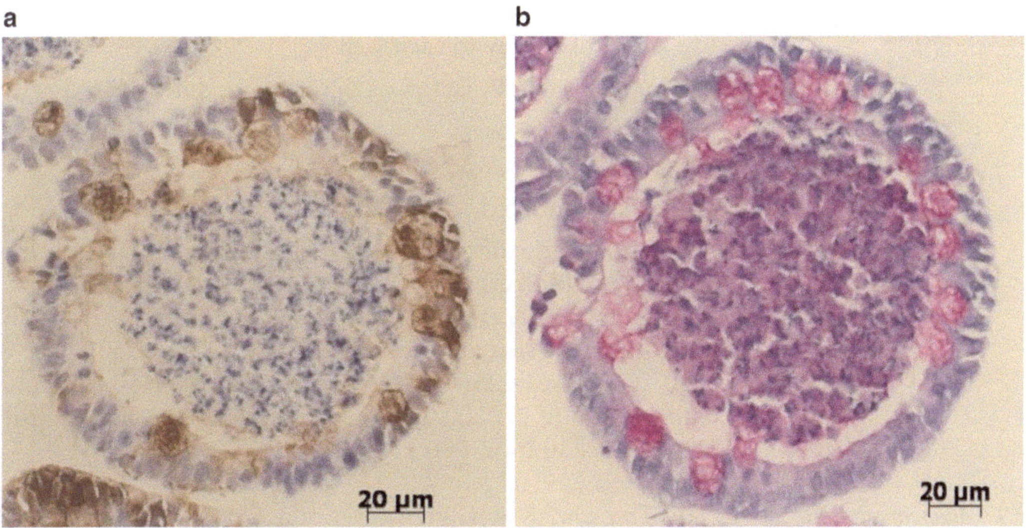

Fig. 2 Immunohistochemistry of embedded *APC 1322 T/+* spheres culterd in vitro. Lysozyme (**a**) and PAS staining (**b**) show the presence of Paneth cells and goblet cells, respectively

human disease, including the metastatic process [19]. Finally, this model can be used to test drug sensitivity. From our different mouse models, we can now transplant cultured spheres generated by deleting *APC*. These can be combined with other tumor suppressor genes/oncogenic mutations that are common to colon cancer such as mutations in *KRas*, *p53*, and *SMAD4* [20].

2 Materials

2.1 Intestinal Regeneration

1. Micropore surgical tape 2.5 cm × 5 m.
2. 10 ml syringe with a pipette tip on the end.
3. 10 % neutral-buffered formalin.

2.2 Intestinal Crypt Culture

1. Crypt culture medium: Add to one bottle of Advanced DMEM/F12 (Invitrogen) 5 ml HEPES (1 M), 5 ml l-Glutamine (100×), 5 ml Pen/Strep (100×). Make aliquots of B27, N2 (both Gibco), and 12.5 % BSA (Sigma) stock and keep at −20 °C. To 50 ml of this, add 1 ml of B27 (50×), 0.5 ml of N2 (100×), and 0.4 ml of 12.5 % BSA/PBS (final concentration of 0.1 %).
2. Growth factors:
   - Mouse R-spondin1 (Rspo1, R&D): Make a stock of 25 μg/ml in PBS; add 250 ng (10 μl) every other day to each well.
   - Noggin (Peprotech): Make stock of 5 μg/ml in 0.1 % BSA/PBS; add 50 ng (10 μl) every other day to each well.
   - Epidermal growth factor (EGF) (Peprotech): Make stock of 1 μg/ml in 0.1 % BSA/PBS, and add 25 ng (25 μl) every other day to each well.
3. Matrigel, growth factor reduced, no phenol red (BD Biosciences).
4. 24-well plate.
5. Phosphate-buffered saline (PBS).
6. 4-Hydroxytamoxifen (Sigma H7904, 25 mg): Make stock of 10 mM in 100 % ethanol. When using 25 mg, dissolve powder in 6.45 ml 100 % ethanol. If problems occur in dissolving the powder, just warm up solution to 37 °C. Aliquot and keep at −20 °C. Make a 250 μM working solution (1:40 dilution), and use 1 μl of working solution per well for the crypt culture (final concentration 500 nM).

2.3 Adenoma Culture

1. 10× Trypsin 2.5 % (Gibco).
2. Same materials as for the crypt culture; use same concentrations of the growth factors as well.

2.4 Orthotopic Injection of Cells/ Spheres into Mice

1. Single cell dissociation:
   (a) TrypLE Express (1×) solution (Gibco).
   (b) DNase I (10,000 U).
   (c) DNase incubation buffer.
2. Subcutaneous injection:
   (a) Immunodeficient mice.
   (b) 27 G needle syringe.
   (c) Matrigel, growth factor reduced, no phenol red (BD).
   (d) Precooled tips.
   (e) Ice.
3. Orthotopic injection:
   (a) Immunodeficient mice.
   (b) 27 G needle syringe.
   (c) Anesthetic.
   (d) Antiseptic solution.
   (e) Sterile forceps and scissors.
   (f) Sterile gauze.
   (g) Sterile PBS.
   (h) Sterile thread.
   (i) Sterile stainless steel clips.
   (j) Precooled tips.
   (k) Ice.

3 Methods

3.1 Intestinal Regeneration

1. Removal and preparation of the intestine.

   For regeneration assays:

   (a) 72 h post-irradiation with 14 Gy the mice are euthanized. Remove the small intestine (SI) up to the appendix by holding the stomach with forceps and then pulling gently to free the intestine (trim connective tissue) so that the SI is removed in one piece.
   (b) After removing the intestine, flush with water or PBS—use a syringe with a pipette tip insert into the end of the intestine and flush (*see* **Note 1**).
   (c) Remove 1 cm from the SI nearest to the stomach and discard. From the following part of the SI, cut ten pieces about 1 cm in length. Place these pieces in a pile on a piece of micropore surgical tape in a “pyramid” (*see* **Note 2**).

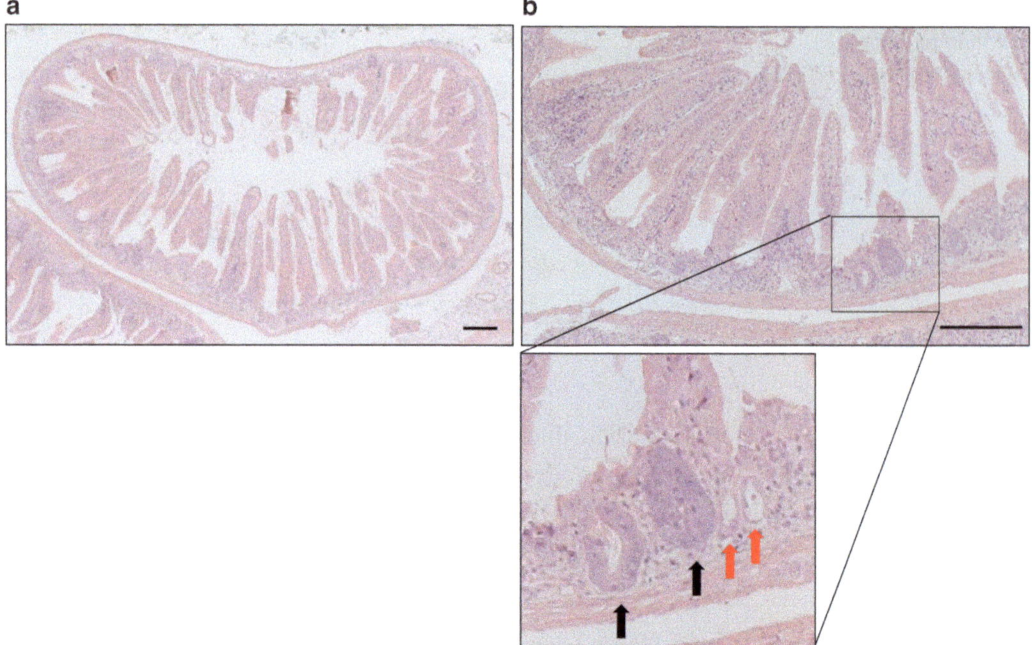

Fig. 3 Hematoxylin and eosin-stained section of 14 Gy irradiated small intestine (SI). (**a**) Overview of one of the ten cross sections of SI that is scored. (**b**) Closeup section of regenerating guts. *Box* shows magnified regenerating crypts (*black arrows*) and dying crypts (*red arrows*). Scale bars represent 200 μm

(d) Wrap the tape around the "pyramid" and seal. Fix in a universal tube filled with formalin overnight at room temperature.

(e) Process the intestines and embed them end on, so histological sections show the circumference of the intestinal parcels (*see* Fig. 3).

(f) Stain one section with hematoxylin and eosin.

2. Scoring regeneration:

(a) Score the regenerating crypts from hematoxylin and eosin-stained sections: for each SI cross section (of which there should be ten), score the number of "regenerating crypts." A regenerating crypt is a crypt containing at least six consecutive cells (Fig. 3).

(b) Record the numbers of regenerating crypts for each roll and calculate the average per mouse (*see* **Note 3**).

3.2 Intestinal Crypt Culture (See Note 4)

1. Thaw an aliquot of Matrigel on ice (~500 μl per 24-well plate = ~20 μl/well) (*see* **Note 4**).

2. The small intestine is dissected and flushed with cold PBS (*see* **Notes 5** and **6**). A piece of 10 cm is sufficient for culture.

3. Open the small intestine longitudinally.
4. Scrape villi using a glass cover slip with little pressure. Keep the piece of SI in cold PBS; move to the tissue culture hood.
5. Cut the small intestine into small pieces (5 mm) with scissors. The pieces should be small enough to fit through the opening of a 10 ml pipette.
6. Wash the pieces of tissue with 10 ml of cold PBS by pipetting up and down. After the pieces have settled, discard the supernatant.
7. Repeat **step 6** until supernatant is relatively clear, typically around five times.
8. Add 25 ml 2 mM EDTA/PBS and place on a roller in a cold room for 30 min.
9. Discard the EDTA; rinse the pieces with cold PBS, to remove remaining EDTA. Discard the supernatant, and add 10 ml of fresh PBS. Pipette the pieces up/down with the 10 ml pipette; the supernatant is your first fraction. Repeat this step three times. Check each fraction under the microscope. Fraction 1 should have more villi and single cell contamination. Fractions 2–4 should contain crypts and no villi structures.
10. Crypt fractions 2–4 are combined, filled up to 50 ml with Advanced DMEM/F12 (ADF), and centrifuged (224 × *g*, 5 min).
11. Resuspend the pellet with 10 ml ADF, and collect in a 50 ml Falcon tube after passing through a 70 μm cell strainer; wash the cell strainer with additional 5 ml. Transfer the 15 ml to a 15 ml Falcon. The tube is centrifuged with lower speed (56 × *g*, 2 min), so that single cells cannot settle down.
12. Repeat washing with ADF until you cannot see single cells in the supernatant. Typically it takes 2–3 washes.
13. During wash step, calculate number of crypts per ml by placing 3 drops of 10 μl of your crypt suspension in an empty well of a 24-well plate each. Take the average of the three numbers, and calculate the total number of crypts according to the volume of your crypt solution.
14. Discard the supernatant as much as possible, and resuspend with thawed Matrigel [at 500 crypts/100 μl Matrigel, i.e., 100 crypts/well]. 500 μl of Matrigel will be sufficient for the whole 24-well plate. Keep on ice.
15. Place a drop of about 20–30 μl of Matrigel-crypt mix on a pre-warmed 24-well plate (*see* **Note 7**). Incubate at 37 °C, 5 % CO_2 for 5 min till the Matrigel drop is solid. Then add 500 μl of the crypt culture medium (incl. N2, B27, BSA) and add Rspo1, Noggin, and EGF.

16. Feed the organoids every other day and change the medium once a week. Viable crypts will round up and form ball-like structures. After 2–3 days protrusions will appear. Passage the crypts once a week.
17. Passaging the organoids is done by mechanical dissociation (*see* **Note 8**). Remove the media from the wells, and scrape the villi off the bottom of the wells with a p1000 pipette. Collect the organoids in a 15 ml tube, add 5 ml of ADF, and spin down with low speed (56 × *g*, 3 min). Discard the supernatant and remaining Matrigel. A pellet of organoids should remain. Use a p200 pipette with 200 μl of ADF and pipette up/down until all the big pieces are broken up. Wash 2× with 5 ml of ADF and spin down (56 × *g*, 3 min). From here, follow **steps 13–15** again (*see* **Note 9**).
18. An established crypt culture allows gene deletion of genes such as c-Myc using tamoxifen-inducible Cre-lox technology in vitro. Use a concentration of 500 nM of 4-hydroxytamoxifen and incubate for 16–20 h with the culture medium. Incubate the control crypts with the same volume of 100 % ethanol.

3.3 Adenoma Culture

1. Cut adenomas/polyps off the small intestine/colon with scissors. It is possible to pool the polyps from one mouse, or use a single adenoma.
2. Cut each adenoma into small pieces by means of scissors, as small as possible.
3. Wash ~3 times with cold PBS to remove debris.
4. Incubate in 5 mM EDTA/PBS for 10 min at RT. Pipette up and down with a 5 ml pipette during 10 min.
5. Wash 2–3 times with PBS to remove EDTA (*see* **Note 10**).
6. Incubate in 3 ml 10× trypsin (5 mg/ml) supplemented with 100–200 U DNase and the appropriate DNase buffer (300 μl of 10× DNase buffer) for 30 min at 37 °C.
7. After trypsinization, add 5 ml of ADF and shake vigorously. The aim is to dissociate the tumor chunks into single cells. Repeat this step about 5–6 times. Fill up to 50 ml with ADF.
8. Spin down at 224 × *g* for 5 min. Discard supernatant and resuspend pellet in 10 ml of ADF. Pass through 70 μm cell strainer. Wash strainer with 5 ml ADF. Transfer to a 15 ml Falcon (easier to spot pellet). Spin down at 224 × *g* 5 min.
9. Optional: count cells. Plate around 10,000–50,000 cells per well. Resuspend in 20 μl Matrigel/well (*see* **Note 11**).
10. Wait 5 min to let the Matrigel solidify. Add crypt culture medium supplemented with EGF and Noggin (same concentration as for crypt culture). Change media every other day.

Once the spheres have formed, add growth factors after each passage (*see* **Note 12**).

11. After approximately 5 days, the cultures must be passaged. Take p1000 to take up the Matrigel and put in a 15 ml tube with around 5 ml of ADF; spin down at 56×*g* for 3 min. Discard supernatant. Mechanically dissociate pellet with a p200. Wash two times with ADF. Repeat **step 9–10**.

3.4 Injection of Cells/Spheres into Mice

1. Single cell dissociation:
   (a) Pre-warm TrypLE Express (Gibco) to 37 °C before use (*see* **Note 13**).
   (b) Resuspend cells in 1 ml of TrypLE Express, 100–200 U of DNase and appropriate volume of DNase buffer.
   (c) Keep in a 37 °C water bath for 30 min or till cells are completely dissociated.
   (d) Add 10 ml of ADF.
   (e) Centrifuge for 5 min at 300×*g*.
   (f) Discard supernatant and repeat **steps d** and **e**.
   (g) Use cells for further analysis.
2. For subcutaneous injection:
   (a) Thaw Matrigel overnight on ice (*see* **Note 14**).
   (b) Keep syringe on ice.

   For the sphere injection:

   - Mechanically dissociate spheres the day before the injection.
   - Count around 50–100 spheres (each sphere should contain around 100 cells).
   - Mix them to 200 μl of Matrigel using precooled tips; keep on ice till the injection.

   For single cell injection:

   - Dissociate the spheres in pre-warmed TrypLE Express solution as described above.
   - Determine viable and total cell counts.
   - Mix cells with 200 μl of Matrigel using precooled tips; keep on ice till the injection.

   (c) Inject the cells subcutaneously in immunocompromised mice using precooled syringes.
3. For orthotopic injection:
   (a) Passage spheres the day before the injection.
   (b) On the day of injection, count around 50–100 spheres.
   (c) Mix them with a maximum of 50 μl Matrigel using precooled tips. Keep the cell suspension on ice till the injection.

(d) Anesthetize the mouse and shave the abdomen.

(e) Disinfect the abdomen with antiseptic solution.

(f) Incise the left abdomen longitudinally for 2 cm with a sterile surgical scissors.

(g) Separate the skin from the muscle layer below with sterile surgical scissors.

(h) Exteriorize the cecum.

(i) Isolate the cecum with sterile gauze. You can keep the cecum moist with warm PBS.

(j) Inject the cells mixed in Matrigel between the mucosa and the muscularis layer of the cecal wall (*see* **Note 15**).

(k) Return the cecum into the abdominal cavity.

(l) Suture the muscle layer.

(m) Close the skin with 3–4 sterile stainless steel clips.

(n) Administer analgesia according to local regulations.

4 Notes

1. If there is a hole in the intestine that makes flushing difficult, flush from the other end (but remember which end is the stomach end). From the proximal to the distal end of the intestine, the number of crypts per circumference of the intestine decreases, and thus it is important to score a similar area of the intestine each time.
2. Try to keep the micropore tape as dry as possible to ensure it seals properly.
3. A minimum of three different mice should be scored.
4. Do not repeatedly freeze/thaw the Matrigel. Use aliquots.
5. The efficiency of surviving crypts depends on the time from sacrifice of the mouse until the end of the procedure. A time frame of 2 h is recommended.
6. When dissecting the gut, flush it with cold PBS and perform all the washings with cold PBS. This will increase the viability of the crypts.
7. It is important that the drop is in the center of the well and does not touch the edge.
8. The organoids will fill up with debris and mucus secreted from the goblet cells. By passaging the organoids, the debris and dead cells will be removed and every protrusion will form a new organoid. The mechanical disruption of the organoids dissociates the protrusions and avoids the separation of the whole organoid into single cells.

9. It is better to plate crypts/organoids more dense than too diluted.
10. Make sure to remove all EDTA before using the trypsin and DNase; the DNase buffer is essential for the DNase.
11. The counting is not too important for the purpose of culturing. Even very dense cells in the Matrigel will grow.
12. Noggin and EGF can be added to enhance the sphere formation after dissociation.
13. When we add TrypLE Express (1×), we find that it is better to mechanically dissociate the cells with a pipette every 10 min. It will help the dissociation.
14. Matrigel has to be thawed on ice and not in the fridge. Be careful; Matrigel solidifies rapidly at room temperature; use precooled tips.
15. During the orthotopic injection, avoid any leakage. You should observe the formation of a bubble under the cecum wall during the injection. Wait a few seconds before removing the needle.

Acknowledgments

This work is supported by the European Union Seventh Framework Programme FP7/2007–2013 under grant agreement number 2785668 and Cancer Research UK.

References

1. He TC, Sparks AB, Rago C, Hermeking H, Zawel L, da Costa LT, Morin PJ, Vogelstein B, Kinzler KW (1998) Identification of c-MYC as a target of the APC pathway. Science 281:1509–1512
2. Korinek V, Barker N, Morin PJ, van Wichen D, de Weger R, Kinzler KW, Vogelstein B, Clevers H (1997) Constitutive transcriptional activation by a beta-catenin-Tcf complex in APC-/- colon carcinoma. Science 275:1784–1787
3. Muncan V, Sansom OJ, Tertoolen L, Phesse TJ, Begthel H, Sancho E, Cole AM, Gregorieff A, de Alboran IM, Clevers H, Clarke AR (2006) Rapid loss of intestinal crypts upon conditional deletion of the Wnt/Tcf-4 target gene c-Myc. Mol Cell Biol 26:8418–8426
4. Soucek L, Whitfield J, Martins CP, Finch AJ, Murphy DJ, Sodir NM, Karnezis AN, Swigart LB, Nasi S, Evan GI (2008) Modelling Myc inhibition as a cancer therapy. Nature 455:679–683
5. Ashton GH, Morton JP, Myant K, Phesse TJ, Ridgway RA, Marsh V, Wilkins JA, Athineos D, Muncan V, Kemp R, Neufeld K, Clevers H, Brunton V, Winton DJ, Wang X, Sears RC, Clarke AR, Frame MC, Sansom OJ (2010) Focal adhesion kinase is required for intestinal regeneration and tumorigenesis downstream of Wnt/c-Myc signaling. Dev Cell 19:259–269
6. Sansom OJ, Meniel VS, Muncan V, Phesse TJ, Wilkins JA, Reed KR, Vass JK, Athineos D, Clevers H, Clarke AR (2007) Myc deletion rescues Apc deficiency in the small intestine. Nature 446:676–679
7. Athineos D, Sansom OJ (2010) Myc heterozygosity attenuates the phenotypes of APC deficiency in the small intestine. Oncogene 29:2585–2590

8. Yekkala K, Baudino TA (2007) Inhibition of intestinal polyposis with reduced angiogenesis in ApcMin/+ mice due to decreases in c-Myc expression. Mol Cancer Res 5:1296–1303
9. Finch AJ, Soucek L, Junttila MR, Swigart LB, Evan GI (2009) Acute overexpression of Myc in intestinal epithelium recapitulates some but not all the changes elicited by Wnt/beta-catenin pathway activation. Mol Cell Biol 29:5306–5315
10. Murphy DJ, Junttila MR, Pouyet L, Karnezis A, Shchors K, Bui DA, Brown-Swigart L, Johnson L, Evan GI (2008) Distinct thresholds govern Myc's biological output in vivo. Cancer Cell 14:447–457
11. Sansom OJ, Clarke AR (2002) The ability to engage enterocyte apoptosis does not predict long-term crypt survival in p53 and Msh2 deficient mice. Oncogene 21:5934–5939
12. Ijiri K, Potten CS (1986) Radiation-hypersensitive cells in small intestinal crypts; their relationships to clonogenic cells. Br J Cancer Suppl 7:20–22
13. Ijiri K, Potten CS (1987) Further studies on the response of intestinal crypt cells of different hierarchical status to eighteen different cytotoxic agents. Br J Cancer 55:113–123
14. Potten CS (1984) Clonogenic, stem and carcinogen-target cells in small intestine. Scand J Gastroenterol Suppl 104:3–14
15. Sato T, van Es JH, Snippert HJ, Stange DE, Vries RG, van den Born M, Barker N, Shroyer NF, van de Wetering M, Clevers H (2009) Paneth cells constitute the niche for Lgr5 stem cells in intestinal crypts. Nature 469:415–418
16. Sato T, Vries RG, Snippert HJ, van de Wetering M, Barker N, Stange DE, van Es JH, Abo A, Kujala P, Peters PJ, Clevers H (2009) Single Lgr5 stem cells build crypt-villus structures in vitro without a mesenchymal niche. Nature 459:262–265
17. Lewis A, Segditsas S, Deheragoda M, Pollard P, Jeffery R, Nye E, Lockstone H, Davis H, Clark S, Stamp G, Poulsom R, Wright N, Tomlinson I (2010) Severe polyposis in Apc(1322T) mice is associated with submaximal Wnt signalling and increased expression of the stem cell marker Lgr5. Gut 59:1680–1686
18. Su LK, Kinzler KW, Vogelstein B, Preisinger AC, Moser AR, Luongo C, Gould KA, Dove WF (1992) Multiple intestinal neoplasia caused by a mutation in the murine homolog of the APC gene. Science 256:668–670
19. Tseng W, Leong X, Engleman E (2007) Orthotopic mouse model of colorectal cancer. J Vis Exp 484
20. Sansom OJ, Meniel V, Wilkins JA, Cole AM, Oien KA, Marsh V, Jamieson TJ, Guerra C, Ashton GH, Barbacid M, Clarke AR (2006) Loss of Apc allows phenotypic manifestation of the transforming properties of an endogenous K-ras oncogene in vivo. Proc Natl Acad Sci USA 103:14122–14127

Chapter 17

Detection of *c-myc* Amplification in Formalin-Fixed Paraffin-Embedded Tumor Tissue by Chromogenic In Situ Hybridization (CISH)

Nataša Todorović-Raković

Abstract

In situ hybridization (ISH) allows evaluation of genetic abnormalities, such as changes in chromosome number, chromosome translocations or gene amplifications, by hybridization of tagged DNA (or RNA) probes with complementary DNA (or RNA) sequences in interphase nuclei of target tissue. However, chromogenic in situ hybridization (CISH) is also applicable to formalin-fixed, paraffin-embedded (FFPE) tissues, besides metaphase chromosome spreads. CISH is similar to fluorescent in situ hybridization (FISH) regarding pretreatments and hybridization protocols but differs in the way of visualization. Indeed, CISH signal detection is similar to that used in immunohistochemistry, making use of a peroxidase-based chromogenic reaction instead of fluorescent dyes. In particular, tagged DNA probes are indirectly detected using an enzyme-conjugated antibody targeting the tags. The enzymatic reaction of the chromogenic substrate leads to the formation of strong permanent brown signals that can be visualized by bright-field microscopy at 40× magnification. The advantage of CISH is that it allows the simultaneous observation of gene amplification and tissue morphology and the slides can be stored for a long time.

Key words *c-myc* amplification, Chromogenic in situ hybridization (CISH)

1 Introduction

c-myc amplification is relatively common in various types of cancer [1], but rigorous studies with consistent methodologies are needed to investigate its potential as a prognostic or predictive marker in clinical cancer practice [2–7]. The widely used methods for evaluation of *c-myc* amplification in cancer are chromogenic in situ hybridization (CISH) and fluorescence in situ hybridization (FISH). Comparative analysis between these two methods was well presented by Hsi et al [8]. Although well correlated, both methods have their distinct advantages. However, CISH is emerging as a practical, cost-effective, and valid alternative to FISH in testing different kinds of genetic alterations [9]. In a study by Rummukainen and colleagues [10], for example, *c-myc* amplification by CISH

Laura Soucek and Nicole M. Sodir (eds.), *The Myc Gene: Methods and Protocols*, Methods in Molecular Biology, vol. 1012, DOI 10.1007/978-1-62703-429-6_17, © Springer Science+Business Media, LLC 2013

yielded a stronger association with survival, in comparison to FISH. The authors suggested that the better survival stratification by CISH was attributed to the high sensitivity of the probe detection, excellent visibility of the gene copy signals, and, in particular, the possibility to restrict the copy number counting to cells that are truly malignant by morphologic criteria.

Here we present a CISH protocol specific for detection of *c-myc* amplification on formalin-fixed paraffin-embedded tissue sections that is adapted, with minor modifications, from the manufacturer's instructions of the Spot Light CISH polymer detection kit (Zymed/Invitrogen). The Zymed Spot Light *c-myc* probe is a double stranded digoxigenin labeled probe in hybridization buffer and binds specifically the *c-myc* gene locus on chromosome band 8q24.12-8q24.13. In this probe, repetitive sequences have been removed by Zymed SPT™ technology, so that such probe does not require repetitive sequence blocking.

2 Materials

1. Deionized water.
2. Xylene.
3. 70, 85, 95 and 100 % ethanol.
4. 30 % Hydrogen peroxide.
5. 100 % Methanol.
6. 50 % Tween 20.
7. 20× SSC buffer: 3 M NaCl, 0.3M sodium citrate.
8. Phosphate buffered saline (PBS): 10 mM PBS, pH 7.4.
9. 0.025 %Tween/PBS: 0.025 % Tween 20 in PBS.
10. Heat pretreatment solution: Tris–EDTA buffer, ready to use reagent from Zymed/Invitrogen.
11. Enzyme: pepsin solution with 0.05 % sodium azide and detergent, ready to use reagent from Zymed/Invitrogen.
12. Formamide.
13. Denaturation buffer: for 100 ml mix 10 ml 20× SSC, 20 ml deionized water and 70 ml formamide.
14. Quenching solution: 3 % hydrogen peroxide in absolute methanol.
15. Superfrost slides.
16. UnderCover slips or coverslips with rubber cement.
17. Spot Light *c-myc* probe (Zymed/Invitrogen).

18. Spot Light CISH polymer detection kit (Zymed/Invitrogen):
    - CAS-block (reagent containing buffer, stabilizer, and 0.01 % sodium azide, ready to use).
    - Mouse antidigoxigenin antibody (containing BSA, buffer, and 0.1 % sodium azide, ready to use).
    - Goat anti-mouse HRP polymer conjugate (containing stabilizer, buffer, 0.005 % gentamicin sulfate, and 0.1 % Proclin 300, ready to use).
19. 20× DAB concentrated chromogen solution.
20. Histomount (mounting solution).
21. Mayer hematoxylin (Invitrogen).

3 Methods

3.1 CISH Procedure (See Note 1)

1. Preparation of formalin-fixed paraffin-embedded slides: tissue sections should be fixed in 10 % neutral buffered formalin up to 48 h prior to paraffin embedding. The optimal tissue section thickness is 4 μm on SuperFrost/Plus microscope slides. Bake slides for 4 h at 60 °C and dry them at 37 °C (or air dry) overnight, prior to deparaffinization.
2. Deparaffinization: Submerge slides in Xylene, 2×10 min (*see* **Note 2**); then put them in 100 % Ethanol, 3×3 min; finally, submerge them in deionized water, 3×3 min.
3. Pretreatment and enzyme digestion: heat the pretreatment solution in a glass container on a hot plate at ≥98 °C and cover it with aluminum foil to prevent evaporation (*see* **Note 3**); submerge slides in boiling solution, cover, and boil for 15 min. After boiling, immerse slides in PBS (or deionized water), 2×3 min. Cover the tissue sections with enough enzyme solution (about 100 μl) and incubate at RT for 10–15 min (*see* **Note 4**). Finally, submerge slides in PBS (or deionized water), 2×3 min.
4. Dehydration: dehydrate slides in graded ethanol series (70, 85, 95, 2×100 %) for 2 min each. Air dry. Store ethanol solutions in a freezer at −20 °C for later use.
5. Denaturation (*see* **Note 5**): denaturation could be done in two ways:
    (a) Concomitant denaturation: add 15 μl probe to the middle of the slide, seal with coverslip, and denature at 94–95 °C in the slide block of a PCR machine or hot plate.
    (b) Prepare a coplin jar with 100 ml of denaturation buffer and put it in a water bath at 78 °C (this temperature is suitable for simultaneous work with four slides maximum).

When the denaturation solution reaches 78 °C, immerse slides and incubate for 5 min. At the same time, immerse tubes with aliquoted probes (15 μl) in a water bath under the same conditions (78 °C for 5 min). After denaturation, put the tubes with probe quickly on ice (to prevent renaturation). Slides should be dehydrated in cold (−20 °C) graded ethanol series (70, 85, 95, 2×100 %) for 2 min each and air dried.

6. Hybridization: add 15 μl probe to the central area of the tissue section or in the middle of the coverslip, then place the coverslip on the appropriate area covering the tissue section. Seal coverslip with rubber cement to prevent evaporation or, if using UnderCover slips, simply stick them to the tissue section and seal. Place slides in a humidifying chamber or in the slide block of PCR machine at 37 °C overnight.
7. Stringent wash (*see* **Note 6**): Prepare two coplin jars with 0.5× SSC buffer, one at room temperature and the other in a water bath at 75 °C. Carefully remove coverslips (or rubber cement). Immerse slides in the jar at room temperature for 5 min and, after that, in the jar at 75 °C, for 5 min. Wash in PBS (or deionized water).
8. Immunodetection using Spot Light CISH polymer detection kit (Invitrogen): immerse slides for 10 min in peroxidase quenching solution. Wash in PBS or in 0.025 % PBS/Tween20, 3×2 min. Add enough CAS-block (blocking reagent for reducing non-specific background staining) to cover the tissue section and incubate for 10 min at room temperature. Blot off the CAS-block reagent. Add approximately 100 μl antidigoxigenin (mouse) antibody to cover the tissue section and incubate for 30–45 min at room temperature. Wash in PBS or in 0.025 % PBS/Tween20, 3×2 min. Add approximately 100 μl of polymerized HRP conjugated goat anti-mouse antibody and incubate at room temperature for 30–45 min. Wash in PBS or in 0.025 % PBS/Tween20, 3×2 min. Add enough DAB chromogen to cover tissue section (approximately 150 μl) and incubate at room temperature for 30 min. Wash under running tap water for 3 min. Cover tissue sections with hematoxylin and incubate them for 5–10 s (*see* **Note 7**). Wash under running tap water for 3 min. Dehydrate slides in graded ethanol series (70, 85, 95, 2×100 %) for 2 min each. Immerse in xylene, 2×2 min. Apply mounting solution before xylene evaporates.

3.2 Evaluation of Results

In CISH, the analysis of results requires bright-field microscopy. It is possible to easily distinguish signals by using 40× magnification, but for better identification of dots/clusters, 100× magnification could be used as well. Typical signals appear as small, brown (DAB stained) discreet dots within the nucleus, or clusters, in which single dots cannot be counted.

Scoring: Normal, diploid cells have usually one or two signals (brown dots); aneuploid cells might have less or more than two signals. The presence of more than five signals is considered to be the cutoff for amplification: the visualization of 6–10 dots or a small cluster per nucleus in >50 % of cells is an indication of low amplification, and more than 10 dots or large clusters in >50 % of cells is a sign of high amplification.

4 Notes

1. General rule: during the procedure, slides should not be let dry, except after alcohol treatment. All reagents should be brought at room temperature before use.
2. Depending on tissue thickness and fixation, xylene treatment in deparaffinization section could be longer, up to 12 h before analysis. Longer xylene treatment is recommended for thicker samples.
3. Optimal heat pretreatment procedure is essential for the test. Inadequate heat pretreatment procedure (i.e., suboptimal time and temperature) could result in a weak or no signal at all. Keep the temperature at ≥98–100 °C for 15 min before starting.
4. Optimal enzyme digestion procedure is essential for the test. Inadequate enzyme digestion procedure (excessive digestion or under-digestion) could result in weak or no signal at all or poor tissue morphology. Depending on tissue thickness (the optimal one is 4 μm), enzyme digestion time should be decreased or increased (up to 15 min).
5. Inadequate denaturation and hybridization time, due to shorter incubation and/or lower temperature, could result in weak or no signal and poor tissue morphology. Always check the temperature and be sure that it is maintained during the whole process. Be sure that humidity in the humidifying chamber is satisfactory. Recommended conditions are 95–98 °C on PCR thermal cycler or 80–90 °C on heating block for 5 min.
6. Stringent wash temperature should be properly adjusted according to the number of slides. Recommended temperature is 72 °C for one slide. For each additionally slide, temperature should be increased of 1 °C, up to a maximum temperature of 80 °C. Stringent wash temperature under 70 °C could cause background staining and stringent wash temperature exceeding 80 °C could cause loss of signal.
7. Excessive hematoxylin counterstaining could result in poor signal evaluation. Best results are obtained when tissue is counterstained for 3–5 s. If you are not sure about time, check under microscope before applying coverslips.

References

1. Deming SL, Nass SJ, Dickson RB, Trock BJ (2000) C-myc amplification in breast cancer: a meta-analysis of its occurrence and prognostic relevance. Br J Cancer 83:1688–1695
2. Masramon L, Arribas R, Tartola S, Perucho M, Peinado MA (1998) Moderate amplifications of the c-myc gene correlate with molecular and clinicopathological parameters in colorectal cancer. Br J Cancer 77:2349–2356
3. Augenlicht LH, Wadler S, Corner G, Richards C, Ryan L, Multani AS et al (1997) Low-level c-*myc* amplification in human colonic carcinoma cell lines and tumors: a frequent, p53-independent mutation associated with improved outcome in a randomized multi-institutional trial. Cancer Res 57:1769
4. Baker VV, Borst MP, Dixon D, Hatch KD, Shingleton HM, Miller D (1990) C-myc amplification in ovarian cancer. Gynecol Oncol 38:340–342
5. Mitani S, Kamata H, Fujiwara M, Aoki N, Tango T, Fukuchi K, Oka T (2001) Analysis of c-myc DNA amplification in non-small cell lung carcinoma in comparison with small cell lung carcinoma using polymerase chain reaction. Clin Exp Med 1:105–111
6. Rodriguez-Pinilla MS, Jones RL, Lambros MBK, Arriola E, Savage K, James M et al (2007) MYC amplification in breast cancer: a chromogenic in situ hybridisation study. J Clin Pathol 60:1017–1023
7. Todorović-Raković N, Nešković-Konstantinović Z, Nikolić-Vukosavljević D (2011) C-myc as a predictive marker for chemotherapy in metastatic breast cancer. Clin Exp Med DOI10. 1007/s10238-011-0169-y
8. Hsi B-L, Xiao S, Fletcher JA (2002) Chromogenic in situ hybridization and FISH in pathology. Methods Mol Biol 204:343–351
9. Madrid MA, Lo RW (2004) Chromogenic *in situ* hybridization (CISH): a novel alternative in screening archival breast cancer tissue samples for HER-2/*neu* status. Breast Cancer Res 6:R593–R600
10. Rummukainen JK, Salminen T, Lundin J et al (2001) Amplification of c-myc oncogene by chromogenic and fluorescence in situ hybridization in archival breast cancer tissue array samples. Lab Invest 811:545–1551

Chapter 18

Cell-Based Methods for the Identification of MYC-Inhibitory Small Molecules

Catherine A. Burkhart, Michelle Haber, Murray D. Norris, Andrei V. Gudkov, and Mikhail A. Nikiforov

Abstract

Oncoproteins encoded by dominant oncogenes have long been considered as targets for chemotherapeutic intervention. However, oncogenic transcription factors have often been dismissed as "undruggable." Members of Myc family of transcription factors have been identified as promising targets for cancer chemotherapy in multiple publications reporting the requirement of Myc proteins for maintenance of almost every type of tumor. Here, we describe cell-based approaches to identify c-Myc small molecule inhibitors by screening complex libraries of diverse small molecules based on Myc functionality and specificity.

Key words c-Myc, MYCN, Small molecules, Cell-based assays, Functional screening

1 Introduction

c-Myc is a member of the Myc family of transcription factors that regulate expression of multiple genes involved in many cellular processes, including promotion of proliferation, enhancement of cellular metabolism, and induction of apoptosis [1]. Among Myc proteins, c-Myc and MYCN are most frequently implicated in tumorigenesis [2]. Moreover, these proteins are structurally and functionally very similar such that replacement of the *c-myc* gene with the *MYCN* gene creates a viable mouse with no developmental abnormalities, suggesting that the c-Myc and MYCN proteins share all critical functions [3]. Because a large number of human tumors exhibit deregulated expression of Myc-family members (more than 50 % of all human malignancies [4]) and because of the high dependency of tumor growth on elevated Myc levels in several experimental systems [5, 6], Myc proteins are attractive targets for cancer chemotherapy. Accordingly, it has been demonstrated recently that whole-mouse genetic inhibition of transactivating properties of c-Myc resulted in rapid regression of incipient and

Laura Soucek and Nicole M. Sodir (eds.), *The Myc Gene: Methods and Protocols*, Methods in Molecular Biology, vol. 1012, DOI 10.1007/978-1-62703-429-6_18, © Springer Science+Business Media, LLC 2013

established tumors, whereas the side effects to normal tissues were well tolerated and completely reversible even over extended time periods of c-Myc inhibition [7]. Thus, the inhibition of Myc appears to be a safe and efficient method to eliminate cancer. Several approaches have been pursued to develop anti-MYC therapeutics [8]; however, no drugs targeting c-MYC or MYCN have reached clinical trials. Therefore, identification of anti-Myc pharmaceutical agents capable of either direct tumor elimination or sensitization of a tumor to conventional chemotherapy is an important goal for anticancer drug development.

2 Materials

2.1 Cell Lines

Grow all cell lines in DMEM medium supplemented with 10 % fetal bovine serum:

1. SHR6-17 (SH-EP human neuroblastoma cell line that expresses low level *C-MYC* and no *MYCN* stably transfected with a MYC-responsive luciferase reporter to follow the effects of library compounds on MYC transactivation).
2. SH-CMV-luc (SH-EP cells with a constitutive luciferase reporter for identification of luciferase inhibitors or general transcription inhibitors).
3. HO15.19 (Rat-1 fibroblasts with both alleles of *c-myc* gene deleted via somatic recombination) [9].

2.2 Lentiviral and Retroviral Vectors

The plasmids used for the cell-based readout system and described in this chapter are as follows:

1. pTZV3 vector (kindly provided by Tranzyme, Inc).
2. pTZV3-eGFP-N3i (*N3i*; shRNA against human *MYCN* sequence GCAGCAGTTGCTAAAGAAA in replication-incompetent lentiviral vector TZV3-*eGFP*).
3. pTZV3-*eGFP-GFPi* (control shRNA).
4. pTZV3-CMV-*hMYCN* (generated by replacement of eGFP in vector pTZV3-eGFP with human *MYCN* cDNA (kindly provided by Dr. William Weiss, University of California at San Francisco, USA)).
5. Lentiviral packaging plasmids (pTRE-gag-pro-RRE-poly A, pCMV-vpr-RT-IN-poly A, pCMV-VSV-G-poly A, pCMV-tetoff-poly A, and pCMV-tat/rev) (kindly provided by Tranzyme, Inc).
6. The Myc-responsive reporter plasmid, pR6mHSP-luc: consisting of a minimal heat shock protein promoter containing six E-box sequences that were cloned from a modified ornithine

decarboxylase (*ODC*) promoter construct (kindly provided by Mary Danks of St. Jude Children's Hospital, USA).

7. pLXSH-FLAG-c-MYCN and pLXSH-FLAG-N-myc (kindly provided by Dr. Michael D. Cole, Dartmouth College, USA).

2.3 Other Reagents

1. Bright-Glo™ Luciferase Assay System (Promega Corporation).
2. Polybrene (Sigma-Aldrich).
3. Propidium iodide (Sigma-Aldrich).
4. Methylene blue (US Biological).
5. Methanol.
6. Sodium dodecyl sulfate (SDS).
7. Phosphate buffer saline (PBS).
8. DMSO.

3 Methods

3.1 Overview

The readout system described here (SHR6-17, *see* **Notes 1–3**) is based on measuring anti-Myc activity by following the effect of small molecules on MYCN-mediated transactivation of an MYC-responsive luciferase reporter in neuroblastoma cells that are transduced with *MYCN* lentivirus prior to the addition of the small molecule library (Fig. 1). In this system, SHR6-17 cells, which express low levels of c-Myc and no MYCN, are transduced with lentivirus for human MYCN. After 24 h, when MYCN levels begin to increase, library compounds are added to the cells. At 48 h, luciferase activity is measured. If a library compound is inactive, the luciferase activity will continue to increase; however, if a compound is active, the luciferase activity will remain low. Strong hits are classified as those that reduce Myc-responsive luciferase activity to levels comparable to wells transduced with *MYCN* shRNA (*N3i*). However, because positive compounds may be quenchers or direct inhibitors of luciferase, "weak" hits are also validated (40–60 % reduction in luciferase activity). The readout cell line is validated by the demonstration of dose-dependent response of the Myc-responsive reporter to *MYCN* lentivirus (Fig. 2a) and a dose-dependent inhibition of MYCN-induced luciferase reporter by *N3i MYCN* shRNA (Fig. 2b), but not by nonspecific shRNA (Fig. 2c). To verify activity, "hit" compounds are passed through a series of filters to eliminate false positives, including dose-dependent effects on *MYC*-mediated transcription, luciferase inhibition/quencher, and general transcription inhibition assays followed by a cell-based assay for specific inhibition of proliferation of HO15.19 *myc*-null cells ectopically expressing mouse *c-myc* or *MYCN* cDNAs. HO15.19 is the only cell line that

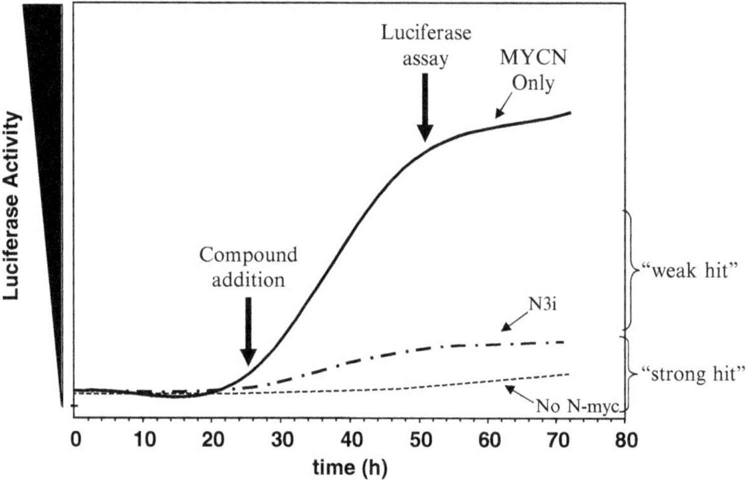

Fig. 1 Schematic representation of MYCN primary screen. SHR6-17 cells with low basal levels of luciferase reporter activity are transduced with *MYCN* lentivirus. 24 h post-transduction, library compounds are added to cells while MYCN levels are still low (~twofold induction), and then luciferase activity of cells is measured ~24 h after incubation with compounds. N3i, an *MYCN* shRNA lentivirus, serves as a positive control for inhibition of MYCN-driven reporter activity. Two categories of hits are obtained: strong hits return luciferase activity back to baseline or to N3i shRNA levels, and weak hits reduce luciferase activity to ~40–60 %

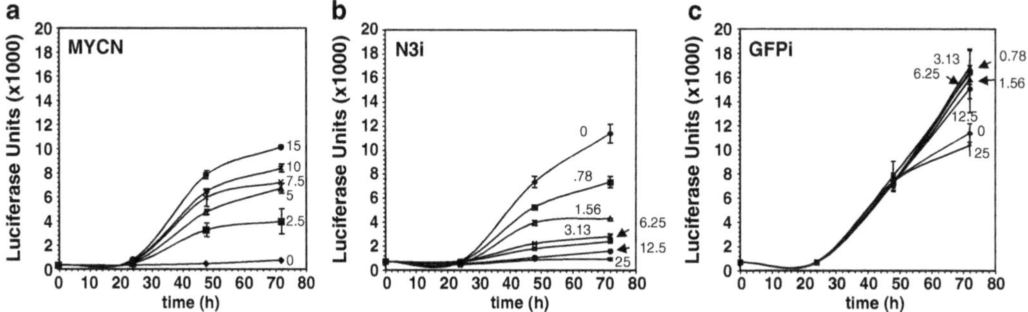

Fig. 2 Induction of luciferase activity by transduction of *MYCN* lentivirus. (**a**) *MYCN* lentivirus induces luciferase activity in a dose-dependent manner. SHR6-17 cells containing the MYC-responsive luciferase reporter were transduced with increasing concentrations of *MYCN* lentivirus for 24, 48, and 72 h in 96-well plates. At each time point, cells were assayed for luciferase activity. *Numbers* next to each curve correspond to the amount of lentivirus (μl) added per well in a total volume of 200 μl. (**b, c**) *MYCN* shRNA lentivirus (N3i, **b**) but not GFP shRNA (GFPi, **c**) block MYCN induction of luciferase activity in SHR6-17 cells. Reporter cells were transduced with *MYCN* lentivirus (10 μl) alone or in combination with increasing amounts of N3i or GFPi lentivirus for 24, 48, and 72 h as described above. *Numbers* next to curves correspond to the amount of shRNA lentivirus used in the experiment

is capable of continuous proliferation, albeit slowly, in the absence of any Myc protein expression, a feature that makes this line a standard for studying Myc-dependent phenotypes [9]. Reconstitution of these cells with ectopically expressed c-Myc or MYCN completely reverses the slow-growth phenotype [9].

3.2 Procedure for Screening Small Molecules for Myc Inhibition

1. Prepare a bulk batch of *MYCN, N3i, and GFPi* lentiviruses in sufficient quantities to cover the entire library screening. Titrate the amount of *MYCN* virus needed for optimal induction of the MYC-responsive reporter and the amount of *N3i* needed to block MYC-mediated transcription. The volumes of each lentivirus applied to the readout cells will depend on this optimization procedure.
2. Divide the viruses into aliquots, with the size of each aliquot being sufficient for 1 day of screening based on the titrations performed above (*see* **Note 4**).
3. For day 1 of screening, seed 7,500 SHR6-17 cells in each well of a 96-well plate in a volume of 100 μl per well.
4. Remove cell medium and add MYCN and/or N3i or GFPi viruses plus 2 μg/ml Polybrene following the template in Fig. 3. The amount of virus required will be determined from **item 1** above, and the volume adjusted to 100 μl with DMEM + 10 % FBS. Each readout plate should include nontransduced cells and cells transduced with only *MYCN* lentivirus as well as positive control, *N3i* and nonspecific shRNA.

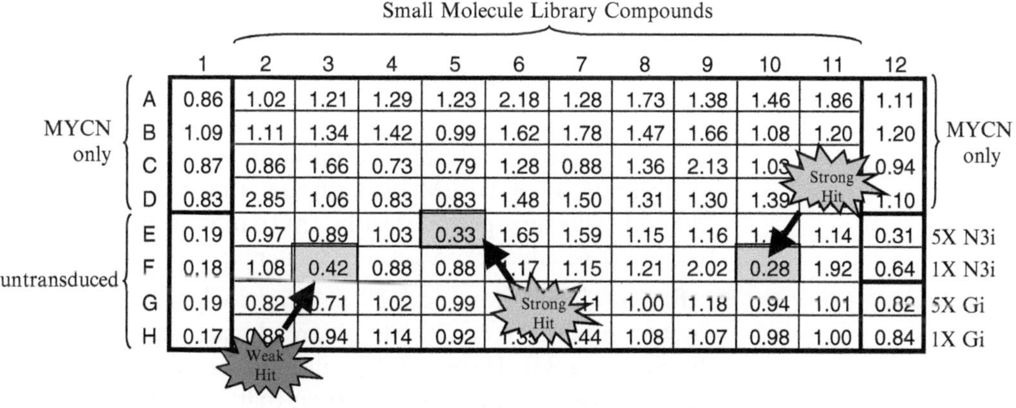

	1	2	3	4	5	6	7	8	9	10	11	12
A	0.86	1.02	1.21	1.29	1.23	2.18	1.28	1.73	1.38	1.46	1.86	1.11
B	1.09	1.11	1.34	1.42	0.99	1.62	1.78	1.47	1.66	1.08	1.20	1.20
C	0.87	0.86	1.66	0.73	0.79	1.28	0.88	1.36	2.13	1.03	[illegible]	0.94
D	0.83	2.85	1.06	0.83	0.83	1.48	1.50	1.31	1.30	1.39	[illegible]	1.10
E	0.19	0.97	0.89	1.03	0.33	1.65	1.59	1.15	1.16	[illegible]	1.14	0.31
F	0.18	1.08	0.42	0.88	0.88	[illegible]	1.15	1.21	2.02	0.28	1.92	0.64
G	0.19	0.82	0.71	1.02	0.99	[illegible]	[illegible]	1.00	1.18	0.94	1.01	0.82
H	0.17	0.88	0.94	1.14	0.92	[illegible]	[illegible]	1.08	1.07	0.98	1.00	0.84

Fig. 3 Representative plate from a small molecule library screening for MYC inhibitors. 80 library compounds were tested per 96-well plate, and the level of luciferase activity in the presence of compound was compared to that of the average of the luciferase activity in cells transduced with *MYCN* lentivirus (MYCN only) to produce an inhibition ratio as presented in the table. Strong hits represent those compounds that reduce the luciferase activity back to basal levels or to a level equivalent to that obtained with the highest *MYCN* shRNA (*N3i*) dose. Weak hits reduce the luciferase levels to 40–60 % of the control. Gi represents wells transduced with an shRNA control virus. Two *strong hits* and one *weak hit* are shown on this plate

5. 24 h after infection, treat cells with library compounds at a final concentration of ~10 μM and incubate for additional 24 h.
6. Prior to determining luciferase activity, view each well under the microscope to identify compounds that are generally cytotoxic, based on the complete rounding up of cells and detachment from the well surface. These molecules are eliminated from the "hit" list as potential false positives.
7. Add 10 μl Bright-Glo Luciferase Assay System reagent to each well and gently tap side of plates to mix the Bright-Glo with cell medium (*see* **Note 5**).
8. Read plates on a luminometer.
9. To identify hits, calculate the inhibition ratio (ratio of luciferase measurement values of test compound divided by that of the average MYCN-only controls on the same plate). Compounds with a ratio of ≤0.6 are classified as hits with compound evaluation prioritized based on their ratio (i.e., strong hits—inhibition ratios <0.4, similar to that of *MYCN* shRNA; weak hits—inhibition ratio 0.4–0.6). An example plate of screening results is presented in Fig. 3.
10. To validate the hits, the above procedure should be performed for selected compounds at three different doses to establish the dose-dependence of each hit compound. This is typically done by taking an aliquot of each putative hit directly from the library and testing it at 0.1, 1 and 10 μM.
11. Hits that demonstrate dose-dependence can be ordered from the library source for further characterization, pending filtering for false positives (*see* Subheading 3.3).

3.3 Filtering for False Positives

3.3.1 Luciferase Inhibition/Quenchers

1. Plate SHR6-17 cells in 96-wells and transduce with *MYCN* lentivirus as described in Subheading 3.2.
2. 48 h after transduction (i.e. when maximum luciferase expression is reached), add "hit" compounds to the cells at a final concentration of 10 μM in duplicate. Incubate at 37 °C for 30 min. This time period is sufficient to inhibit the enzyme but too short to affect expression of luciferase.
3. After the incubation, add 10 μl Bright-Glo Luciferase Assay System reagent to each well, mix, read plate, and analyze as in Subheading 3.2. If the results still indicate inhibition of luciferase activity for a compound, then that compound is considered a false positive due to quenching of the signal or direct inhibition of luciferase. It is not necessary to distinguish between those two conditions.

3.3.2 General Transcription Inhibition

This filter can be performed with any cell line containing a luciferase reporter under the control of a constitutive promoter. As an example, we use SH-CMV-luc cells, which are SH-EP human neuroblastoma cells that contain a luciferase reporter driven by the CMV promoter.

1. Seed 10,000 SH-CMV-luc cells in each well of a 96-well plate.
2. The next day, add "hit" compounds to the cells at a final concentration of 10 μM in duplicate. Incubate at 37 °C for 24 h.
3. After the incubation, add 10 μl Bright-Glo Luciferase Assay System reagent to each well, mix, read plate, and analyze as in Subheading 3.2. If the results still indicate inhibition of luciferase activity for a compound, then that compound is considered a false positive due to general transcription inhibition. Those compounds that are not toxic during screening and pass previous filtering are deemed validated hits and proceed to specificity testing as described below to evaluate their anti-Myc properties.

3.4 Cell-Based Assays for Inhibition of Endogenous Myc-Dependent Phenotypes

Analyze structures of confirmed hits and divide compounds into classes when two or more compounds share significant structural similarity. Based on the dose-response data from MYCN transactivation assay described in Subheading 3.2, derive IC50 values (the inhibitory dose resulting in 50 % decrease in Myc-specific luciferase activity) for each compound. Rank compounds within each chemical class by their IC50 value for MYCN transactivation and identify, where possible, several best compounds within each class. The best of each of the structural classes and compounds with unique structures are subjected to the Myc specificity filter that uses HO15.19 *myc*-null cells transduced with pLXSH–vector, pLXSH–c-Myc, or pLXSH–N-Myc.

3.4.1 Proliferation Assay

1. Seed 4,000 HO15.19–vector cells and 2,000 pLXSH–c-Myc or pLXSH–N-Myc cells in each well of a 96-well plate.
2. The next day, prepare two fold serial dilutions of the compounds selected in Subheading 3.3 to achieve a range of concentrations (e.g. 0.08–20 μM). Add the compounds to the cells and incubate at 37 °C for 72 h.
3. Remove medium from plates and add 100 μl of 0.5 % methylene blue/50 % methanol in water, incubate for 30 min, wash three times with water, and air dry. Add 100 μl of 1 % SDS in PBS per well for 10 min, measure optical density of each well at 650 and 540 nm, and then subtract the background at 540 nm from the absorbance at 650 nm (*see* **Note 6**).
4. Calculate an IC50 value based on the above measurements (a dose of the compound that inhibits proliferation by 50 % under the above experimental conditions). These calculations can be done using GraphPad Prism software (Fit Spline–Lowess) or equivalent program.

3.4.2 The Cell Cycle Distribution Assay

From the above data, a MYC Index can be calculated for each compound by dividing the IC50 value of a compound in HO15.19–vector cells by the average of the IC50 of that compound in HO15.19–N-myc cells and HO15.19–c-Myc cells (active compounds possess a MYC Index >1). Rank compounds within each chemical class by their MYC Index (the higher the MYC Index, the better the compound). The best of each of the structural classes and compounds with unique structures are subjected to filtering based on the cell cycle distribution of cells treated with the compound.

1. Seed 7,000 HO15.19–vector cells or 3,000 pLXSH–c-Myc or pLXSH–N-Myc cells in each well of a 96-well plate.
2. The next day, add vehicle (DMSO) or the compounds to the cells at concentrations equal to the IC50, 0.5 × IC50 and 0.25 × IC50. Incubate at 37 °C for 48 h.
3. Remove medium, trypsinize the cells, and subject them to standard propidium iodide FACS analysis.
4. Calculate the percentage of cells in G0/G1, S, and G2/M phases of the cell cycle in treated and untreated populations. Based on the above measurements, calculate a Cell Cycle Index: divide an average ratio between the proportion of G0/G1 cells in treated versus untreated HO15.19–c-myc and HO15.19–N-myc cells by a similar ratio in HO15.19–vector cells (*see* **Note 7**).

4 Notes

1. This type of system can be set up in any cancer cell line of interest. As a general rule, however, the cell line used for the readout should be adherent, transduced easily (i.e., ~100 % transduction efficiency), and should form a single cell suspension following trypsinization (i.e., not a cell line prone to clumping) so that it can be uniformly delivered across 96-well plates. These characteristics will enable a more reproducible readout across days of screening.
2. The readout cell line should be selected as a single cell clone rather than a population of cells such that there is a significant differential in luciferase activity between cells prior to transduction with *c-myc* or *MYCN* and after (ideally five- to tenfold). In our experience, the Myc reporter in the SHR6-17 cells was increasingly silenced the longer it was maintained in culture (i.e., drop in fold activation of reporter over time) so it is important to maintain a substantial collection of early passages in liquid nitrogen.
3. For some cell lines, prolonged incubation with viruses can be toxic. If this is the case, virus containing medium can be

removed and replaced with virus-free medium prior to addition of compounds.

4. Do not freeze-thaw the virus.
5. We titrated the amount of Bright-Glo required to give optimal signal in our system (10 μl). This may vary with different cell lines and from batch to batch so it is important to purchase sufficient Bright-Glo to cover the screen and to optimize for particular batches. Aliquot the Bright-Glo as repeated freeze-thaw cycles can affect activity.
6. In the event that a plate reader does not have a 650 nm filter, a 595 nm filter can be used. The background corrected values will be lower but should not affect the overall results as long as the controls (i.e., vehicle control treated wells) are approaching confluence at the time the assay is harvested.
7. The procedures described in Subheading 3.4 utilize HO15.19 cells and their derivatives, since these have for a long time been considered as the standard in the field. With the obvious goal to develop c-Myc-targeting therapeutic agents, the next step should include testing of the compounds in the transformed cells. However, due to the extreme variability in the phenotypes caused by genetic inhibition of Myc in different tumor cell lines [10], it is becoming increasingly difficult to provide a detailed protocol for testing the compounds beyond the described systems. We therefore suggest further characterization of specific phenotypes caused by the compounds and vis-à-vis genetic inhibition of c-Myc using siRNA technology in the cell line of choice. This may include comparison of changes in global cellular transcription or the ability to induce a specialized form of proliferation arrest such as differentiation or senescence or cell death such as apoptosis or mitotic catastrophe.

Acknowledgments

This work was supported by NIH R01 CA120244 and ACS RSG-10 121-01 grants to M.A.N., and by CINSW and NHMRC grants to M.H. and M.D.N.

References

1. Dang CV, O'Donnell KA, Zeller KI, Nguyen T, Osthus RC, Li F (2006) The c-Myc target gene network. Semin Cancer Biol 16:253–264
2. Dang CV (2012) MYC on the path to cancer. Cell 149:22–35
3. Malynn BA, de Alboran IM, O'Hagan RC, Bronson R, Davidson L, DePinho RA, Alt FW (2000) N-myc can functionally replace c-myc in murine development, cellular growth, and differentiation. Genes Dev 14:1390–1399
4. Hermeking H (2003) The MYC oncogene as a cancer drug target. Curr Cancer Drug Targets 3:163–175
5. Arvanitis C, Felsher DW (2006) Conditional transgenic models define how MYC initiates and maintains tumorigenesis. Semin Cancer Biol 16:313–317

6. Schulte JH, Lindner S, Bohrer A, Maurer J, De Preter K, Lefever S, Heukamp L, Schulte S, Molenaar J, Versteeg R, Thor T, Künkele A, Vandesompele J, Speleman F, Schorle H, Eggert A, Schramm A (2013) MYCN and ALKF1174L are sufficient to drive neuroblastoma development from neural crest progenitor cells. Oncogene 32(8):1059–65
7. Soucek L, Whitfield J, Martins CP, Finch AJ, Murphy DJ, Sodir NM, Karnezis AN, Swigart LB, Nasi S, Evan GI (2008) Modelling Myc inhibition as a cancer therapy. Nature 455:679–683
8. Vita M, Henriksson M (2006) The Myc oncoprotein as a therapeutic target for human cancer. Semin Cancer Biol 16:318–330
9. Mateyak MK, Obaya AJ, Adachi S, Sedivy JM (1999) Phenotypes of c-Myc-deficient rat fibroblasts isolated by targeted homologous recombination. Cell Growth Differ 8:1039–1048
10. Wang H, Mannava S, Grachtchouk V, Zhuang D, Soengas MS, Gudkov AV, Prochownik EV, Nikiforov MA (2008) c-Myc depletion inhibits proliferation of human tumor cells at various stages of the cell cycle. Oncogene 27:1905–1915

Index

Laura Soucek and Nicole M. Sodir (eds.), *The Myc Gene: Methods and Protocols*, Methods in Molecular Biology, vol. 1012, DOI 10.1007/978-1-62703-429-6, © Springer Science+Business Media, LLC 2013

C

D

E

H

N

S

T

MIX
Papier aus verantwortungsvollen Quellen
Paper from responsible sources
FSC® C105338

If you have any concerns about our products,
you can contact us on
ProductSafety@springernature.com

In case Publisher is established outside the EU,
the EU authorized representative is:
Springer Nature Customer Service Center GmbH
Europaplatz 3, 69115 Heidelberg, Germany

Printed by Libri Plureos GmbH
in Hamburg, Germany